VIII All-Russian (with International Participation) Conference on Low Temperature Plasma in the Processes of Functional Coating Preparation 2016

Journal of Physics: Conference Series Volume 789

Kazan, Russia
6 - 9 November 2016

ISBN: 978-1-5108-3686-0
ISSN: 1742-6588

Printed from e-media with permission by:

Curran Associates, Inc.
57 Morehouse Lane
Red Hook, NY 12571

CURRAN ASSOCIATES INC. proceedings.com

Some format issues inherent in the e-media version may also appear in this print version.

Printed by Curran Associates, Inc. (2017)

Institute of Physics
Dirac House, Temple Back
Bristol BS1 6BE UK

Phone: 44 1 17 929 7481
Fax: 44 1 17 920 0979

techtracking@iop.org

Additional copies of this publication are available from:

Curran Associates, Inc.
57 Morehouse Lane
Red Hook, NY 12571 USA
Phone: 845-758-0400
Fax: 845-758-2633
Email: curran@proceedings.com
Web: www.proceedings.com

Table of contents

Volume 789

VIII All-Russian (with international participation) Conference on Low Temperature Plasma in the Processes of Functional Coating Preparation

6–9 November 2016, Kazan, Russian Federation

Accepted papers received: 7 December 2016
Published online: 10 February 2017

Preface

Papers

Under the chairmanship of a member-correspondent of the Academy of Sciences of RT, Doctor of Technical Sciences, Professor N F Kashapov , in 2016 from 6 to 9 November in the Academy of Sciences of the Republic of Tatarstan, Kazan Federal University held the scientific-technical conference "Low-temperature plasma in a functional coating process" (with international participation).

The modern development of technology requires a broad introduction to the industry of new efficient processes based on innovative achievements in the field of science and technology. One of the areas, significantly expands technological possibilities of material processing process, is the use of concentrated energy flows, which are prepared using a low-temperature plasma generators. In order to create conditions for the unification of Russian and foreign scientists, engineers and technologists in the solution of fundamental and applied problems in the fields of research processes applying functional coatings, plasma dynamics, plasma modification of materials of low-temperature plasma flows.

Topics of the conference were oriented in the following directions:

1. Gas discharge physics.
2. Physical processes in low-temperature plasma in the deposition of functional coatings.
3. Thermodynamic, portable and optical properties of low-temperature plasma in the coating process.
4. Plasma chemistry. Electrochemistry.
5. Combustion and heat and mass transfer in multi-phase media.
6. Diagnostics of low-temperature plasma.
7. Dusty plasma.
8. Interaction of plasma and high-energy ion beams with a solid state. Laser plasma.
9. Characterization of coatings produced using low-temperature plasma.
10. Use of ion-plasma coatings.

Presented posts set high scientific level of the conference. The reports correspond to the world level of science in the field of plasma physics. The conference was productive for the participants both in terms of familiarization with the new results, often from related fields, and in terms of establishing and strengthening contacts between scientists from different research centers. A large number of young participants testifies to the continuity of scientific traditions represented at the conference of scientific schools, the present and the relevance of these research schools.

Peer review statement

All papers published in this volume of *Journal of Physics: Conference Series* have been peer reviewed through processes administered by the proceedings Editors. Reviews were conducted by expert referees to the professional and scientific standards expected of a proceedings journal published by IOP Publishing.

Gas flow influence on streamer-to-leader transition in surface barrier discharge in air at atmospheric pressure

Y Akishev[1,2], V Karalnik[1], M Medvedev[1], A Petryakov[1], N Trushkin[1] and A Shafikov[1]

[1]SRC RF TRINITI, 108840, Russia, Moscow, Troitsk, Pushkovykh Str., vlad. 12
[2]National Research Nuclear University MEPhI, 115409, Russia, Moscow, Kashirskoe sh.-32

E-mail: akishev@triniti.ru

Abstract. The experimental results on study of a sinusoidal surface barrier discharge (SBD) in airflow at different velocities are presented. Influence of gas flow velocity and its orientation on plasma structure of SBD is established and the conditions providing transition of SBD from the streamer mode to the mode with surface leaders (SL) are found out as well. It is shown the formation of the SL in SBD happens owing to thermal effects associated with a local gas heating in both the bright current spots disposed at the sharp edge of high-voltage electrode and thin current channels (streamers) originated from these spots. It was revealed that gas flow with a high velocity directed against the propagation of surface streamers leads to destruction of the bright anode spots and elimination of the SL (but not the streamers) on a barrier surface. However the gas flow with the same velocity but opposite direction leads to regular disposition of the bright current spots at the edge of high-voltage electrode and regular spatial structure of the thin current channels (streamers and leaders) originated from the current spots. It was found out that gas flow of any orientation leads to diminishing the gas temperature in the plasma channels compared to that in SBD in gas at rest.

1. Introduction

The sinusoidal SBD is an alternating current gas discharge which generates a thin layer (or sheet) of non-thermal plasma tightly adjoining the barrier surface. In fact, the SBD is a capacitor with a variable capacitance, and one of the capacitor plates of this capacitor is a thin non-stationary plasma sheet periodically forming on a barrier surface. The spatial configuration of a thin plasma sheet formed at each half-cycle depends on the specific experimental conditions. For instance, plasma sheet can be homogeneous (diffusive regime of SBD) or consist of many thin current filaments (streamer regime of SBD). The streamer regime goes over into so called leader regime, if the electric power feeding the SBD increases. In this regime, plasma sheet is heavily non-uniform - only a few bright current channels (surface leaders) are observed on the barrier. The both diameter of leader and its brightness exceed analogous streamer parameters. The formation of surface leaders in SBD is not always a desirable process. At present the level of knowledge about the processes forming the surface leaders is obviously insufficient. It is known only that the formation of low-current surface leader in SBD differs strongly from the formation of high-current volume leader in the spark [1-3]. The alternating current gas discharge is characterized by that the plasma processes happened in preceding half-cycle can influence the processes in subsequent half-cycle. An alternating current gas discharge is characterized by the fact that the plasma processes which happened in the previous half-cycle can influence plasma processes in the subsequent half-cycle. This dependence can be called as a short "memory" of SBD which is determined by time of decaying the plasma structures created in the previous half-cycle. Both the current spots on HV electrode and non-stationary current channels on barrier provide a strongly inhomogeneous energy deposition into gas adjoining the barrier and into barrier as well. This circumstance leads to non-uniform and gradual (during many periods) warming up of the barrier, HV electrode and gas. The decay time of thermal structures formed in SBD is defined by heat conductivity processes which are slower in comparison with a recombination of plasma structures. This is a reason why thermal structures in SBD keep themselves for many periods and influence slow evolution of plasma structures in subsequent periods. The influence of long-living thermal structure on the surface plasma structure in SBD can be called as a long "memory". This report presents the results on experimental investigation of plasma structures (streamers and leaders) of SBD in air at atmospheric pressure and influence of gas flow on the long memory of SBD.

2. Experimental setup

The sinusoidal SBD influenced by gas flow was organized on a ceramic plate (dielectric permittivity ε=9) with sizes of 60x60x1.5 mm^3. The electrodes were made of aluminum foil with thickness of 50 μm. High-voltage (HV) electrode looks like a rectangular strip with sizes of 35x35 mm^2, the angles of which are rounded. HV electrode has been pasted on top side of the barrier plate. The strip was oriented perpendicularly to a gas flow and disposed at the edge of the plate upstream. The HV electrode was partly covered with a dielectric so that only narrow part (3x35 mm2) of its surface is open. The grounded electrode with sizes of 35x35 mm^2 has been pasted on bottom side of the barrier plate. The upper edge of the grounded electrode was shifted by 2 mm downstream relatively the edge of HV electrode. The grounded electrode was fully covered with a dielectric.

The subsonic jet of air was flat and had the cross-section of 48x2 mm^2, i.e. the wide stream of air was able to cover all plasma area on a barrier. The flat air jet was directed either along a barrier surface or at angle 90^0 and 45^0 to the barrier surface. In the first case, the jet can be directed in two directions - along the movement of the plasma channels (streamers and leaders) and in opposite direction. The magnitude of gas velocity in jet was measured by thermo-anemometer Testo-510. In the experiments gas velocity was varied within limits of 0-80 m/s.

SBD was powered with sinusoidal voltage of 100 kHz in a form of the wave packets of variable duration from 0.1 to 2 s. The voltage amplitude was varied from 2 to 12 kV. The current and voltage waveforms of SBD were recorded by Tektronix TDS-520 and Tektronix DPO-2024 oscilloscopes. Based on these data the electric power of SBD was calculated. The image of plasma structure of SBD averaged over many periods of sin voltage was taken by digital camera Canon EOS 550D. Dynamics of fast development of SBD in a single half-cycle was studied with use of the multi-frame and fast camera LV-03, the frames of which with exposure time of 100 ns were synchronized with current and voltage of SBD. The optical spectrum of SBD plasma radiation was registered by a fiber-optical spectrometer AvaSpec-2048FT with the spectral resolution of 0.1-0.2 nanometers. The translational and vibration temperatures of nitrogen in the SBD plasma sheet were determined from processing of spectra of the second positive system of N_2.

3. Experimental data and discussion

The set of photos presented in figure 1a shows slow evolution of plasma structure formed on a barrier of high-frequency SBD in air at rest. This evolution is traced from the very beginning of the wave packet till its end in 0.5 s. The exposure time of each picture is equal to 16 ms. It means that the image on each photo is formed by overlapping of 3200 instant images of SBD in the half-cycles following one after another. The behavior of SBD current in a positive and negative half-cycle is shown in figure 1b by example of four periods from the whole wave packet. Note that the average electric power of SBD is high enough and equal to 210 W. Due to that the formation of the surface leaders roaming on a barrier surface happens quickly, as a rule, in 30-50 ms (see figure 1a). However the stabilization of spatial localization of the electrode spots and bright surface leaders originated from these spots takes place approximately in 200 ms after application of the wave packet to the SBD. Figures 1c and 1d show instant images of long-lasting SBD in the beginning of negative and positive half-cycle respectively. Exposure time of each shot is 100 ns. Figures 1e and 1f show images averaged over single positive and negative half-cycle respectively. In this case exposure time of each shot is 5 µs. All images are synchronized with appropriate oscillograms of current and voltage shown below.

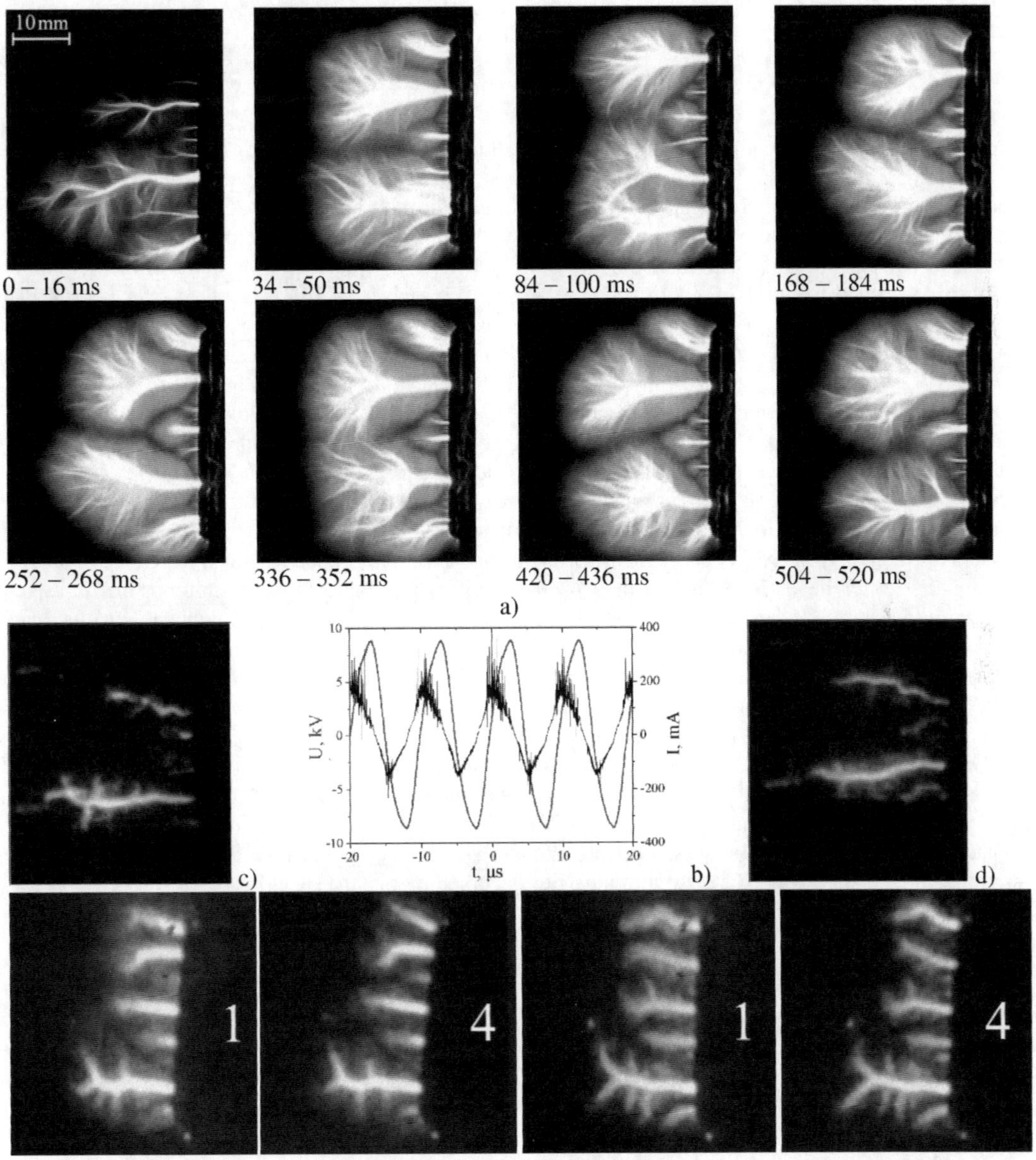

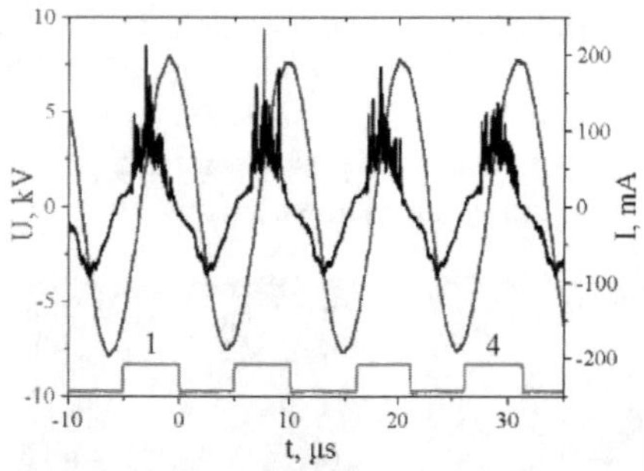

e)

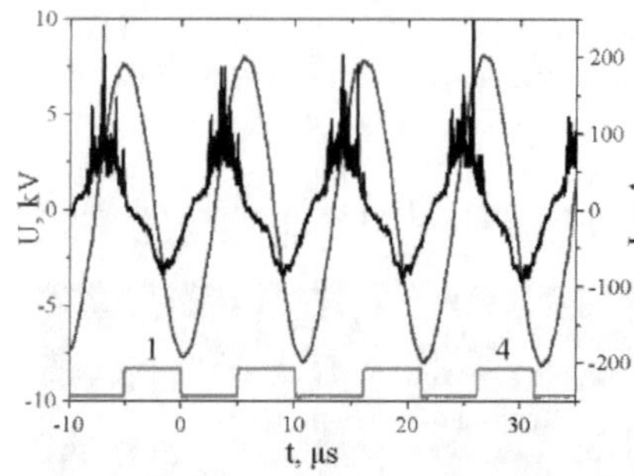

f)

Figure 1. a) Evolution of the SBD plasma structure in ambient air at rest. Each image is the overlapping of 3200 instant images of SBD in the positive and negative half-cycles following one after another. Time is counted from the beginning of imposing of a sin wave packet. The time interval corresponding to each photo is given below. b) The current and voltage waveforms of SBD. c) and d) The instant images of SBD with the exposure time of 100 ns; the pictures were taken in the beginning of positive (c) and negative (d) half-cycles. e) and f) The images of SBD averaged over appropriate positive (e) and negative (f) half-cycle respectively; the exposure time of each shot is 5 μs; all images are synchronized with I, U oscillograms; the appropriate marks numbered as 1 and 4 are shown below.

Close examination of the images presented in figure 1 leads to the conclusion that in the case of long-lasting SBD the localization and configuration of bright plasma structures (anode and cathode current spots on HV electrode and surface leaders on a barrier) approximately repeat each other in every positive and negative half-cycles through many periods of a sin voltage. This feature can be called as long memory. However, the numerous thin and less bright surface streamers surrounding the leaders are subject to more frequent change of their configuration, especially those which are localized around the leader head. Note that leaders in each half-cycle do not appear simultaneously: at the beginning the large leaders develop from the most bright electrode spots and then the smaller ones arise. Another important feature of steady sin SBD sounds as follows: due to existence on a barrier of the residual plasma formed at the previous half-cycle (short memory) there is no necessity to involve any additional ionization mechanisms like photoionization of gas in order to provide the propagation of positive surface streamers and leaders.

Gas flow appreciably influences the plasma structure of SBD. For example, fast airflow directed against movement of the plasma channels breaks large anode and cathode current spots. In this case, numerous and irregular small anode spots and regular small cathode spots happen in positive and negative half-cycles respectively (see pictures 2a and 2b). Hereat the surface leaders disappear (there are only surface streamers). Total area of a barrier occupied by plasma sheet becomes smaller.

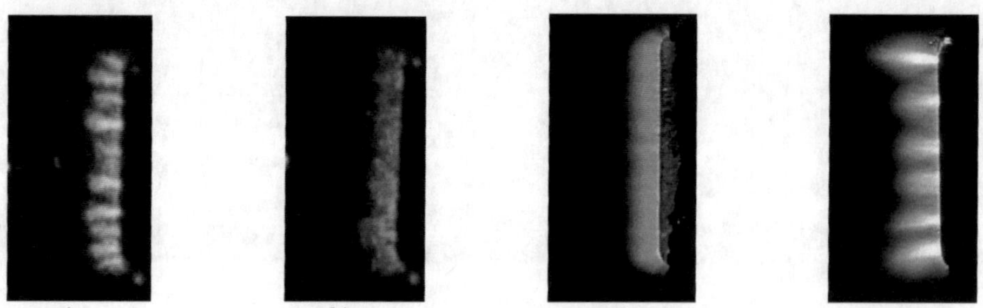

a) b) c) d)

Figure 2. The images of SBD in fast airflow at different direction of gas flow velocity. V=70 m/c.
a) and b) airflow is directed against movement of the plasma channels; negative (a) and positive (b) half-cycles; exposure time is 5 μs. Airflow is directed against (c) and along (d) movement of the plasma channels; exposure time is 16 ms.

Figures 2c and 2d present the images of SBD in fast airflow directed against (c) and along (d) the movement of the plasma channels. These photos were taken with long exposure time (15 ms) and their images were formed due to overlapping of many instant images of SBD. In the (c) case, small electrode current spots and streamers originated from these spots are not localized and slowly and chaotically migrate on a surface within limit of their small spatial period. This is a reason why plasma sheet in picture 2c looks like a homogeneous discharge but in fact this is not the case (see, for instance, the instant image 2a). In the (d) case, the leaders are not destroyed. The number of cathode spots and leaders are less and they are regularly distributed with a large spatial period. Besides, the chaotic migration of these spots and leaders in transverse direction is slow and happens within the limits which are smaller compared to their spatial period. This is a reason why the periodical structure of SBD in pictures 2c and 2d is kept even if they are taken with a long exposure time. In this case there is a long memory which is not destroyed by even fast airflow directed along the plasma channels. Short memory in spatial structure of SBD exhibits itself in negative half-cycles. This memory is determined by cathode spots and therefore cannot be destroyed by airflow of any direction.

Gas flow influences total length and translation temperature of streamers and leaders. The dependence of total length of streamers and leaders on airflow velocity is shown in figure 3a. Curves 1 and 2 in this figure correspond to velocity directed along (1) and against (2) the movement of the plasma channels. In the first case there is a non-monotonic dependence on velocity. At high velocities, the higher velocity, the shorter plasma sheet. There is a correlation between gas temperature in streamer and its possibility to transform into leader. The threshold for appearance of leader in SBD corresponds to the amplitude of sin voltage $U \approx 8$ kV. At this amplitude, the gas temperature in streamer sharply grows up approximately to 1200 K and streamer transforms into leader which appreciably increases own length (see figure 3b).

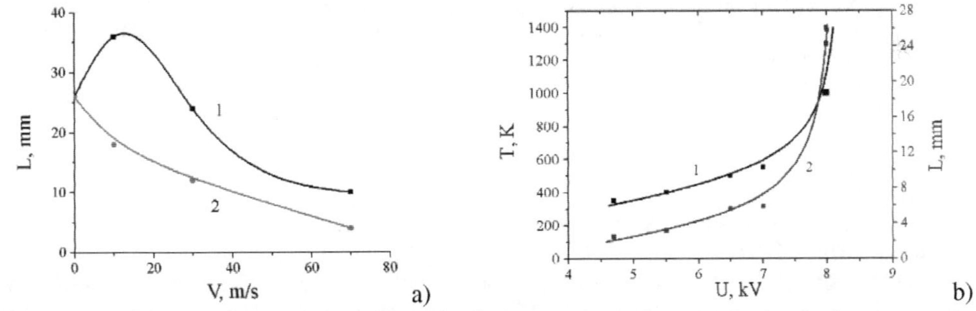

Figure 3. a) The dependence of total length of plasma channels on airflow velocity; curves 1 and 2 correspond to velocity directed along (1) and against (2) the movement of the plasma channels.
b) The dependence of gas temperature in plasma channels (1) and total length of plasma channels (2) on the amplitude of sin voltage for SBD in ambient air at rest.

Acknowledgement
This work was financially supported by Russian Foundation of Basic Research (grant №14-02-01067).

References

[1] Joussot R, Boucinha V, Weber R, Hong D, 2011 Negative spark leaders on a surface DBD plasma actuator – *IEEE Transactions on Plasma Science*, **39** (11):2194-95.

[2] Akishev Yu, Aponin G, Balakirev A, Grushin M, Karalnik V, Petryakov A, Trushkin N, 2013 Stepwise expansion of a surface dielectric barrier discharge as a result of alternation in formation of streamers and leaders.- *J. Phys. D: Appl. Phys.*, **46** 135204.

[3] Baseliyan E M, Raiser Yu P, *Spark discharge*, (Moscow: MIPT), p.320

The probe measurements of the electron density in deuterium plasma created by the electron beam of moderate energy

Y Akishev[1,2], V Karalnik[1], M Medvedev[1], A Petryakov[1], N Trushkin[1] and A Shafikov[1]

[1]SRC RF TRINITI, 108840, Russia, Moscow, Troitsk, Pushkovykh Str., vlad. 12
[2]National Research Nuclear University MEPhI, 115409, Russia, Moscow, Kashirskoe sh.-32

E-mail: akishev@triniti.ru

Abstract. The results of the probe measurements of the electron density in deuterium plasma at gas pressure P=1 Torr are presented. Plasma was created in a quartz tube by stationary beam of electrons with moderate energy up to 10 keV. The electron beam of 35 mm in diameter was generated by the steady-state obstructed glow discharge in a narrow inter-electrode gap. The electrical probe comprises two large metallic plates oriented in parallel to each other with a space gap between them of 15 mm. The electron beam passes through this gap in parallel to the plates and creates plasma in the gap. The plasma density was determined from comparison between the calculated volt-ampere characteristics (VACs) and VACs measured in the experiment. The dependence of plasma density generated by e-beam on energy of electrons in the beam is found out.

1. Introduction

The great advantage of the e-beams of moderate energy (up to 25 keV) is that they do not generate a hard X-radiation which is very dangerous to health. The e-beams of moderate energy (EBME) are environmentally friendly object and therefore they are widely used for generation of non-thermal reactive plasma which is of great interest for different applications such as surface modification of various materials, biomedicine, pharmaceutics, laser techniques, etc [1-2]. One of the promising applications of EBME is the super-high charging of the dust particles in plasma by e-beam. This approach opens the opportunity to create a small-scale neutron source for biomedical applications [3]. In this source the EBME provides two functions: it makes the super-high charging of the dust particles and creates a non-thermal plasma around these particles. Positive ions of plasma accelerate in a strong electric field of highly charged particle and gather energy; after that they strike the target and initiate the reactions of nuclear fusion accompanied by release of neutrons. We have done the numerical calculations of both the super-high charging of the dust particles by e-beam and the appropriate reactions of nuclear fusion [3]. It was shown that the plasma-forming gas, the most suitable for this task, is a hydrogen isotope – a deuterium. In this paper the experimental results on the probe measurements of the electron density in plasma created by EBME are presented. To do that we used not the classical probe method but developed new approach.

2. Experimental setup

The e-beam with energy of electrons up to 25 keV was generated by the obstructed glow discharge in the narrow inter-electrode gap filled with deuterium at low pressure (P = 0.5 - 2 Torr). The electrode system of the e-beam generator (figure 1) comprises the solid metallic cathode (stainless steel) of 24 mm in diameter and two meshy anodes with geometrical transparency of 60% and 70% located at a distance of 1.5 and 6 mm from the cathode. This electrode system forms two gas discharges - basic and auxiliary. Every gas discharge has own power supply electrically isolated from each other. Low-current auxiliary discharge between cathode and distant meshy anode serves for preionization of gas in the gap of the high-current basic discharge between cathode and close-in meshy anode. Basic gas discharge is powered by high voltage and operates in the obstructed glow regime. The electrode system is placed inside quartz tube of 110 mm in diameter and 300 mm in length. This triple electrode system enables to generate a pulsed e-beam and steady-state EBME as well. In the case of a pulsed regime it is possible to get the e-beam with energy of electrons up to 25 keV and electron current up to 150 A. In steady-state regime the e-beam with energy up to 10 keV can be generated.

To measure the density of plasma created by e-beam we used a moveable probe which comprises two large plane electrodes oriented in parallel to each other with a space gap between them of 15 mm. Each electrode is a thin copper foil pasted on a dielectric plate. Sizes of the foil are 15x15 mm^2 with 50 μm in thickness. The VAC of a double electrode probe was measured by the electric scheme which has no electrostatic connection with the "ground" of both the basic and auxiliary discharges. This circumstance ensure the absence of the electric current leakage from the pointed discharges into the probe electric circuit.

Propagation of the e-beam through deuterium in the reactor is accompanied by optical emission of the plasma forming gas. The visible longitudinal structure of self-consistent system "plasma + e-beam" was registered by Canon EOS 550 digital camera. The transverse structure visualization of the e-beam was done with use of the moveable plate of phosphor (oxysulfide of the yttrium activated by terbium). All experiments were done under slow gas blowing through the reactor to keep the initial purity of deuterium (99.99 % D_2).

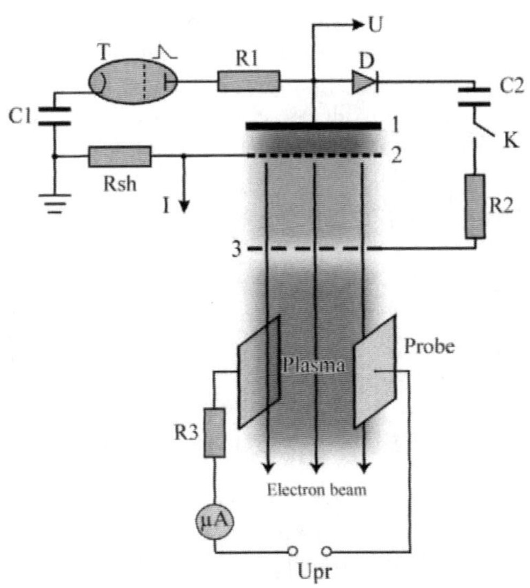

Figure 1. Scheme of the experimental setup. 1- the cathode; 2,3 – the meshy anodes; R1 = 56 Oм, R2 = 1 MOм, R3 = 1 кOм – the ballast resistors, R_{sh} = 0.024 Oм – the current shunt; D – the HV diode; C1 =12.5 nF, C2 =100 μF – the pulsed capacitors; K – the switch to charge the capacitor of auxiliary discharge; T – the thyratron 1000A/25kV; U_{pr} – the power source feeding the probe scheme. For generation of steady-state e-beam the thyratron is removed and capacitors C1, C2 are replaced by the appropriate steady-state power sources.

3. Results and discussion

The photos presented in figure 2 show the spatial distribution of optical emission from low pressure deuterium (P=1 Torr) excited by e-beam propagating through the reactor from right to left. The image in figure 2a corresponds to the case of high-current pulsed e-beam (energy of electrons is 20 keV, the electric current of e-beam is 50 A). The image in figure 2b corresponds to the case of low-current steady-state e-beam (energy of electrons is 5 keV, the electric current of e-beam is 2.2 mA). These photos are taken at different exposures. The exposure time in figure 2b much more exceeds exposure time in figure 2a. Due to that the visible high intensity of emission at the exit of e-beam gun in steady-state regime is a trick - in fact the intensity of optical emission from reactor in the pulsed regime of e-beam gun is much higher than that in steady-state regime. One can see in figure 2 that the e-beam diameter at the exit of the gun is the same (≈35 mm) for both the pulsed and steady-state regimes. However, the spatial structure of the e-beam in a long distance from a gun exit depends on the current and energy of e-beam. High-current pulsed e-beam constricts itself after some distance from the e-gun. Contrariwise, the propagation of low-current steady-state e-beam is accompanied with its monotonic divergence with the angle of disclosure about 10^0. Inside the reactor there is a block made of two dielectric rings which are located at a distance of 40 mm from exit of the e-gun (see figure 2b). This block supports the double electric probe in the region where the e-beam is transversely homogeneous.

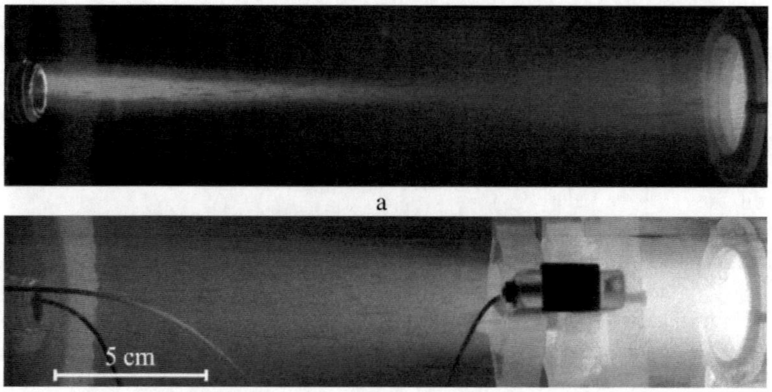

Figure 2. Two photos showing the spatial distribution of optical emission from low pressure deuterium (P = 1 Torr) excited by e-beam propagating through the reactor from right to left.
a) high-current pulsed e-beam, energy of electrons is 20 keV, the electric current of e-beam is 50 A, duration of the e-beam is 200 ns; b) low-current steady-state e-beam, energy of electrons is 5 keV, the electric current of e-beam is 2.2 mA, the exposure time is 200 ms.

The plasma density created by e-beam was determined from comparison between the calculated and measured VACs. Numerical calculations were performed with the use of 1-D non-stationary diffusive-drift model which takes into consideration the continuity equations for electrons and positive ions and Poisson equation for the electric field strength. Poisson equation took into account not only the space charge of electrons and ions of plasma but space charge of the e-beam. The processes of electron-ion recombination as well as ionization of deuterium by plasma electrons and e-beam were included in the model. This approach differs from the classical one used at the processing of VACs of Langmuir's electrical probes. In fact, our model simulates the non-self-sustained gas discharge supported in the gap by e-beam.

As an example, figure 3 presents the calculated distributions in the gap of the number densities for electrons and positive ions and for electric field strength as well. Plasma forming gas is D2, P=1 Torr. Important that the applied voltage across the gap is lower compared to the normal cathode drop for self-sustained glow discharge. One can see that practically all gap is filled with a quasi-neutral plasma except the narrow layer of a positive space charge in the vicinity of a negative electrode (something like cathode layer). Only in this layer there is an ionization by electrons of plasma. However, owing to a small size of this area it does not influence density of the plasma created by the e-beam electrons.

The experimental VACs of a double plane probe measured at different energies of the steady-state e-beam are given in figure 4. Herein, the results of calculations executed for the conditions of experiment are presented as well. One can see the calculated data well coincide with experimental parametrical dependence that proves the applicability of the developed model for processing of the probe VACs.

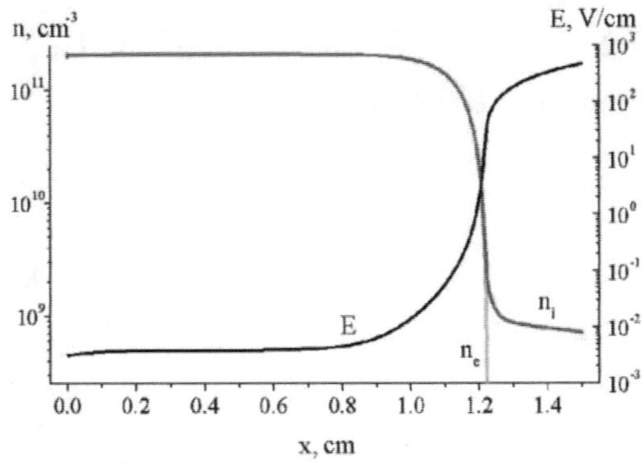

Figure 3. Spatial distribution of electrons, ions and electric field strength in the probe gap. Anode at the left, cathode on the right. The applied voltage across the gap is 75 V. Energy of steady-state e-beam is 5 keV. D_2.

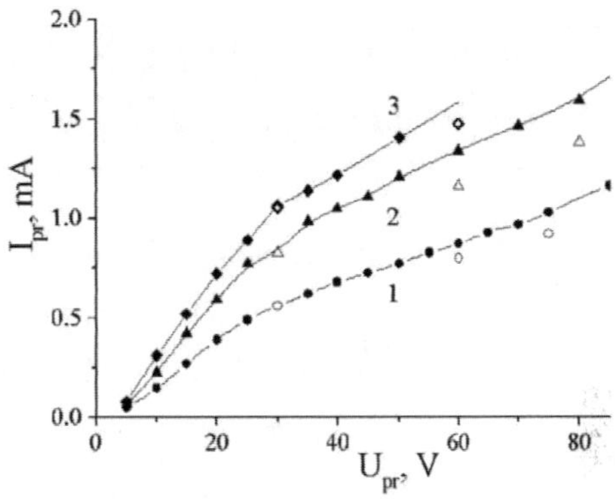

Figure 4. The probe VACs at different energies of the steady-state e-beam. D_2, P=1 Torr. Solid marks are the experimental data; open marks are the calculated data. E-beam energies: 1 - 5 keV; 2 - 7.5 keV; 3 - 10 keV.

The dependences of the experimental e-beam current and calculated plasma density vs energy of the steady-state e-beam are presented in figure 5. One can see that density of plasma created by steady-state e-beam increases monotonically with energy and current of the beam.

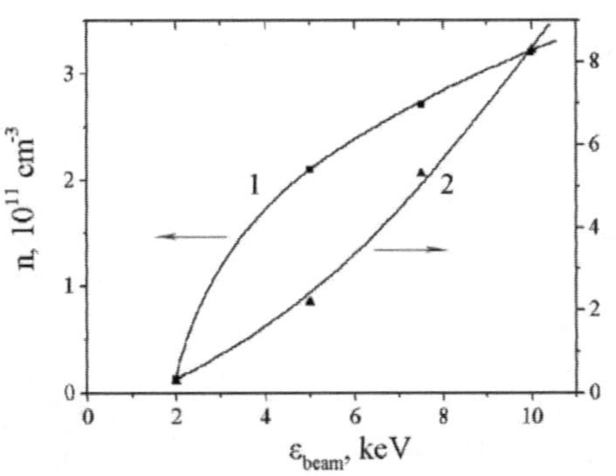

Figure 5. The dependences of the calculated plasma density (1) and the experimental e-beam current (2) vs energy of the steady-state e-beam. D_2, P = 1 Topp.

Acknowledgement
The research is completely executed at the expense of a grant of the Russian Scientific Foundation (project № 16-12-10458).

References

[1] Ozur G E, Proskurovsky D I, Rotshtein V P and Markov A B 2003 *Laser & Particle Beams* **21** 157–74
[2] Burdovitsin V A, Klimov A S, Medovnik A V and Oks E M 2010 *Plasma Sources Science and Technology* **19** 20–26
[3] Akishev Yu S, Karal'nik V B, Petryakov A V, Starostin A N, Trushkin N I and Filippov A V 2016 *Plasma Physics Reports* **42** 14–24

IOP Conf. Series: Journal of Physics: Conf. Series **789** (2017) 012003 doi:10.1088/1742-6596/789/1/012003

The magnetic field application for the gas discharge plasma control in processes of surface coating and modification

T Ya Asadullin, I G Galeev

General Physics Department, Kazan National Research Technical University – KAI, 420111, K. Marx St, 10, Kazan, Russia

E-mail: tyaasadullin@kai.ru

Abstract. In this paper the method of magnetic field application to control the gas discharge plasma effect on the various surfaces in processes of surface coating and modification is considered. The magnetic field directed perpendicular to the direction of electric current in the gas discharge plasma channel is capable to reject this plasma channel due to action of Lorentz force on the moving electrically charged particles [1,2]. The three-dimensional spatial structure of magnetic field is created by system of necessary quantity of the magnets located perpendicular to the direction of course of electric current in the gas-discharge plasma channel. The formation of necessary spatial distribution of magnetic field makes possible to obtain a required distribution of plasma parameters near the processed surfaces. This way of the plasma channel parameters spatial distribution management is the most suitable for application in processes of plasma impact on a surface of irregular shape and in cases when the selective impact of plasma on a part of a surface of a product is required. It is necessary to apply automated computer management of the process parameters [3] to the most effective plasma impact.

1. Low-temperature plasma of gas discharges has been widely applied for a long time in various technological processes effecting the surface of materials: cleaning, modification, coating, plasmochemical reactions in a near-surface layer etc. Impact of gas-discharge plasma on the surface of substance is actively applied. The plasma impact on the processed surface can be various: bombing of a surface of substance by the fast charged particles; radiation by the electromagnetic radiation of a various spectral band, for example ultra-violet radiation; heating of material due to Joule heat at course of electric current through surface material, including induction currents in case of high frequency discharges; chemical impact on the processed surface with the substances which are formed during course of plasmochemical reactions in gas-discharge plasma, etc.

Various ways of the organization of the discharge can be applied to implementation of plasma impact on the surface of material. The simplest way is when the surface of one of electrodes, for example, of the cathode, is processed. By such method it is possible to make, for example, cleaning of a surface of the cathode of oxides or other pollution. The disadvantage of this way is its limitation on available values of parameters. It is possible to use only near electrode layers of gas-discharge plasma. Besides rather often there is an intensive thermal emission in the near electrode layers that also limits opportunities of application, for example, for coating.

Besides the near electrode areas of discharge it is possible to use for impact on surfaces, for example, the area of the positive column of a glow discharge. However this area in case of normal, diffusion burning of discharge, especially with the low pressure, occupies the free space in the discharge camera therefore to organize the plasma influence on processed surface is rather complicated because putting the proceeded material into the plasma space will strongly change the plasma parameters. This way of influence is suitable to process fabrics, nonwoven fabrics and other high porous dielectric materials. But in this case there is a complexity for the organization of a gas flow which is used for the gas mixture cooling and updating, and for the stabilizing of discharge. The cross flow (blow-in) of gas can be used for offset of the discharge area closer to the processed surface as well, for example, in plasmatrons.

It is possible to use the creeping discharge to put discharge area directly near the processed surface, but pulse and unstable character of this discharge makes its usage inconvenient for processing of surfaces.

2. B. Timerkaev and co-workers have considered in recent works [1,2] a way of pressing of the plasma gas-discharge channel to the surface of fuel oil by means of magnetic field. It has been investigated for plasmochemical processing of heavy hydrocarbonic raw materials. Pressing of plasma was carried out under the influence of Lorentz force. It is known that electric current in gases is caused by drift of electrons and ions. Since oppositely charged particles in electric field move in opposite directions, Lorentz force for all of them will be directed in the same direction. Selecting the corresponding directions of electric and magnetic fields it is possible to achieve the discharge channel inclination towards the surface of liquid hydrocarbonic raw materials.

In works [1,2] they have performed an estimated calculation of influence of electric and magnetic fields on the movement of electrons and ions in plasma of the glow discharge near the surface of liquid hydrocarbon raw materials for pressure of 10 Tor, temperature 500 K and induction of magnetic field 0,1 T. According to the results of calculations the free path length for electrons has been obtained about 0,074 mm, the average speed of an electron on free path length (i.e. drift speed in the direction of electric field) has been obtained equal 0,16 10^6 m/s, and the radius of the Larmor movement has been obtained about 1,5 microns. Therefore at such magnetic field range the electron will drift in the direction, perpendicular as to electric, and to magnetic fields. At the same time at such values of magnetic field the Larmor radius for ions is about 1 mm. Thus magnetic field cannot strongly influence the ions movement on the distance of ions free path length. Consequently electrons will participate in the complicated drift movement in the direction, perpendicular to the directions of both electric and magnetic field. And plasma ions will drift towards the electrons.

The experimental setup described in works [1,2], together with system of the organization of several mutually perpendicular magnetic fields provides a possibility of management of a inclination of the plasma channel from its conditional axis. It can be used for processing of surfaces of irregular shape, concentrating plasma area near various sites of a surface during necessary time. In this way it is possible to process, for example, internal surfaces of dielectric tubes. In this case when this technique of the plasma channel inclination management is used for tubes processing it is easy to organize a longitudinal gas flow.

Gas discharge plasma tends to various types of instability and pulsations. It is a serious problem for implementation of surface technological processes. Certainly it produces additional difficulties when the magnetic field is used to manage the gas discharge plasma channel position for the surfaces processing. It is recommended to use a computer based automated control for both electrical discharge parameters and magnetic field distribution of the technology experimental setup to enhance the efficiency, quality and stability of the plasma influence on the surfaces. The method of stability increase of the gas discharge with the use of computer based automated system is described in [3].

References

[1] Timerkaev B A, Sofronitskiy A O, Andreeva A A 2015 The glow discharge creeping along the surface and its applications. ISSN 2078-6255. Vestnik of A.N. Tupolev KSTU (Kazan: Kazan National Research Technical University Press) vol 3 pp 5-9
[2] Timerkaev B A, Sofronitskiy A O Andreeva A A 2016 Carbon nanotubes formation in the

decomposition of heavy hydrocarbons creeping along the surface of the glow discharge. Journal of Physics: Conference Series 669 012062 4 pages

[3] Asadullin T Ya, Galeyev I G, Timerkayev B A 2016 Way of stability increase of the glow discharge. Journal of Physics: Conference Series 669 012062 3 pages

Capacitive coupled RF discharge: modelling at the local statement of the problem

I B Badriev[1], V Yu Chebakova[1] and V S Zheltukhin[2]

[1]Kazan Federal University, 18 Kremlevskaya Street, Kazan, 420008, Russia
[2]Kazan State Technological University, 68 Karl Marx Street, Kazan, 420015, Russia

E-mail: vchebakova@mail.ru

Abstract. In this paper a mathematical model of capacitively coupled RF discharge at atmospheric pressure is constructed, a method of numerical realization of the model is developed, and the numerical calculations are carried out. Comparison of the results of the numerical experiments with the data of other authors, in particular, with the experimental data, is demonstrated as well a model adequacy as effectiveness of the numerical method. A results of calculations of the model problem at pressure of 760 Torr, frequency of generator of 13.76 MHz and interelectrode distance of 20mm, in local approximation are presented.

1. Introduction

Low-temperature non-equilibrium plasma is widely used in various fields of science and technology [1-7]. Low-temperature plasma is applied for material treatment aimed to increase service life and reliability of machine-building products, to create light-weight and service-strong polymeric composite materials, polyethylene-plastic materials, nanodispersed metal powders and compounds, etc. Various types of discharges are used to create plasma and important role in this process is assigned to RF-discharges, and particularly, to CCRF-discharges. Argon as the plasma gas is often usedin the plasma systems. Analysis of the literature revealed a number of disadvantages that are inherent to the using models of the capacitively coupled RF (CCRF) discharges at high pressures. As a rule, a simplified scheme of plasma chemical reactions in the drift-diffusion approximation is taken into account. This scheme is quite adequate for argon at low pressure and contains a reaction for electrons, atomic ions, neutral and excited atoms.

However, studies of microwave discharges showed that the ratio of concentrations of atomic and molecular ions depends on the gas temperature [8]. The present work is a continuation of previously published [9,10]. Our calculations for simplified kinetic scheme which is typically used at low pressures have shown that there is a significant change of the gas temperature in the interelectrode space at atmospheric pressure, and the boundary conditions of the first kind (i.e. the gas temperature at the boundary is constant and equal to the temperature of the electrodes) while a gas temperature in the interelectrode space is practically constantat low pressures. Changes of the gas temperature profile in the interelectrode space is less pronounced at the third kind boundary conditions (i.e. with free heat exchange). As a result, we include the molecular ion and argon dimerto the kinetic scheme.

2. Mathematical model

We have constructed a self-sustained mathematical model, which is described a capacitive coupled RF discharge between two plane-parallel electrodes, one of them is grounded and the other is connected to

RF generator, wherein the interelectrode distance is smaller than the electrode size. In this case the electric field is close to the potential field and the discharge is uniform along the electrode. It allows us to use one-dimensional approximation [11]. Estimates of the time and the distance of the electrons lossesof an energy acquired from the electric field, are showed that a local approximation can be used for the simulation of an RF discharge at atmospheric pressure, when the parameters of the electron components of the plasma (such as diffusion coefficients, mobility, the average energy, constants of the plasma-chemical reactions, etc.) depend on the local value of the reduced electric field (ratio E/N of the electric field E to the concentration of heavy particles N) [12].

Self-sustained model of the CCRF discharge at atmospheric pressure in the local approximation contains the balance equations for electrons, metastable atoms, molecular and atomic ions, the kinetic equations for argon dimer and neutral atoms, the Poisson equation for the electric fieldpotential, as well as stationary atomic heat equation ion gas with the boundary conditions of heat transfer, calculated on the average for the period of oscillation of the field parameters. Rate coefficients for processes taking place in electron impact depends on the local value of the reduced electric field and the degree of ionization taking into account electron-electron collisions. A simplified diagram of argon atom wherein the 4 lowest nearby electronically excited states is described as the single level excited atoms of density. The following plasma-chemical reactionsis considered in the model:

$Ar+e \rightarrow Ar^{+}+2e$; $Ar^{*}+Ar^{*} \rightarrow Ar+Ar^{+}+e$; $Ar^{*}+e \rightarrow Ar^{+}+2e$; $Ar^{+}+e \rightarrow Ar+h_{v}$; $Ar^{+}+2e \rightarrow Ar+e$; $Ar+e \rightarrow Ar^{*}+e$; $Ar^{*} \rightarrow Ar+h_{v}$; $Ar^{*}+Ar \rightarrow 2Ar$; $Ar^{*}+e \rightarrow Ar+e$; $Ar^{*}+Ar^{*} \rightarrow Ar_2^{+}+e$; $2Ar+Ar^{+} \rightarrow Ar_2^{+}+Ar$; $Ar_2^{+}+e \rightarrow Ar+Ar^{+}+e$; $Ar_2^{+}+e \rightarrow Ar+Ar$; $Ar_2^{*} \rightarrow 2Ar+h_{v}$; $Ar^{*}+2Ar \rightarrow Ar_2^{*}+Ar$; $Ar_2^{*}+Ar_2^{*} \rightarrow e+2Ar+Ar_2^{+}$; $Ar_2^{*}+Ar^{*} \rightarrow e+Ar+Ar_2^{+}$; $e+Ar2^{*} \rightarrow 2Ar^{*}+e$; $Ar^{*}+2Ar \rightarrow 3Ar+h_{v}$; $Ar_2^{+}+Ar \rightarrow 2Ar+Ar^{+}$.

Constants of the processes are taken from [13-20].

3. Numerical method for solving the problem

The mathematical model is described by a system of boundary and initial-boundary problems, which are characterized by several features complicatinga numerical solution: different time scales of variation of the main characteristics of the steady discharge; the presence of large gradients of the density of charged particles, the electric field and the electron temperature near the electrodes; the system is nonlinear, both on separate equations and as whole. existing methods of nonlinear problems solving (see.,eg, [21-26]) were unusable in this case. Therefore,we used an approximate method for solving nonlinear system of boundary and initial-boundary problems, which is considered in this paper. The metod is based on a preliminary finite-dimensional approximation of the problem by using finite-difference schemes with a subsequent application for its implementation of the iterative process.

The finite-difference approximation of the convection-diffusion equations of charged particles is constructed by an integro-interpolation method with a method of directed differences to approximate first-order derivative terms. An implicit Euler scheme was used to solve the Cauchy problems. A Newton linearization is usedto approximate the quadratic nonlinearities in the right side of the equations. Density flows for the ion and electron gas are calculated by the Scharfetter-Gummel scheme [27].

The numerical algorithm is based on moving the non-linearity in coefficients (diffusion coefficients, the mobility and the constants of plasma-chemical reactions) to the low layer. The linearization of the system is carried out by the method of the Seidel type. The linearization of the heat equation for the atom temperature was carried out by Jacobi method, in which the initial approximation is the value of the atomic-ion temperature on the previous period. The calculations were carried out until the steady state of the full charge balance in the interelectrode spaceis reached, i.e., charges which is brought to the electrode by electrons andpositive ions for the period of the field oscillation are exactly compensated.

4. The results of numerical experiments

The numerical simulation is carried out by a software package based on theMatlab. Numerical calculations performed at atmospheric pressure, interelectrode distance of 2mm, and frequency of the high-frequency generator of 13.56 MHz is compared with results of [18] and is showed a good correspondence.

Additionally, the model problem was solved at atmospheric pressure, the interelectrode distance of 20mm, and theRF generator frequency of 13.56 MHz. It is found that in the last case the gas temperature isabout 2000 K in the discharge center and about 930 K near the electrodes. This difference between the gas temperatureamounted to 53.5%, is effected to significantly on the charged particlesdensities in the interelectrode space. The densities is more convex at 20 mm than at 2 mm. The most significant changes are observed in the distribution of molecular positive argon ions. The graph of argon molecular positive ions is substantially constant at the distance is 2 mm, while there is a local maximum of ones near of the electrode at the distance of 20 mm, and the local minimum is observed in the center of discharge area. The ratio of the minimum of the density of molecular positive argon ions to the maximum ones is about 0.43, i.e, the molecular ions density falls more than 2 timesin the middle of the discharge gap. At the same time, there is the preservation of quasi-neutrality of the plasma in the ambipolar field. The ratio of dimer to the excited atoms also changes: the maximum is about 0.78at the distance 2 mm, while ones is about 0.38when the distance is 20 mm. Thus, if the dimers density at the distance of 2 mm can be assumed constant, whileones is decreased at 20 mm, when the gas temperature is maximum.

5. Conclusion

The mathematical model of low-temperature CCRF plasma in argon, eliminating the drawbacks of existing models is constructed. The model includes drift–diffusion equations for electron and ions, Poisson's equation for the potential of the electrical field, balance equations for metastables and grounded state atoms as well as a steady state equation for atomic-ion gas conductivity. A numerical method is proposed for model solution. The solution for the model with non-ideal conductor is obtained. The results of simulations are shown that in case of a significant gas heating aswell at higher pressures, as in the case of large interelectrode distances at low pressures, the processes of formation and destruction of particles, and consequently influences on the distribution and balance of charged (electrons, atomic and molecular ions) and excited particles in the discharge gap.

Acknowledgments
The work is performed according to the Russian Government Program of Competitive Growth of Kazan Federal Universityandby Russian Science Foundation, project no. 16-11-10299.

References

[1] Fridman A 2008 *Plasma chemistry* (Cambridge: Cambridge Univ. Press)
[2] Raiser Yu P 2009 *Physics of gas discharge* (Dolgoprudny: Publishing house "Intellect")
[3] Abdullin I Sh, Zheltukhin V S and Kashapov N Ph 2000 *Radio-frequency plasma jet treatment under low pressure. Theory and practice in the application* (Kazan: Publishing house affiliated with Kazan University)
[4] Savinov V P 2013 *Physics of a high-frequency capacitive discharge* (Moscow: PHYSMATLIT)
[5] Riser YuP, Shneider MN and Yatsenko NA 1995 *Righ-frequency capacitive discharge. Physics. Experimental procedures. Attachments*(Moscow: Publishing house affiliated with MPhTI)
[6] Chebert P, Braithwaite N 2011 *Physics of radio-frequency plasmas* (Cambridge: Cambridge Univ. Press)
[7] Abdullin I Sh, Zheltukhin V S, Uglov A A and Khusainov I G 1992 The effect of non-equilibrium low-temperature plasma flux on germanium and silicon*Fizika i Khimiya Obrabotki Materialov* **6** 82-85
[8] Martinez E Castonos, Kabouzi Y, Makasheva K and Moisan M 2004 Modeling of microwave-sustained plasmas at atmospheric pressure with application to discharge contraction *Physical review E70* 066405

[9] Dubrovin V T, Chebakova V Ju and Zheltukhin V S 2016 Radio-Frequency Discharge at Low Pressure: a Non-Local Problem Statement Approach *Procedia Engineering* **150** 1041–5

[10] Chebakova V Ju, Zheltukhin V S and Dubrovin V T 2016 Problem of radio-frequency discharge at atmospheric pressure in local statement *Applied Mathematical Sciences***10**(21)1013-22.

[11] Riser YuP, Shneider MN and Yatsenko NA 1995 *Righ-frequency capacitive discharge. Physics. Experimental procedures. Attachments*(Moscow: Publishing house affiliated with MPhTI)

[12] Kudryavtsev A A, Smirnov A S and Tsendin L D 2010 *Physics of glow-discharge* (Saint Petersburg: Publishing house "Lan'")

[13] Hagelaar G J M, Pitchford L C 2005 Solving the Boltzmann equation to obtain electron transport coefficients and rate coefficients for fluid models *Plasma Sources Sci. Techn.***14** 722–33.

[14] Ferreira C M, Loureiro J and Ricard A 1985 Populations in the metastable and the resonance levels of argon and stepwise ionization effects in a low-pressure argon positive column, *Journal of Applied Physics***57** (82) 82–90

[15] Lymberopoulos D P, Economou D J 1993 Fluid simulations of glow discharge & Effect of metastable atoms in argon, *J. Appl. Phys.***73** (8) 3668–79

[16] *Moravej M, Yang X, Hicks R F, Penelon J and Babayan S E 2006 A radio-frequency nonequilibrium atmospheric pressure plasma operating with argon and oxygen J. Appl. Phys.* ***99*** *Art. 093305*

[17] Bora B, Bhuyan H, Favre M, Wyndham E and Chuaqui H 2011 Diagnostic of Capacitively Coupled Low Pressure Radio Frequency Plasma: An Approach through Electrical Discharge Characteristic *International Journal of Applied Physics and Mathematics***1** (2) 124–8

[18] Balcon N P, Aanesland A, Hagelaar G J M., Boswell R and Boeuf J P 2007 Atmospheric pressure RF discharge in argon: optical diagnostic, fluid model and applications, *28th ICPIG, July 15-20, 2007, Prague, Czech Republic* 957–60

[19] Diatko N A, Ionokh Yu Z, Meshanov A B and Napartovich A P 2005 Study of dark study of development of positive column of glow discharge in argon *Plasma Physics Reports***31** (10) 939-53

[20] Epstein I L, Gavrilović M, Jovircević S, Konjević N, Lebedev Yu A and Tatarinov A V 2014 The study of a homogeneous column of argon plasma at a pressure of 0.5 torr, generated by means of the Beenakker's cavity *Eur. Phys. J. D***68** (334)

[21] Badriev I B, Zadvornov O A 2003 A decomposition method for variational inequalities of the second kind with strongly inverse-monotone operators *Differential equations.***39** (7) 936-44

[22] Badriev I B; Zadvornov O A and Lyashko A D 2004 A study of variable step iterative methods for variational inequalities of the second kind *Differential equations.* **40** (7) 971-83

[23] Badriev I B, Banderov V V 2014 Numerical method for solving variation problems in mathematical physics *Mechanical components and control engineering III.* **668-669** 1094-7

[24] Badriev I B, Zadvornov O A, Saddek A M 2001 Convergence Analysis of Iterative Methods for Some Variational Inequalities with Pseudomonotone Operators*Differential equations***37** (7) 934-42

[25] Badriev I B 1983 Difference-schemes for linear-problems of the filtration theory with discontinuous law *Izvestiya Vysshikh Uchebnykh Zavedenii Matematika.* **5** 3-12

[26] Badriev I B 2013 On the solving of variational inequalities of stationary problems of two-phase flow in porous media Source of the Document *Applied Mechanics and Materials* **392** 183-187

[27] Kulikovsky A A 1995 A More Accurate Scharfetter-Gummel Algorithm of Electron Transport for Semiconductor and Gas Discharge Simulation *Journal of Computional Physics***119** 149–55

Experimental study of the polymer powder film thickness uniformity produced by the corona discharge

Marsel Fazlyyyakhmatov
Kazan (Volga Region) Federal University, 18 Kremlyovskaya str., Kazan 420008, Russian Federation

E-mail: mfazlyjy@kpfu.ru

Abstract. The results of an experimental study of the polymer powder film thickness uniformity are presented. Polymer powder films are produced by the electrostatic field of corona discharge. Epoxy and epoxy-polyester powder films with thickness in the range of 30-120 microns are studied. Experimentally confirmed possibility of using these coatings as protective matching layer of piezoceramic transducers at frequencies of 0.5-15 MHz.

1. Introduction

In recent years polymer powder coatings are widely used in various industries. Major application area is obtaining protective, functional and decorative coatings in the finishing industry [1].

Several ways of producing powder coatings are existing. The most often used at present is a method of applying powder coatings in the electrostatic field [2]. Powder coating guns are widely used in industry for manual and automatic operations. Two types of gun are available: the triboelectric [3,4] and the corona gun [5]. Both techniques have some advantages and disadvantages.

In recent decades, electrostatic spraying techniques are used in such fields as pharmaceutical [6] and food [7]. Corona discharge plasma is widely used in different branches of industry and manufacturing, in particular for the surface modification [8].

Electrostatic powder coating was first introduced to the finishing industry in 1962. Since then are developed many types of guns: conventional corona guns, low-voltage (EGD) guns, electrokinetic gun, cyclo gun, edge effect guns [9]. All these guns can be classified into two types: with external and inner powder charging. The most perspective and have the greatest advantages is conventional corona guns.

Conventional corona guns are versatile and cater for a wide variety of workpieces of different shapes and sizes. An air flow, which must be carefully regulated, conveys powder via a diffuser and through a region rich in ions from a corona electrode. Ion attachment occurs to a proportion, but by no means all, of the particles so that the grounded workpiece is bombarded by charged particles, ions and even some uncharged particles. Negative corona charging is normally used [9].

There are two types of polymer powder coatings: thermosets and thermoplastics. For obtaining protective coatings in recent years in most cases, are used thermosetting polymer powders. These coatings are formed by fusion of the powder particles and further chemical transformations.

By type of film former the most common thermosetting compositions are epoxy, polyester, epoxy-polyester, polyurethane and polyacrylate [10,11].

In [12,13] was proposed to use powder coatings for the protection and recovery of medical ultrasound sensors.

Ultrasound is the most popular method of non-destructive testing and diagnostics in various fields including medicine. In [14,15] are shown results of research and development of new methods and equipment for non-destructive testing multiphase and biological objects.

The main source of ultrasonic vibrations are electromechanical transducers based on piezoelectric materials. Despite the appearance of new piezoelectric composite materials and various polymer piezoelectric films the most common materials remain barium titanate ($BaTiO_3$) and lead zirconate titanate (PZT) ceramics. The main problem of these materials is the brittleness.

In the process of diagnosis liquid and solid objects the close contact of the ultrasonic transducer with the test object is necessary. This requires to use different protective coatings and prisms. They are preventing the damage transducer and protect it from aggressive environments. Particularly in the medicine copolymers of vinyl chloride and ethyl acrylate are applied. However, the protective coatings are abraded and are damaged, resulting in a failure of the transducer. To solve this problem, the applying of polymer powder coating in the electrostatic field are proposed [16].

2. Materials and methods of research

For most tasks of ultrasound diagnostics and probing the frequency range 0.5-15 MHz is used. Ultrasonic vibrations in this frequency range provide sufficient resolution in the low level vibration damping. In previous paper [16] was shown that the thickness of the matching layer the piezoceramic at these frequencies is 10-300 um.

The studies [12] of properties the samples of the coatings of polyether paint PD510226 («BECKER»), epoxy-polyether paint EP110022G («BECKER») showed the suitability of using this type of coating to the surface of the ultrasonic transducers. Such parameters as impact strength, flexibility, adhesion, gloss, hardness and wear resistance were evaluated.

In this paper we investigated the epoxy-polyether powder coating EP111095G («BECKER»). This powder coating also corresponds safety requirements due to the absence of organic solvents and heavy metals.

The substrate samples for studies are chosen according to ISO 1514:2016. The quartz glass panels with dimensions of 10x5x0.2 sm are used in this studies. The surface of the glass is polished. It should be noted that quartz glass is a dielectric. Therefore, powder coating by electrostatic spraying is difficult. To solve this problem, on the glass surface previously suffered by the magnetron sputtering an aluminum oxide layer about 200 nm thick. The surface preparing process are discussed in [13]. It should be noted that surface must be clean and dry.

Spraying the coating is carried out using a spray gun START-50 with voltage 30 kV on the corona discharge electrode. The measurement of the potential electrostatic field during application polymer powder coatings on metallic substrates using different speeds of micro-particles and the distance to the substrate was carried out [17].

Spraying is carried out in the chamber of coating polymer powder materials at the samples. The chamber designed to prevent scattering and collection powder coating particles are not deposited on the painted product.

Experimental studies in [16] was shown optimum operating parameters of the system of homogeneous coating polymer powder coatings. Based on these data the following parameters are chosen: time of spraying – 6 sec., distance from gun to substrate – 15 sm., air pressure – 200 kPa.

Baking was carried out in a SNOL 350 drying chamber at 200 °C for 10 min.

3. Experimental results and discussion

The experiments to measure uniformity of the polymer powder film thickness produced by DektakXT Stylus Profiler. Was chosen following settings: Scan type – Standard Scan, Range – 65.5 um, Profile – Hills and Valleys, Stylus Type – Radius 12.5 um, Stylus Force – 1 mg, Duration – 100 sec, Length – 30 mm. Experimental result are shown at Figure 1.

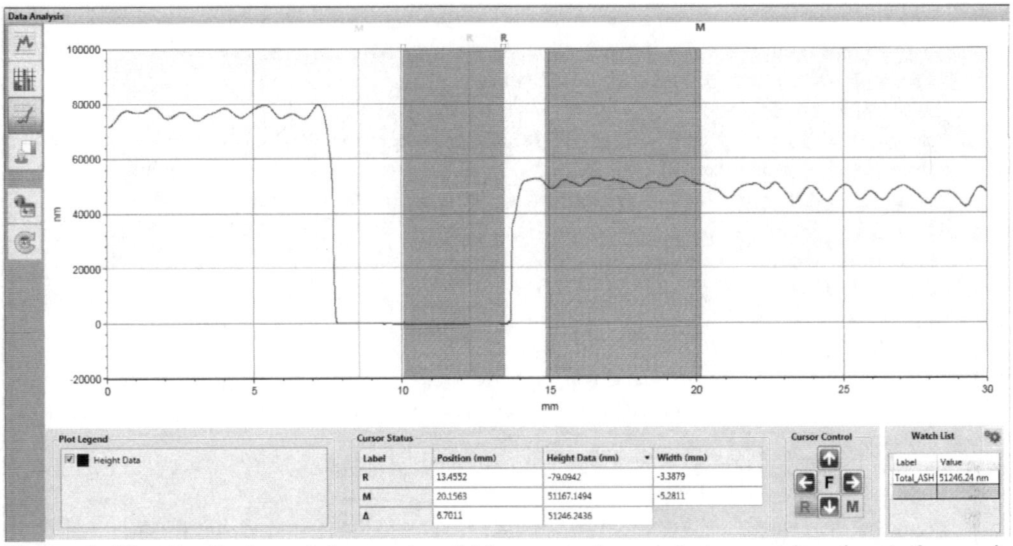

Figure 1. The distribution of the coating thickness on the substrate: spraying time t=6 sec, air pressure P=200 kPa

Measurement carried out from left to right along the central axis of substrate. The length of the scanned surface was 3 cm. Spraying center is located at around 1.5 cm. o the left from the center the mark from the high-temperature adhesive tape width of 5 mm was applied. It is necessary to reference the ground from the surface.

Figure 1 shows that the surface within the 1.5 cm is homogeneous with small dips in the range of 5 microns. It can be noted that the left from the mark coating thickness at 25-30 microns larger than the right. Apparently this is explained by the fact that the charged powder particles did not deposited to the dielectric layer adhesive tape and carries away with the air flow to the left from the center of the substrate and the deposited there. In the next time will be necessary to carry out measurements with a careful selection of the location of reference mark.

Earlier, in [18] were produced coating thickness measurements of polyether powder PD810119G («BECKER») sprayed in various modes. The measurement of thickness the obtained coatings was carried out symmetrically relative to the centre of the substrate to the left and to the right in increments of 5 mm using an ultrasonic thickness gauge PosiTector 200 B/Adv. It was shown that the coating is heterogeneous with a maximum thickness at the centre, followed by a decrease and an increase closer to the substrate edge.

Comparison with the published data [19,20] simulation of spraying process was produced. Conducted calculations [20] showed that during application the aerosol to a flat surface oriented normal to the axis spraying in an electrostatic field, the solid phase is applied nonuniformly and its density increases towards the free edge. At free edge the solid phase reaches the highest value. This is defined bigger intensity of the electrostatic field in this area. The experimental results confirm these calculations.

Optimal modes to obtain uniform coating was chosen. Result of this process are shown in Figure 1. The powder coating in chosen mode was sprayed at the spherical piezoceramic PZT26 [21].

4. Conclusions

Results of experimental studies the distribution coating thickness are presented. The modes to obtain polymer powder coating of ultrasonic transducers with different geometric forms is proposed. Experimentally confirmed possibility of using these coatings as protective matching layer of piezoceramic transducers at frequencies of 0.5-15 MHz.

References

[1] Dastoori K, Makin B and Telford J 2001 *J. Electrostat.* **51-52** 545-551
[2] Wu S 1976 *Polym.-Plast. Technol. Eng.* **7(2)** 119-220
[3] Kleber W and Makin B 1998 *Part. Sci. Technol.* **16(1)** 43-53
[4] Boncza-Tomaszewski Z and Penczek P 2002 *Macromol. Symp.* **187** 417-425
[5] Kleber W 1993 *J. Electrostat.* **30** 393-402
[6] Prasad L K, McCinity J W and Williams R O 2016 *Int. J. Pharm.* **505** 289-302
[7] Khan M K I, Schutyser M A I, Schroen K and Boom R M 2012 *J. Food Eng.* **111** 1-5
[8] Liston E M, Martinu L and Wertheimer M R 1993 *J. Adhes. Sci. Technol.* **7(10)** 1091-1127
[9] Bailey A G 1998 *J. Electrostat.* **45** 85-120
[10] Bosshard G 1982 *Prog. Org. Coat.* **10** 205-212
[11] Weiss K D 1997 *Prog. Polym. Sci.* **22** 203-245
[12] Gavrilova V A, Kashapov N F and Kashapov R N 2011 *Biomedical Engineering* **45** 198-200
[13] Gavrilova V A, Fazlyyyakhmatov M G and Kashapov N F 2013 *Repair, Rebuild, Modernization* **12** 13-16
[14] Denisov E S, Temyanov B K, Sagdiev R K and Fazlyyyakhmatov M G 2014 *IOP Conf. Series: Mater. Sci. Eng.* **69** 012014
[15] Sagdiev R K, Denisov E S, Evdokimov Yu K, Fazlyyyakhmatov M G and Kashapov N F 2014 *IOP: Conf. Ser. Mater. Sci. Eng.* **69** 012012
[16] Gavrilova V A, Fazlyyyakhmatov M G and Kashapov N F 2013 *J. Phys.: Conf. Series* **479** 012010
[17] Gavrilova V A, Fazlyyyakhmatov M G and Kashapov N F 2014 *Russian Physics Journal* **57 (3/3)** 119-123
[18] Gavrilova V A, Fazlyyyakhmatov M G and Kashapov N F 2014 *J. Phys.: Conf. Series* **567** 012023
[19] Ye Q, Steigleder T, Scheibe A and Domnick J 2002 *J. Electrost.* **54** 189-205
[20] Ye Q. and Domnick J 2003 *Powder Technol.* **135-136** 250-260
[21] Fazlyyyakhmatov M and Kashapov N 2014 *High Temperature Material Processes* **18(4)** 273-279

Dynamics of coalescence and spreading of liquid polymeric particles during coating formation

V L Fedyaev[1,2], E R Galimov[2], N Ya Galimova[2], I R Gimranov[3], A R Siraev[1]

[1] Institute of mechanics and mechanical engineering of Kazan scientific centre of Russian academy of science, Kazan, Russian Federation
[2] Kazan national research technical university named after A.N. Tupolev, Kazan, Russian Federation
[3] Gazprom Transgaz Kazan[3], Kazan, Russian Federation

E-mail: kstu-material@mail.ru

Abstract. Processes of agglutination, coalescence and spreading of polymeric powder particles during coating formation are considered using methods of mathematical modeling. The relationships to evaluate time of particles agglutination, velocity of coalescence and spreading of material on the surface of a treated body are given. Recommendations on intensification of the given technological stages are presented.

Particles agglutination occurs in the deposition process of polymeric powder particles on a treated surface and after completion of this technological stage, in the following stage – coalescence, spreading, and finally, coating formation [1, 2]. Peculiarity of agglutination stage is that in the course of its implementation the certain connections are set between particles of the same materials in contact. In adhesion theory this case is called autohesion [3]. It is assumed that material of the particles in the contact region is not mixed; boundary surface of particles is retained.

In the next stage – coalescence stage of powder particles provided that their material is in the fluid (liquid) state the boundary surface of the particles unnoticeable. Over time, in the absence of external factors the two merging particles under the action of capillary forces form one particle close in shape to a sphere.

It should be noted that these two processes can occur, primarily, between softened particles of thermosetting materials as well as between fritted or melted thermoplastics particles.

1. Agglutination of polymeric powder particles

In order to mathematically describe the agglutination process of powder particles in an applied layer let us move from real porous medium of this layer to fictitious [4] and such that its voids are identical gaseous cavities of spherical shape uniformly distributed over the volume. It is easy to assure that in case of rhombohedral laying of spherical particles the diameter of these pores

$$d_{\Pi} = d_p (6\sin^2 \vartheta \sqrt{1+\cos\vartheta} / (\pi(1+\cos\vartheta)) - 1)^{1/3}, \qquad (1)$$

where d_p – particles diameter, ϑ – angle characterizing the peculiarity of laying of particles (balls) ($60^\circ \le \vartheta \le 90^\circ$). Approximately

$$d_{\Pi} \sim d_p (1 - 0.22\cos\vartheta). \qquad (2)$$

Accordingly, the initial porosity of the medium

$$m_0 \sim 0.5 - 0.33\cos\vartheta\,. \qquad (3)$$

We emphasize that these spherical pores are equivalent to real cavities of complex shape.

Following Frenkel Ya. I. [1] to describe the time history of equivalent radius of the spherical pore (bubble) $a = a(\tau)$ we use the equation:

$$da / d\tau = -3\text{щ}\sigma / (4\eta)\,. \qquad (4)$$

Here щ – correction factor, σ – surface tension coefficient (surface tension) at the contact boundary of the gaseous medium of the bubble and the polymeric material, *N/m*; η – Viscosity of this material, *kg/(m s)*; τ – time, *s*.

Integrating (4) provided that for a time τ parameters σ, η are close to their average values σ_0, η_0 we find:

$$a = a(\tau) = a_0 - K_0\tau\,,$$

where a_0 – initial equivalent radius of the spherical pores, *m*; $K_0 = 3\text{щ}\tau_0 / (4\eta_0)$.

Hence time of full agglutination of the polymeric powder particles (collapse of pores) will be:

$$\tau_{c\Pi} = a_0 / K_0\,.$$

If at the initial time $\tau = 0$ laying of spherical particles of the applied layer of the powder with a radius r_p is rhombohedral, then $a_0 = 0.5d_\Pi$ (see. (3)),

$$\tau_{c\Pi} \sim 1.33(1 - 0.22\cos\vartheta)\eta_0 r_p / (\text{щ}\sigma_0)\,. \qquad (5)$$

Slightly different evaluation of the agglutination dynamics of polymeric powder particles can be obtained by considering an ideal model [4] of the layer instead of the named above fictitious model. At the initial time using the Kozeny formula we define the channel radius of an ideal porous medium by the formula:

$$r_{\Pi 0} = 2m_0 r_p / (3(1 - m_0))\,, \qquad (6)$$

or by the relationship:

$$r_{\Pi 0} \approx 2.83\sqrt{k / m_0}\,, \qquad (7)$$

where k – permeability coefficient of the applied layer of the medium [4, 5].

Further, assuming that dependence (4) where the bubble radius a is replaced by the channel cross-section radius r_Π acts in a circular cross-section of the pore channel we find:

$$r_\Pi = r_\Pi(\tau) = r_{\Pi 0} - K_0\tau\,. \qquad (8)$$

According to (6), (8) time of the full agglutination of the polymeric powder particles

$$\tau_{c\Pi} = r_{\Pi 0} / K_0 = 8m_0\eta_0 r_0 / (9\text{щ}\sigma_0(1 - m_0))\,. \qquad (9)$$

Substituting dependence (3) in the equation (9), we approximately obtain:

$$\tau_{c\Pi} \sim 0.89(1 - 0.66\cos\vartheta)^2 \eta_0 r_p / (\text{щ}\sigma_0)\,. \qquad (10)$$

At an average value $\cos\vartheta = 0.25$ according to formula (5) time $\tau_{c\Pi} \sim 1.26\eta_0 r_0 / (\text{щ}\sigma_0)$ then it follows from (10) $\tau_{c\Pi} \sim 0.74\eta_0 r_p / (\text{щ}\sigma_0)$. It can be seen that usage of model of the ideal porous medium results in less time of full agglutination of particles than model of the fictitious medium. Probably, since its structure is more favorable for the agglutination of the material. In general, the time of full agglutination of powder particles can be decreased by reducing the particles size, changing their laying (decreasing angle ϑ), reducing the polymeric material viscosity, increasing the surface tension coefficient at the polymer-gas interface.

2. Coalescence and spreading of material of polymeric particles

The coalescence process of the polymeric powder particles is typical, in our opinion, for the particles the material of which is in slow-flowing state or close to it, for example, in a plastic state. In this case with respect to the processes occurring in the applied powder layers we will assume that particles coalescence is the stage that follows the agglutination when as a result of heating of material up to a certain temperature $T_T < T_*$ (T_* – limiting temperature, probably, decomposition temperature of the polymeric material) the surface tension, viscosity decrease and its fluidity increases.

Considering a layer of liquid polymeric particles on the body surface we can identify, firstly, a layer of particles in contact with the body; secondly, layer of particles in contact with the ambient gaseous medium and, finally, a plurality of particles located in the inner region of the layer. For the latter it is typical that each of them is surrounded by other particles.

It should be expected that coalescence processes of liquid particles in the selected areas of the applied layer will be slightly different from each other. The near-border particles, their coalescence will of course be under a certain influence of both the solid body and ambient gaseous medium (air).

Let us consider further as the most interesting from a practical point of view coalescence of polymeric powder particles that are located on the surface of a treated body. Considering the regularities of spreading liquid on the solid surface [6-8] we adopt the following assumptions as the initial ones. Let us assume that liquid and gas media are incompressible; process of deformation, form changing of particles, pores is slow (quasi-stationary), isothermal; pressure in material of particles p_p does not change over time ($p_p = const$) however in both opened and closed pores pressure p_g can change. Loads from side of the overlying layers of particles, mass forces are not taken into account, movement of the media in the vertical to the body surface section, passing through the mass centers of the selected neighbouring polymeric particles is plane-parallel (two-dimensional).

Fig. 1 shows the possible change over time of periodicity cell initially representing a half circle of the spherical particle cross-section with radius r_p, in case of wettable solid body surface.

Here AB – upper limit of the cell, $A'B'$ – its middle position, $h_k = h_k(\tau)$ – cell height, PQ – interface between polymeric material and gas medium of the cavity, $BQ(PP')$ – conventional border between the considered cell and cell of the neighbouring particle symmetric to it, θ – limiting wetting angle. We emphasize that the border PQ close to the arc of a circle is equivalent to the real border that has a more complex shape (fig. 2).

Thus the problem is to define the changing over time radius of the curvature border PQ $R = R(\tau)$, height $h_k = h_k(\tau)$, ordinate of point Q $y_Q = y_Q(\tau)$, distance from point P to the coordinates origin $x_P = x_P(\tau)$. Following dependencies are used to solve it. According to the Laplace's formula on the interphase boundary PQ

$$R = \sigma / \Delta p, \tag{11}$$

where $\Delta p = p_g - p_p$ – pressure difference in the gas (p_g) and material of the particles (p_p).

At the point P on the solid body surface equilibrium condition of the interface between the liquid and gaseous phase (Young's equation) is as follows:

$$\cos\theta = \Delta\sigma_s / \sigma \tag{12}$$

Here $\Delta\sigma_s = \sigma_{sg} - \sigma_{sl}$ – jump of the surface tensions on a solid body; σ_{sl}, σ_{sg} – surface tension on the interface between solid body – liquid, solid body – gas, respectively.

Due to the fact that at coalescence of adjacent particles the tangent EQ (fig. 3) must be parallel to the axis Ox center of the circle C is located on the line DQ we get:

$$\theta = \pi - \gamma, \tag{13}$$

where angle $\pi / 2 \le \gamma \le \pi$.

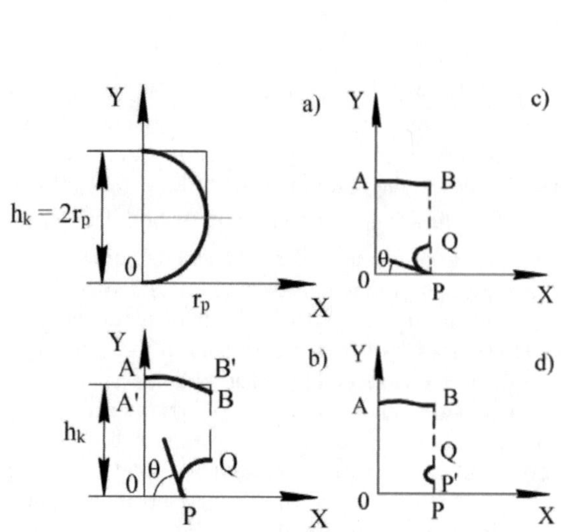

Figure 1. Dynamics of coalescence of liquid particles of polymeric powder on the wetted surface.

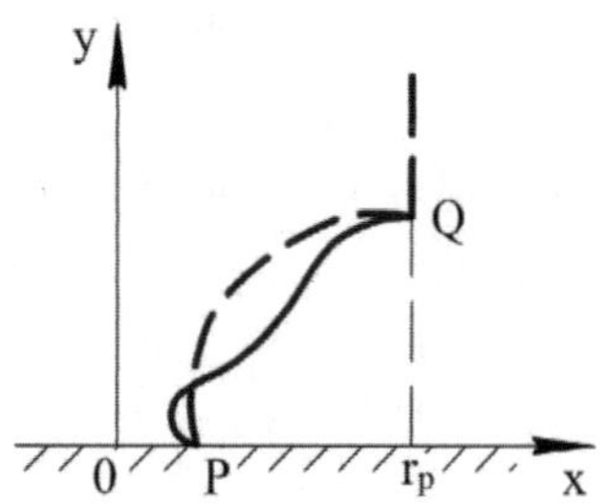

Figure 2. Approximate shape of the interface between liquid medium and gas at particles coalescence.

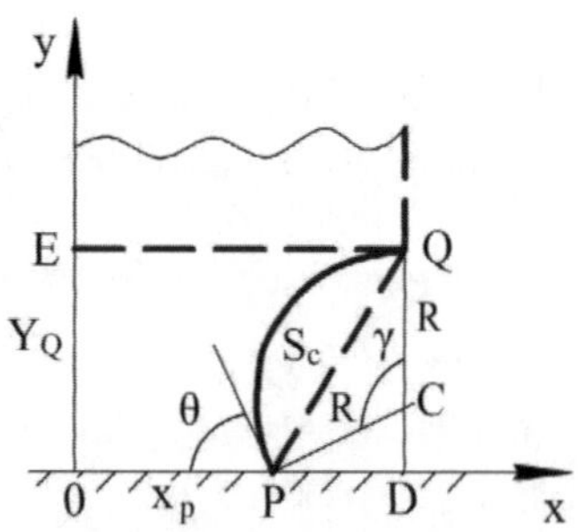

Figure 3. Part of the periodicity cell with an equivalent interface.

The condition of liquid conservation in the cross section $OPQBA$ in the process of particles coalescence is represented by the dependence:

$$S_p = S_{QB'A'E} + S_{OBQE} - S_C = 0.5\pi r_p^2 = r_p h_k - 0.5(r_p - x_p)(y_Q - R) - 0.5\gamma R^2 . \quad (14)$$

Finally, from the triangle PDC (fig. 3) it follows that

$$y_Q = R + \sqrt{R^2 - (r_p^2 - x_p^2)} . \quad (15)$$

Since it is assumed that at coalescence, shifting of the points P and Q the line PQ remains an arc limiting wetting angle θ must change over time $\theta = \theta(\tau)$, accordingly, $\zeta = \cos\theta = \zeta(\tau)$. Indeed [8] since limiting wetting angle depends on the properties of liquid, solid medium, structure of the body surface, presence of adsorbed vapors and surface-active substances on it, as well as on temperature, mass forces, velocity of movement of the interface between the liquid and gaseous media along the body, it is quite possible that when shifting the point P the value of angle θ will be changing. Considering this fact we can write:

$$\zeta = \zeta(\bar{x}_p) = (\bar{x}_p + \bar{R} - 1) / \bar{R}, \tag{16}$$

where $\bar{x}_p = x_p / r_p$, $\bar{R} = R / r_p$. Further taking into account
Ошибка! Источник ссылки не найден., we find:

$$\bar{x}_p = \bar{x}_p(\tau) = 1 - \delta_s \bar{R}, \tag{17}$$

where $\delta_s = 1 - \Delta\sigma_s / \sigma$.
In accordance with **Ошибка! Источник ссылки не найден.** the dimensionless radius of curvature

$$\bar{R} = \bar{R}(\tau) = 1/((1-\tilde{p})\bar{p}_g) \approx (1+\tilde{p})/\bar{p}_g \sim 1/\bar{p}_g. \tag{18}$$

Here $\tilde{p} = \bar{p}_p / \bar{p}_g$ – ratio of pressures p_p and p_g ($\tilde{p} \square 1$); $\bar{p}_p = r_p p_p / \sigma$, $\bar{p}_g = \bar{p}_g(\tau) = r_p p_g / \sigma$ – dimensionless pressure in the material of particles of a gas cavity, respectively.

Taking into account (17) the dimensionless coordinate of the point Q (see **Ошибка! Источник ссылки не найден.**) is given by the formula:

$$\bar{y}_Q = y_Q / r_p = \bar{R}(1 + \sqrt{1 + \delta_s(1 - 2/\bar{R})}). \tag{19}$$

If the value $\delta_s(1 - 2/\bar{R})$ is small then approximately

$$\bar{y}_Q \sim \bar{R}(2 - 0.5\delta_s(1 - 2/\bar{R})).$$

Passing in (14) to the dimensionless variables, we get the evaluation of the height of the periodicity cell:

$$\begin{aligned} \bar{h}_k = \bar{h}_k(\tau) = h_k / r_p \approx 0.5\pi + 0.5(1 - \bar{x}_p)(\bar{y}_Q - \bar{R}) + 1.2\bar{R}^2 \square \\ \sim 0.5\pi + 0.5\delta_s \bar{R}^2(1 - 0.5\delta_s(1 - 2/\bar{R})) + 1.2\bar{R}^2. \end{aligned} \tag{20}$$

In this case for the sake of simplicity, the angle γ is replaced by its average value of 2.355.
In the limit when $\bar{R} \to 0$ accordingly $\bar{y}_Q \to 0$ (see **Ошибка! Источник ссылки не найден.**), $\bar{x}_p \to 1$, height $\bar{h}_k = 0.5\pi$ and area of the rectangle $OPBA$ is equal to S_p.

It should be noted that in the proposed model of powder particles coalescence, spreading of polymeric material on the body surface the main reason of particles transformation is the change of gas pressure p_g in the pores over time. In turn the change of pressure p_g is possible in the following cases. If the pores are open and connected with an external environment then pressure p_g changes in accordance with the changes of the external pressure $p_a = p_a(\tau)$. In the case of closed (dead-end) pores the pressure p_g will change by the action of external loads on the particles trying to change the pores size when heating or cooling of powder particles. Despite the fact that we are not taking into account the direct influence of external factors on the considered particles we will assume that their influence has effect indirectly through the pressure p_g change.

For further analysis of the obtained solutions let us assume that the dimensionless gas pressure in the pores depends on time linearly:

$$\bar{p}_g = щ_0 + щ_1\tau.$$

If $\bar{p}_g$ changes over time relatively small then the radius of curvature

$$\bar{R} = \bar{R}(\tau) \sim (1 - k_p\tau)/щ_0 \ (k_p = щ_1/щ_0), \tag{21}$$

decreases over time according to the law which is close to the linear one.

Further considering that when the angle θ changes from $\pi/2$ to 0 parameter δ_s decreases from 1 to 0, we can write:

$$\delta_s = \delta_s(\tau) = 1 - \tau/\tau_p, \tag{22}$$

where τ_p – characteristic time of spreading of polymeric material on the body surface, for example, time during which the point P reaches the position D.

Taking into account dependencies (21), (22), we define the velocity of spreading of material of particles (velocity of point P movement):

$$v_P = v_P(\tau) = dx_p/d\tau = \text{ш}_p(\text{ш}_{p0} - 2\bar{\tau}). \tag{23}$$

Here $\text{ш}_p = r_p k_p/\text{ш}_0$, $\text{ш}_{p0} = 1 + 1/(\tau_p k_p)$, $\bar{\tau} = \tau/\tau_p$.

From formula (23) it follows that at the beginning of the spreading ($\bar{\tau} \sim 0$) point P velocity

$$v_P = v_{P0} = (1 + \tau_p k_p) r_p/(\tau_p \text{ш}_0).$$

Since $\text{ш}_0, \text{ш}_1 > 0$ (gas pressure increases over time) $v_{P0} > 0$. By the end of shifting of the point P it reaches the position D, when time $\bar{\tau} = \bar{\tau}_*$, velocity

$$v_P = v_{P1} = (1 + \tau_p k_p(1 - 2\bar{\tau}_*)) r_p/(\tau_p \text{ш}_0).$$

In order that point P would actually reach D at the time $\bar{\tau} = \bar{\tau}_*$, the polymeric material completely wetted the body surface, the condition $v_{P1} \geq 0$ must be satisfied. Hence it follows that

$$\tau_p k_p \leq 1/(2\bar{\tau}_* - 1). \tag{24}$$

At fixed time $\bar{\tau}_*$, initial dimensionless gas pressure ш_0 from relationship (24) we find that the parameter ш_1 characterizing the intensity of the increase of gas pressure in the cavity must satisfy the condition:

$$\text{ш}_1 \leq \text{ш}_0/(\tau_p(2\bar{\tau}_* - 1)) = r_p p_{g0}/((\sigma\tau_p)(2\bar{\tau}_* - 1)), \tag{25}$$

where p_{g0} – gas pressure at $\tau = 0$. If this condition is not satisfied the pressure in the cavity increases at a higher rate, then by the time $\tau = \tau_*$ the cavity will not close, gas bubbles are formed on the surface of a treated body. On the other hand when parameters $\text{ш}_0, \text{ш}_1$ are specified the time τ_* is defined from the equation $x_p(\tau = \tau_*) = r_p$:

$$\tau_* = 1/k_p = \text{ш}_0/\text{ш}_1. \tag{26}$$

Therefore, time of spreading of the polymeric material and hence the stages of particles coalescence can be decreased by lowering the initial gas pressure p_{g0}, increasing the intensity of increase of p_g. It is also necessary to keep in mind that according to (23) over time τ velocity of spreading v_P decreases. Slowing down the movement of the point P along the surface of the solid body is, apparently, caused by counteraction of the gaseous medium, pressure of which in the cavity of pores increases over time.

In conclusion we note that for intensification of impregnation of capillary-porous bodies by the liquid at which the movement processes of liquid media in the pores are very similar to those considered in this paper, the following techniques are used [9-11]: vacuumization of the bodies before immersion to the liquid, increasing the pressure after the loading, replacement of air that fills the pores on the gases easily soluble in liquid.

References

[1] Yakovlev A D 1987 *Powder paints* (Leningrad: Khimiya) 216

[2] Galimov E R, Zverev E V, Tukbaev E E, Galimova N Ya, Kuryntsev S V and Mukhin A M 2012 *Polymeric powder coatings for special purposes* (Kazan: Ofset servis) 164
[3] *Encyclopedia of Polymers* 1972 Editor in chief Kargin V A V.1 (Moscow: Soviet encyclopedia) 1224
[4] Polubarinova-Kochina P Ya 1977 *Theory of motion of ground waters* (Moscow: Nauka) 664
[5] Belov S V 1981 *Porous metals in mechanical engineering* (Moscow: Mashinostroenie) 247
[6] Bertenev G M and Zelenov Yu V 1976 *Course of polymer physics* (Leningrad: Khimiya) 288
[7] Stromberg A G and Semchenko D P 1973 *Physical chemistry* (Moscow: Vysshaya shkola) 480
[8] Adamson A 1979 *Physical chemistry of surfaces* (Moscow: Mir) 568
[9] Kogan V B 1977 *Theoretical foundations of unit operations of chemical engineering* (Leningrad: Khimiya) 592
[10] Kashapov L N, Kashapov N F and Kashapov R N 2013 J. Phys.: Conf. Ser. 479 012003
[11] Saifutdinov A I, Saifutdinova A A, Kashapov N F and Fadeev S A 2016 *J. Phys.: Conf. Ser.* **669** 012045

Mathematical modeling of processes occurring during deposition of sprayed particles of polymeric powder

V L Fedyaev[1,2], E R Galimov[2], N Ya Galimova[2], M S Takhaviev[3], A R Siraev[1]

[1] Institute of mechanics and mechanical engineering of Kazan scientific centre of Russian academy of science, Kazan, Russian Federation
[2] Kazan national research technical university named after A.N. Tupolev, Kazan, Russian Federation
[3] Gazprom Transgaz Kazan[3], Kazan, Russian Federation

E-mail: kstu-material@mail.ru

Abstract. The deposition of polymeric powder particles on a surface of a treated body and a layer of particles deposited previously is considered using mathematical modeling. Basic provisions of the impact theory are used. The relationships to evaluate the characteristic parameters of the processes under study are given, the results of their analysis and recommendations to improve the deposition efficiency are presented.

When applying the polymeric powder coatings their operational parameters are greatly influenced by technologies and the equipment; used consumables; shape and surface conditions of a treated body; external conditions in which the appropriate technologies are implemented; technological modes and other circumstances [1, 2]. In case of a jet spraying of polymeric powder materials during coating formation process there are three main stages: deposition of powder particles on a treated body surface, formation of a porous dispersed layer; film formation accompanied by agglutination, coalescence, sintering and spreading of particles; structuring of the material of the formed coating. In turn the following phases can be distinguished as a refinement of these stages: deposition of a polymeric powder on a substrate surface; deformation of particles under impact, action of other factors, their agglutination; coalescence, sintering of particles in the sprayed layer; degassing, gas discharge from the forming layer; wetting of the substrate surface; film formation, material curing; molecular and supramolecular structuring, material polymerization of the formed coating; coating adhesion to the solid.

Each of the named phases of the polymeric coating formation in a certain way influences on its operational parameters. In order to evaluate this influence it is necessary to analyze the phases taking into account the peculiarities of the processes preceding to these phases [3]. The present article discusses the first phase – deposition of polymeric powder particles on a treated solid body surface.

Separate powder particles, probably, clusters moving in the jet of air and flue gases firstly fall on the body surface and then on the layer of already deposited particles. Particles behavior at the moment of impact is largely dependent on the particles size, their shape, conditions of the particles material (solid, fritted, melted), its mechanical characteristics; on conditions of substrate surface, its roughness, temperature; on incidence angle of the particles, velocity of fall, impact force and etc. Extensive materials on the processes occurring at the impact of solid and liquid particles (droplets) with a barrier are given in works [4-10].

For an impact interaction of material bodies it is typical that for a short time of the impact τ_0 the velocity of colliding bodies or one of them changes considerably.

Following [7-10] when analyzing the impact of polymeric powder particles with a treated body surface and layer of previously deposited particles we distinguish the following groups of main parameters.

A. Indicators of particles approaching the surface: sizes, shape, volume concentration of particles in the gas jet approaching a treated body.
B. Impact parameters: the particles mass at impact, impact velocity, incidence angle of the particles.
C. Parameters characterizing the particles material: density, Young's modulus, Poisson's ratio, tensile and compression waves velocity, melt viscosity, surface tension.
D. Indicators of the treated body (barrier, substrate): thickness, curvature, surface roughness, material density, Young's modulus, Poisson's ratio, compression and shear waves velocity, velocity of body movement and rotation.
E. Parameters of the sprayed layer: thickness, curvature and roughness of the external surface of the layer; density, porosity, Young's modulus, Poisson's ratio of the material layer, compression and shear waves velocity, melt viscosity.

When denoting the indicators of powder particles we shall use the subscript «*p*», the sprayed layer– «*н*», the coating – «*c*», the body parameters – index «*m*». In order to distinguish the phase state of materials: solid, liquid, gaseous we shall use indexes «*s*», «*l*», «*g*», respectively.

1. Fall of powder particles on the body surface

Let us consider preliminary a unit volume of the gas jet containing a certain amount of powder particles N_p. For the sake of simplicity we assume that all particles have a shape close to a sphere, their radius r_p. When modeling the particles, in particular, there are distinguished the 50% diameter d_{p1} which can be found from condition that 50% of particles are particles with smaller diameter the remaining 50% – particles of larger diameter and the dominant diameter d_{p0} defined as the particles diameter that is characteristic for the maximum amount of particles,

$$d_{p0} = (6V_p / (\pi N_p))^{1/3},$$

where $V_p = \sum_{i=0}^{N_p} V_{pi}$, V_{pi} – volume of i-th powder particle.

Let us also introduce the volumetric particle size distribution

$$E_d = V_d / V_p,$$

where V_d – the volume of those particles whose diameter is equal to d or close to it. Hence the number of particles with diameter d that are contained in the considered unit volume will be:

$$N_d = 6V_p E_d / (\pi d^3).$$

The total number of particles $N_p = \sum_d N_d$.

Further let us assume that the treated body surface is stationary and its curvature is small. We distinguish on it a unit area S_m with outward normal n. If in the unit volume of the jet adjacent to this area all particles with diameter d move to the surface at the same velocity $\vec{v}_d$ then number of particles that have fallen on this area in a time τ

$$N_{dm} = v_{dn} N_d \tau / l,$$

where v_{dn} – the projection of the velocity vector $\vec{v}_d$ on the normal n, l – Length equal to one. Accordingly the number of impacts of all particles in a time τ will be equal to:

$$N_m = \sum_d N_{dm} .$$

We define the intensity of particles deposition on the body surface I_m as the thickness of the material of particles deposited on it per unit time. Under condition that the number of impacts of the particles on the area $S_m = l^2$ per unit time is equal to N_m / τ we find:

$$I_m = \frac{1}{l^3} \sum_d (v_{dn} V_d) .$$

In the case where the sizes of all particles are identical ($d_{p0} = d$) velocity $v_{dn} = v_n$ we obtain:

$$N_m = v_n N_p \tau / l, \; I_m = \pi d^3 N_m / (6\tau l^2) .$$

2. Evaluation of stresses in the body material with multiple impacts of sprayed particles

In addition to the above assumptions we assume that all particles are spherical, of the same size, when approaching to the body are distributed uniformly have the same velocity of fall its magnitude is relatively small (stress in the body material caused by the impact of a particle has time to significantly decrease before the fall of the next particle) all powder particles located close to the considered part body fall on it.

The powder particles falling at first on the body surface and then on the previously deposited particles exert an impact action on the barrier material. In the one-dimensional approximation an impact pressure corresponds to the pressure arising upon the collision of a semi-infinite flat body with a rigid substrate and is described by the dependence [7]:

$$p = \rho_p C_p v_n ,$$

where ρ_p – material density of the particles, C_p – sonic velocity in this material (shock-wave velocity).

Taking into account the mechanical characteristics of the barrier material we obtain:
at impact with the substrate

$$p = \rho_p C_p v_n / (1 + \rho_p C_p / (\rho_m C_m)) ,$$

at impact with the deposited layer of the particles

$$p = \rho_p C_p v_n / (1 + \rho_p C_p / (\rho_H C_H)) .$$

Here ρ_m, ρ_H ; C_m, C_H – density, sonic velocity in the substrate material and sprayed layer of particles, respectively.

According to the theory of Tayruvengadam specific intensity of the energy absorbed by the material of the treated body at multiple impacts of particles is defined as follows:

$$I_e = \alpha_e \zeta(\tau)(p / (R + R_H))^2 / (\rho_p C_p) ,$$

where α_e – proportionality coefficient, $\zeta(\tau) = 1 - \exp(-(\tau / \tau_0)^\beta)$, parameter β depends on the properties of the substrate material and amount of load, R – depth of the substrate material layer, $R_H = R_H(\tau)$ – layer thickness of the deposited particles. Under condition that $R_H / R \ll 1$, energy intensity:

$$I_e \approx \alpha_e \zeta(\tau)(1 - 2R_H(\tau) / R)(p / R)^2 / (\rho_p C_p) .$$

Since at $\tau \to 0$ $\zeta(\tau) \to 0$, and at $R_H(\tau) \to 0.5R$ function $1 - 2R_H(\tau) / R \to 0$, it should be expected that in the given range of parameter variations the energy intensity will be maximum. As a result of a number of simplifications and transformations we find time τ_* at which energy intensity I_e reaches its maximum,

$$\tau_* \approx \tau_0(-\ln(\tilde{R} / (\tilde{\tau}_1 + \tilde{R})))^{1/\beta} ,$$

where $\tilde{R} = 2R_{H1} / R$, $R_{H1} = \dot{R}_H \ (\tau = 0.5\tau_1)$, $(\dot{\ }) \equiv \partial(\ )/\partial\tau$, τ_1 – spraying time, when $R_H(\tau) = 0.5R$; $\tilde{\tau}_1 = 2\beta(\tau_1/(2\tau_0)^\beta/\tau_1$.

3. Impact of solid deformable particle with the barrier surface

Let us assume, as before, that before the impact the falling particle has a spherical shape, it is a ball. Two stages are distinguished at direct impact of a deformable ball with a surface of a solid stationary body. During the initial stage the velocity of the mass center of a particle $v_p = v$ decreases to zero, the ball deforms, all the initial kinetic energy converts into its internal potential energy. Then the ball starts to move in the opposite direction reaching the velocity u. In this case part of mechanical energy is consumed to impart the residual deformations to the material of the ball and on its heating. By introducing a parameter $k = u/v$ called coefficient of restitution, we find the average reaction force of the barrier directed along the outer normal to it,

$$P_{cp} = (1+k)mv/\tau_0 \tag{1}$$

Here $v = -v_n$, v_n – projection on the normal of the velocity of fall of the body $\vec{v}$, $m = m_p$ its mass.

In case when during the impact a particle is under the influence of external forces, for example, gravity force $\vec{F}_g = m\vec{g}$ ($\vec{g}$ – acceleration of gravity), electrostatic force $\vec{F}_e$ (if a charged particle is in an electrostatic field), force of a particle adhesion to the substrate material $\vec{F}_c$, impact force of the following particle on the previous one $\vec{F}_x$, rebounding velocity u will depend on both its deformation described by the coefficient of restitution k, and on the action of external force $\vec{F} = \vec{F}_g + \vec{F}_e + \vec{F}_c + \vec{F}_x$. Accordingly,

$$u = \bar{k}v,$$

where $\bar{k} = k + k_f$, k_f – coefficient characterizing the influence of force $\vec{F}$ ($k_f = 0$, if $\vec{F} = 0$) on velocity u. Considering this fact by analogy with (1) we can write:

$$P_{cp} + F_{cpn} = (1+\bar{k})mv/\tau_0.$$

Hence it follows that the rebounding velocity will be minimum ($\bar{k} \sim 0$) if the projection of the external force average during the impact time is

$$F_{cpn} \sim mv/\tau_0 - P_{cp}.$$

Thus the most favorable situation to decrease the particle rebounding velocity will be the case when gravity force, electrostatic force are directed along the normal to the side of the substrate; surface layer is softened (slightly melted); substrate surface is rough, adhesive (glue) layer is on it; particle following the first one hits it at the end of time period τ_0 (at the beginning of the backward motion of the first particle).

It is known that shock-waves propagate in the material of both the particle and the barrier when particle hits the barrier surface. Basic relationships on the shock-wave are the mass conservation equation and momentum conservation law. As a result of the experimental data analysis it is found that shock-wave velocity $\tilde{v}$ and velocity of the particle at impact v_p are related by [7]

$$\tilde{v} \approx v_p + \delta\Delta v.$$

Here δ – empirical parameter ($0 < \delta < 2$); Δv – particle velocity jump at the moment of impact. Hence we can obtain the dependence (shock adiabat) for pressure on the shock-wave $\tilde{p}$:

$$\tilde{p} \approx \rho_0 \tilde{v} \Delta v,$$

where ρ_0 – initial material density in which the wave propagates. If the barrier is at rest then initial velocity of the barrier particles is equal to zero and impacting particle – v_p. Upon particles contact at the moment of impact their velocities and pressures on both sides in accordance with the consistency condition will be v_*, p_*. Thus at the initial contact step the velocity jump of the falling particle is $\Delta v = \Delta v_p = v_p - v_*$ and in the barrier – $\Delta v = v_*$.

Pressure p_* (impact stress) that has arisen in the particle and the barrier in the beginning of the impact is spreads along the material of the particles, reaches its opposite surface and is reflected. When shock-wave is approaching the contact surface the velocity of the falling particle and pressure in it decrease rapidly. If the acoustic impedance of the particle characterized by the product of the density of its material on the velocity is small the particle stops the movement inward the barrier. When impedance is high (high density, velocity of the particle) it continues to penetrate deep into the barrier.

A detailed solution to the problem of the impact of spherical body with isotropic elastic barrier is presented in the work of Timoshenko S.P. [6, 8]. Let us give the final calculated ratios not going into details. Provided that the contact duration of the ball and the barrier is high according to the Hertz's law the impact force will be:

$$P = \delta_{pm} s^{3/2}. \quad (2)$$

Here $\delta_{pm} = 4\sqrt{r_p} / (3\pi(k_p + k_m))$, $k_p = (1 - v_p^2) / \pi E_p$, $k_m = (1 - v_m^2) / \pi E_m$; $E_p, E_m; v_p, v_m$ – Young's modulus, Poisson's ratio of the particle and barrier material, respectively; s – convergence of the ball and the barrier, caused by the local compression of the body at the point of contact. Assuming that the barrier is semi-infinite and stationary we find the maximum value of the deformation (the closest convergence of the particle and the barrier)

$$s_0 = (5m_p v_p^2 / 4\delta_{pm})^{2/5}.$$

Hence, according to (2), the maximum impact force

$$P_* = \delta_{pm}^{2/5} (5m_p v_p^2 / 4)^{3/5},$$

and the maximum radius of the contact area

$$a_* = \sqrt{r_p} (5m_p v_p^2 / 4\delta_{pm})^{1/5}.$$

It is known that in the considered problem the pressure distribution along the area has the form

$$p = p(r) = p_0 (1 - r^2 / a_*^2)^{1/2}, \quad (3)$$

where $p_0 = p_0(\tau)$ – surface pressure in the area center, $r = \sqrt{x^2 + y^2}$; x, y – Cartesian coordinates on an area with a center in the impact point of the particle. It is easy to show

$$p_0 = 3P / (2\pi a_*^2). \quad (4)$$

At the moment when deformation stops when $\dot{s} = 0$, the radius of the contact area $a = a_*$, force $P = P_*$, pressure

$$p_0 = p_{0*} = \frac{3\delta_{pm}}{2\pi r_p} \left(\frac{5m_p v_p^2}{4\delta_{pm}} \right)^{1/5}.$$

In order to evaluate the change of function $s(\tau)$ and also $a(\tau)$, $p(\tau)$, $P(\tau)$ depending on time it is necessary to integrate the equation that follows from Newton's second law

$$\dot{s}^2 = v_p^2 - 4\delta_{pm} s^{5/2} / (5m_p).$$

After a number of transformations, integration we get

$$\tau = \frac{s_0}{v_p} \int_0^{\bar{s}} \frac{dt}{(1 - t^{5/2})^{1/2}}, \quad (5)$$

where $\bar{s} = s / s_0$.

In particular, for $\bar{s} = 1$ from (5) it is possible to find the impact time:

$$\tau_0 = 2.94(5m_p / (4\delta_{pm} v_p^{1/2}))^{2/5}.$$

If $\bar{s} < 1$, accordingly, integration variable $t < 1$ we obtain:

$$1/(1-t^{5/2})^{1/2} \sim 1 + 0.5t^{5/2}.$$

Therefore,

$$\tau \sim (s_0 / v_p)(1 + \bar{s}^{5/2} / 7),$$

or

$$s \sim v_p \tau \quad (0 \le \tau \le \tau_0) \tag{6}$$

Substituting s from the dependence (5) into (2), we find the impact force $P = P(\tau)$, the radius of the contact area

$$a = a(\tau) = (3\pi(k_p + k_m) r_p P(\tau) / 4)^{1/3}, \tag{7}$$

By the formula(4), (3), respectively, pressure $p_0 = p_0(\tau)$ and $p = p(r, \tau)$.

Fig. 1 shows the changes of function $s(\tau)$ over time τ characterizing the convergence of the particle and the barrier, the radius of the contact area $a(\tau)$ at $r_p = 8\times10^{-8}$ m, $\rho_p = 900$ kg/m^3, $v_p = 15$ m/s, $E_p = 1{,}9\times10^8$ Pa, $\nu_p = 0{,}43$ (high pressure polyethylene); $E_m = 2\times10^{11}$ Pa, $\nu_m = 0.27$ (steel). It can be seen that in the beginning of the impact dependence $s = s(\tau)$ is close to linear whereas $a = a(\tau)$ is substantially non-linear. However at $\tau \to 0.1$ s both functions asymptotically tend to the limiting values.

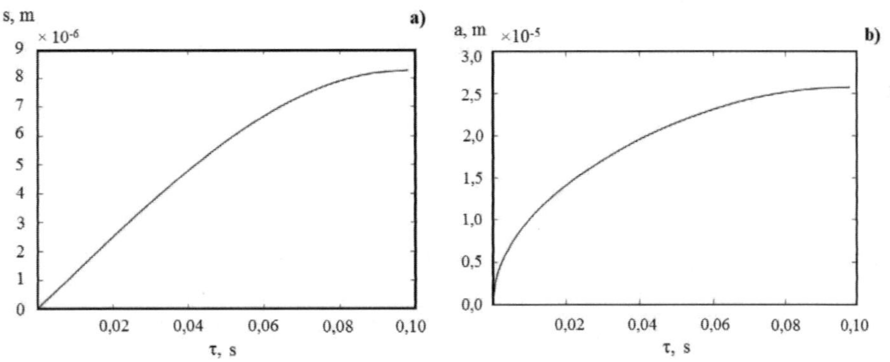

Figure 1. Convergence of the particle and the barrier (a), change of the radius of the contact area (b) over time.

It is also necessary to take into account that with a high degree of accuracy [8]

$$s = s_0 \sin(\pi\tau / \tau_0).$$

For small τ / τ_0 approximately

$$s \sim \pi v_p \tau / 2.94 \cong 1.1 v_p \tau,$$

that, in general, is consistent with (6).

It should be noted that relationships (2)-(7) written for the case of particle impact with the treated body surface can, in principle, be extended on the impact of the particles with previously deposited ones, changing parameters E_m, ν_m on Young's modulus E_H, Poisson's ratio ν_H of the sprayed layer.

Thus analytical dependencies to evaluate the intensity of particles deposition, absorption of the impact energy by the material including taking into account previously deposited powder particles on the treated body were obtained at impact interaction of solid polymeric powder particles with the barrier. Relationships to calculate the contact time of deformable particles with the barrier, the radius of the contact area, pressure distribution along it, the maximum impact force and other indicators are presented.

References

[1] Yakovlev A D 1987 *Powder paints* (Leningrad: Khimiya) 216
[2] Galimov E R 2012 *Polymeric powder coatings for special purposes* (Kazan: Ofset servis) 164
[3] Fedyaev V L, Siraev A R and Martines Markes L 2016 *Current problems of Continuum Mechanics*. (Kazan: Publishing house «FEN» AS RT) 452
[4] Targ S M 1966 *Short course of theoretical mechanics* (Moscow: Nauka) 450
[5] Landau L D and Lifshitz E M 1965 *Mechanics* (Moscow: Nauka) 204
[6] Timoshenko S P and Gud'ir Dzh N 1975 *Theory of elasticity* (Moscow: Nauka) 576
[7] *Erosion* 1982 edited by K Pris (Moscow: Mir) 464
[8] Zukas Dzh A, Nikolas T, Svift Kh F, Greshchuk L B and Kurran D R 1985 *Impact dynamics* (Moscow: Mir) 296
[9] Batuev G S, Golubkov Yu V, Efremov A K, Fedosov A A 1977 *Engineering methods, stuadies of impact processes* (Moscow: Mashinostroenie) 240
[10] Kashapov L N, Kashapov N F and Kashapov R N 2013 *J. Phys.: Conf. Ser.* **479** 012003

Investigation of ring plasmatron for thermal purifacation of the dismantled pipes of an oil assortment from asphalt-resin-paraffin sediments

Az T Gabdrakhmanov, A T Galiakbarov, A D Samigullin, Al T Gabdrakhmanov

Department of Energy and Information, Kazan Federal University, Naberezhnye Chelny, Russian Federation

E-mail: ATGabdrahmanov@kpfu.ru

Abstract. This paper proposes a method of the thermal purification and removal paraffin, without using of mechanical purification of the oil wells and the oil pipeline.

1. Introduction

It's a new question about cost-effective and concurrently a reasonable the raw materials extraction in Russian conditions, there is a necessity of oil transportation from places of extraction to places of consumption. In these conditions the most effective way is the fluid-transmission system. In the operation process of the oil pipeline on the internal surface of the pipes there is an accumulation of asphalt-resin-paraffin sediments , which essentially influence the effectiveness and reliability of their work (reduction of bandwidth, increasing the pressure, absence of reliable diagnostic data). To prevent the catastrophic consequences of pipeline blockage (discontinuities or cracks) should be timely carry out repair and maintenance work. Before carrying out the complex repair works, need to purify the internal surface of the pipes from sediments, which formed on it during in the operation. The task of combating asphalt-resin-paraffin sediments of the pipes remains one of the most important for the industry.

The technology for the thermal purification of the dismantled pipes of an oil assortment from asphalt-resin-paraffin sediments consists by using concentrated annular plasma, which allows uniformly heat the pipe from with sides at high speed. The ring plasmatron, moving by means of clamping wheels and mini gear motor with frequency converter, heats the cleaned pipe with asphalt-resin-paraffin sediments from the outside, thereby melting the asphalt-resin-paraffin sediments due to high thermal conductivity of the tube material. The pipe with asphalt-resin-paraffin sediments is set on the operating position at an angle of 20°C to the horizontal surface. The bulk of asphalt-resin-paraffin sediments are not melted in the center of the heated area of the cleaned pipe, due to low thermal conductivity of asphalt-resin-paraffin sediments. The gradual heating of the cleaning tube, starting from the bottom, causes a progressive melting of the wall layer of asphalt-resin-paraffin sediments along the cleaned pipes, which is provided free of exit gases and the melting products of asphalt-resin-paraffin sediments, without causing explosions during gas formation and thermal expansion of the wall layer of asphalt-resin-paraffin sediments in a closed space. The melted wall layer of asphalt-resin-paraffin sediments capable act as a lubricant at the exit from the cleaning pipe melted solid mass by force of gravity.

2. Experimental studies

The plasmatron of a special design was designed and developed for creation of ring plasma [1]. The plasmatron comprises ring electrodes, a frame, chamber of the working gas and a device for controlling the interelectrode gap. The electric arc under the influence of electromagnetic forces runs through the electrodes with great speed, at the same time heating the mixture of working [2-3].

In the course of experimental studies on the ring plasmatron were obtained following the energy and thermal characteristics (Figure1):

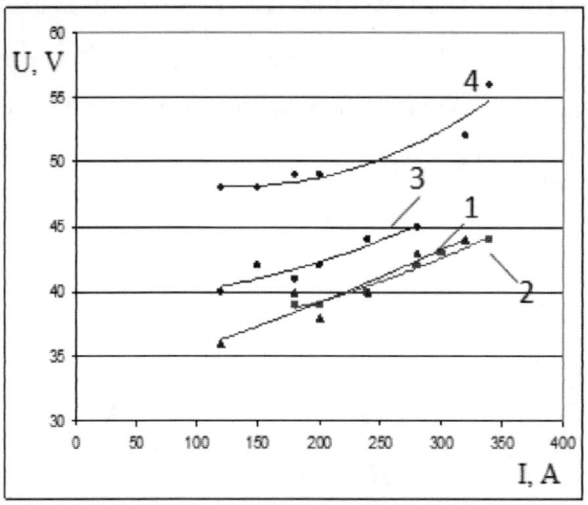

a)

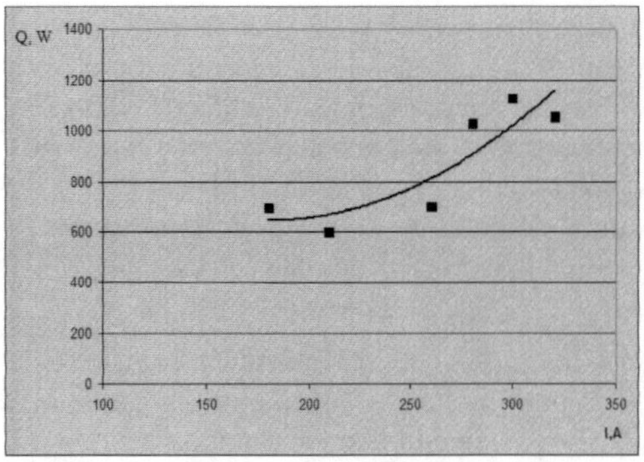

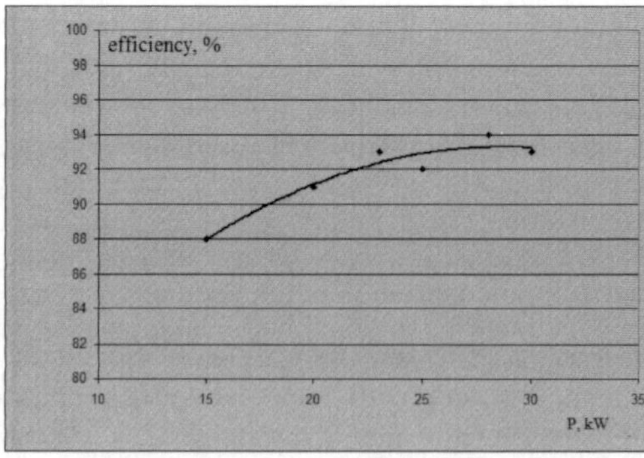

b) c)

Figure 1. Energy and thermal characteristics: a) current-voltage characteristics of the plasmatron for different diameters of electrodes: 1– d_E=4mm, 2– d_E=6mm, 3– d_E=12mm, 4– d_E=14mm (d_E - diameter electrodes), b) the dependence of the heat losses in the electrodes and the coolant on the current strength, c) the dependence of the efficiency on power.

For current of 150-300A and voltages of 100-150V current-voltage characteristic increasing [4]. This is the result of compression the arc of magnetic fields and limitations the arc of electrodes. The arc burns in the smallest interval to increase the arc length necessary to increase the voltage, but it is limited by the power supply voltage, which cannot be increased. In the course the experiments high-speed photography showed that a moving electrical charge is drawn into the interelectrode space and the length of the arc is the magnitude of the interelectrode gap.

The diagram (Figure 1 b) shows that with increasing current, the losses increase. This is connected with increase of the current density in the reference arc spots on the electrodes.

The diagram of relationship the efficiency of the injected power (Figure 1c) shows that the curve has a distinct maximum efficiency, which corresponds to the optimal operating mode of plasmatron of these electrical and geometrical parameters. The ring plasmastron developed by with inter-electrode gap of 5 mm, with power of about 28 kW has an efficiency of 94%.

The plasmatron produces a large volume of plasma in the form of a ring, which eliminates the need to rotate a part because the part immediately processes the uniformly on all sides.

The design of the plasmatron is envisaged the regulation of the interelectrode gap, which changes the power, pumped into the electric arc, and thereby regulates the temperature of the plasma. The essential brightness of the plasma temperature for heating the pipe obtains by regulating of the plasma gas flow [5-7].

3. The conclusion

The results of the experimental researches show the stability of the characteristics of the plasma technological complex so that get a steady the plasma flow [8-14] and carry out purification of the pipe from asphalt-resin-paraffin sediments.

An analysis of the characteristics shows that the optimum values of control limits of electrical parameters to ensure the effectiveness of plasma technological complex are: a current intensity 150...300 A, the voltage of 100...150 V, the total plasma gas flow of 1...1.5 g/s, the interelectrode gap 3...5 mm.

The result of this work is the increase of the performance heat pipes by replacing a linear plasmatron on the ring, and also the reduction of energy consumption by stabilizing the specified energy characteristics of plasma technological complex and as a consequence quality purification of pipes from asphalt-resin-paraffin sediments.

References

[1] Israfilov I.H., Galiakbarov A.T., Israfilov. Z.H., Israfilov D.I. *Plasmatron* // Patent № 2363119. Application number 2006123860 dated July 3, 2006

[2] Gabdrakhmanov Az.T., Israfilov I.H., Galiakbarov A.T., Samigullin A.D. *Calculation of the parameters of the annular plasma torch using dimensionless criteria of similarity* // Collection of Materials of interregional scientific-practical conference «III Kamskie read" part 3 - Nab. Chelny: INEKA, 2011.- p. 42-44.

[3] Zhukov M.F. *The generation of a low-temperature the plasma and the plasma technologies*: Problems and prospects. -Novosibirsk, 2004. p. 246-253.

[4] Gabdrakhmanov Az.T. *Features of the current-voltage characteristics of pulsed plasma generator* / Online electronic scientific and technical journal "Socio-economic and technical systems." №2 (59) - Nab. Chelny: INEKA, 2011.

[5] Israfilov I.H., Saubanov R.R., Rakhimov R.R. *Prospective application of highly concentrated energy for the surface heat treatment products* // Socio-economic and technical systems: research, design, optimization. № 1, 2011, p. 25-30.

[6] Israfilov I.H., Galiakbarov A.T., Gabdrakhmanov Az.T., Nugumanova A.I. *The Improving of the efficiency the plasma processing plant for heat treatment of metals* // Interuniversity scientific book "Designing and research of technical systems", Naberezhnye Chelny, 2008, Issue 12, s.101-105.

[7] Zvezdin V.V., Galiakbarov A.T., Nugumanova A.I., Gabdrakhmanov Az.T., Saubanov R.R.

The investigation of the influence the parameters the pulsed plasma generator on quality process // Bulletin KSTU. AN Tupolev. Number 2, 2010, s.50-52.
[8] Denisov D.G., Kashapov N.F., Kashapov R.N. *The appearance of shock waves in the plasma electrolytic processing* // Iop conference series: materials science and engineering – 2015. – №012005.
[9] N.F. Kashapov, A.G. Luchkin. *The use of low-temperature plasma deposition of hardening coatings on plastics* // Proceedings of the higher educational institutions. Physics. 2014. T. 57. № 3-3. p. 160-163.
[10] N.F. Kashapov, R.N. Kashapov. *A study of plasma-electrolytic process for different ratios of the anode space to the cathode* // Proceedings of the higher educational institutions. Physics. 2014. T. 57. № 3-3. Pp 168-170.
[11] Kashapov L N, Kashapov N F and Kashapov R N 2013 *J. Phys.: Conf. Ser.* **479** 012011
[12] Kashapov L N, Kashapov N F and Kashapov R N 2014 *J. Phys.: Conf. Ser.* **567** 012025
[13] Zaripov R G, Kashapov N F, Tkachenko L A and Shaydullin L R 2016 *J. Phys.: Conf. Ser.* **669** 012053
[14] Denisov D G, Kashapov N F and Kashapov R N 2015 *IOP Conference Series: Materials Science and Engineering* **86** 012005

Preparation of metal surfaces for application of functional coatings

Az T Gabdrakhmanov, I H Israphilov, A T Galiakbarov, Al T Gabdrakhmanov

Department of Energy and Information, Kazan Federal University, Naberezhnye Chelny, Russian Federation

E-mail: ATGabdrahmanov@kpfu.ru

Abstract. This paper proposes an efficient method of plasma purification of metal surfaces before applying coatings or the product finishing treatment. The plasma purification is performed in a steam-gas discharge with an aluminum anode and a liquid cathode.

1. Introduction

During the production process (manufacturing, processing, storage) on the surface of metal parts appear contaminations or foreign substances, whose presence is undesirable or harmful.

The purification of metal surfaces is a process of removing contaminations from the metal surface to a certain level of purity. The purification is produced by various methods - mechanical, physical, chemical, physicochemical and thermochemical.

Mechanical purification of removing contaminations occurs due to their mechanical destruction by rubbing, scraping, milling, under the impact of water jet, solid particles (cast-iron and grit, glass spheres, etc.). The increase in productivity of mechanical purification is achieved by use of mechanized tools (brushes, milling cutters) with electric or pneumatic actuator by increasing of pressure jets to 5-63 MPa. The advantages of the process of mechanical purification are low energy capacity, the versatility, the removing of various contaminations, the simplicity of utilization, the disadvantage is the application of manual labor.

Mechanical physical purification consists in dissolving contaminations in various solvents and its removal from the surface of cleaned product. The intensification of the purification process is achieved by introducing into the zone of purification of ultrasonic oscillations, and also the application of a jet stream dousing and solvent vapors. The advantages of physical purification method are a large purification speed and its high quality, non-waste production, the possibility of the mechanization and the automation of purification processes. At the same time, this method's disadvantage involves the fact that occupational hazard, a difficult of waste disposal, the application for a small group of contaminations.

Physicochemical purification method consists in dissolving, emulsification, and chemical destruction of contaminations (application of a dissolving-emulsifying means with rinsing in solutions of synthetic detergents). The possibility of increasing the speed and quality of purification consists the movement (vibration, rotation) of the object which cleaning during the purification process. The positive aspect of the physicochemical method is high speed of purification and its high quality, low energy capacity of the purification process, moderate temperature of the authority process (20-500C), the possibility of the mechanization and the automation of the process, and negative is the application for a small group of contaminations, occupational hazard and the presence of waste products.

Thermochemical method consists of chemical destruction (combustion) of contaminations in a flame or in an alkaline melt at high temperature (400-4500C), and also in volumetric and structural changes of contaminations. The improving of purification productivity is possible by optimizing the composition of alkaline melt and the automation of the process. The advantages of this method is high speed of purification and its high quality, and also the ability to automate the process, disadvantages - the application for a small group of contaminations, high energy capacity of the purification process, the deformation and the failure of parts.

2. Experimental studies

This paper proposes an efficient and non-polluting method of the purification of metal surfaces in the steam-gas discharge [1-3]. The discharge area is located in the air space between the liquid electrolyte and the aluminum anode [4]. The plasma stream expires from the discharge area in the surrounding atmosphere. The electrolyte stream circulates inside the cathode unit, thus, creating favorable conditions to reduce of heat losses at the cathode. The cleaning detail used as the cathode.

The structures and the current-voltage characteristics of the discharge between the aluminum anode and the electrolytic cathode under atmospheric pressure at different inter-electrode distances were experimentally studied (Figure 1).

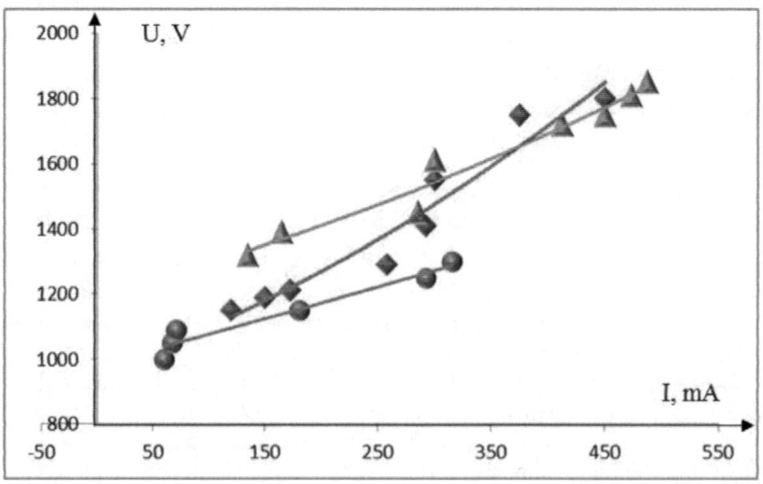

Figure 1. The current-voltage characteristics of the discharge electrode, diameter d = 4.5 mm, the liquid electrod- distilled water; ● - L (electrode gap) of 2 mm; ♦ -L = 4 mm; ▲ - L = 6 mm;

The pictures of the discharge at different regimes were obtained during the experimental research (Figure 2). Multichannel discharge between the aluminum anode and electrolytic cathode has the shape of a truncated cone under atmospheric pressure with sharply defined borders, where threadlike lines chaotically move [5]. With growth of interelectrode distance these lines break up into separate, clearly delineated channels. The lower, larger base is the cathode spot and the top-anodic. As a result of influence of the discharge, the electrolyte evaporates. Its vapors enter the positive pole through the cathode spot. The strength of the vapor several distorts the cathode spot, so the liquid surface is not flat. Its convexity faces the depth of the electrolyte.

a) b) c)

Figure 2. Pictures of the discharge, diameter of the electrode d=4.5 mm, the liquid electrode - distilled water; a) U=1150B, I=120mA; b) U=1210B, I=172 mA; c) U=1800V, I=450 mA;

The samples before treatment and after treatment were presented for evaluation the effectiveness of the purification detail (Figure 3) [6].

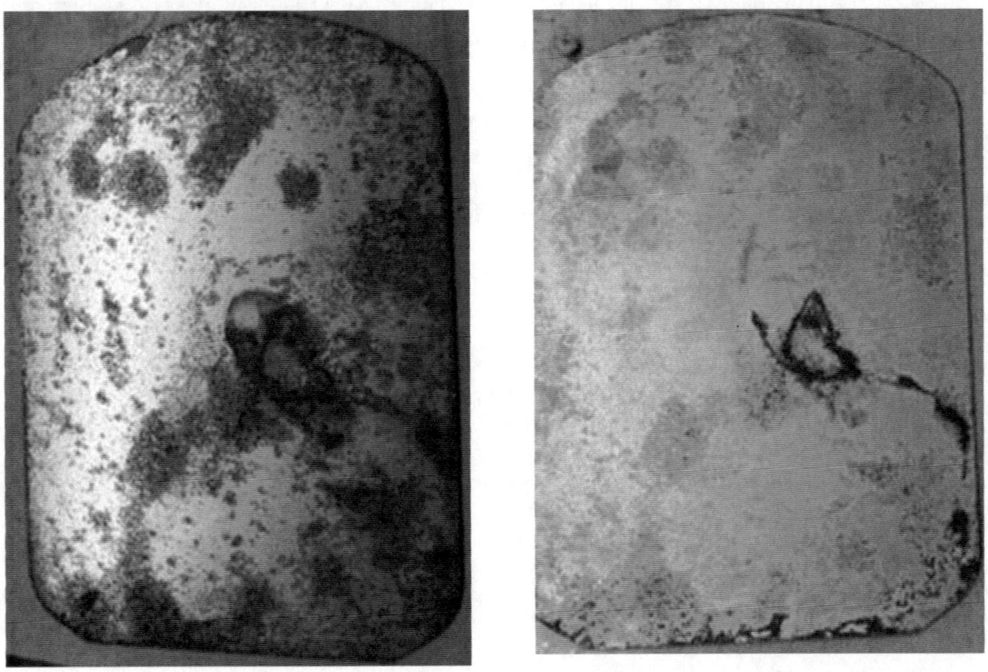

a) b)

Figure 3. The samples (st20, the thickness of 0.5mm 100x150 mm) a) before treatment; b) after treatment, diameter of the electrode d=4.5 mm, the liquid electrode - distilled water; L=4 mm, U=1200V, I=150mA;

3. The conclusion

The results of these studies show that the proposed method provides high-performance and high-quality purification of the surfaces from any contaminations – mineral and organic greases, rust, scale, and remnants of the old galvanic and lacquer coating, enamel insulation with an electric wires[7-12].

References

[1] Hafiz A.A., Valiev R.I., Shakirov Y.I. *Optimize cleaning and deburring process with the surface of the plasma products electrothermal installation with a liquid cathode* // Proceedings of the international scientific conference: (XVIII Benardosovskie reading) "Status and prospects of development of electrotechnology" .2015. -Izdatelstvo: Ivanovo State Power Engineering University named after V.I. Lenin. p. 254-256.

[2] Denisov D G, Kashapov N F and Kashapov R N 2015 *IOP Conference Series: Materials Science and Engineering* **86** 012005

[3] Denisov D.G., Kashapov L.N, Kashapov N.F. Kashapov R.N. *The use of gas discharge with an electrolytic cathode for cleaning products selective laser melting* // Materials of the II International Conference. - Moscow: All-Russian Scientific Research Institute of Aviation Materials, 2016. - p. 10.

[4] Gabdrakhmanov Az.T., Gabdrakhmanov Al.T., Galiakbarov A.T. *The study discharge characteristics of the combined-cycle with an aluminum anode and liquid cathode* // Online electronic scientific and technical journal "Socio-economic and technical systems: research, design, optimization." №3 (70) - Nab. Chelny: KFU, 2016. - p. 35-41.

[5] Tazmeev G.H., Timerkaev B.A., Tazmeev H.K., Arslanov I.M. The research gas discharge with liquid electrolyte cathode near its critical thermal conditions // Journal "Applied Physics", 2016, №1. - p.72-76.

[6] Plasma cleaning method of surface metal / Israfilov I.H., Rakhimov R.R., Saubanov R.R. - Socio-economic and technical systems: research, design, optimization. 2010. T. 56. № 3. p. 1-6.

[7] N.F. Kashapov, A.G. Luchkin. *The use of low-temperature plasma deposition of hardening coatings on plastics* // Proceedings of the higher educational institutions. Physics. 2014. T. 57. № 3-3. p. 160-163.

[8] Luchkin A G, Kashapov N F and Luchkin G S 2013 *J. Phys.: Conf. Series* **479** 012019

[9] N.F. Kashapov, R.N. Kashapov. *A study of plasma-electrolytic process for different ratios of the anode space to the cathode* // Proceedings of the higher educational institutions. Physics. 2014. T. 57. № 3-3. Pp 168-170.

[10] Israfilov I.H., Galiakbarov A.T., Gabdrakhmanov Az.T., Nugumanova A.I. *Thermophysical characteristics process of the pulsed plasma treatment steels* // Izvestiya of the Tula State University. Vol. 6. Part 1. - TSU Publishing House, 2013. - S.253-260.

[11] Kashapov L N, Kashapov N F and Kashapov R N 2014 *J. Phys.: Conf. Ser.* **567** 012025

[12] Zvezdin V.V., Galiakbarov A.T., Nugumanova A.I., Gabdrakhmanov Az.T., Saubanov R.R. *The investigation of the influence the parameters the pulsed plasma generator on quality process* // Bulletin KSTU. AN Tupolev. Number 2, 2010, s.50-52.

Surface thermohardening by the fast-moving electric arch

Az T Gabdrakhmanov[1], L N Shafigullin[1], E R Galimov[2], A R Ibragimov[1]

Department of Energy and Information, Kazan Federal University, Naberezhnye Chelny, Russian Federation[1]

Kazan National Research Technical University named after A.N.Tupolev - KAI (KNRTU-KAI)[2]

E-mail:ATGabdrahmanov@kpfu.ru

Abstract. This paper describes the technology of modern engineering- plasma hardening steels and prospects of its application. It gives the opportunity to manage the process without using of cooling media, vacuum, special coatings to improve the absorptive capacity of hardened surfaces; the simplicity, the low cost, the maneuverability, a small size of the process equipment; a possibility of the automation and the robotization of technological process.

1. Introduction

The majority of details and units of machines from bearing to the cutting tool- works in friction conditions, therefore, are subject to wear out.

The statement is substantially fair that the quality of machines and constructions depend on the surface layer of the material from which they are made. The durability of the machine operation depends from how wear out friction surfaces as arise and develop cracks, especially under alternating loads, i.e. the durability of parts without experiencing contact loads also depends on the quality of the surface detail layer. Therefore, methods of surface hardening are widely used in mechanical engineering.

An effective method of surface hardening of iron-carbon alloys is the treatment of concentrated streams of energy (CSE) – laser, electron, plasma are exposed to the high-speed heating of the surface metal layer and its rapid cooling as a result of the heat transfer to deep layers the material of detail [2-6]. This causes a change in the structure of the surface layer due to the phase transformation – hardening to martensite.

Among the methods of surface treatment using highly concentrated sources of energy hardening by plasma heating has a special place as the most simple and technological operation [2-12]. It is shown that hardening with the plasma heating of the surface of carbon steels leads to increase their abrasion resistance and wear resistance during the friction of metal on metal.

The purpose of plasma hardening is making of details and tools with hardened surface layer a thickness from several millimeters at constant overall chemical composition of the material and saving in the internal layers of the initial properties of the original metal.

IOP Conf. Series: Journal of Physics: Conf. Series **789** (2017) 012010 doi:10.1088/1742-6596/789/1/012010

It is necessary in some cases to harden large flat or cylindrical surfaces with high productivity inaccessible for ordinary plasma installation.

2. Experimental studies

This paper proposes an effective way of thermo-hardening of large surfaces by using fast-moving an electric arc. At this treatment one of the electrodes is a rod electrode made of different materials (steel, copper, etc.), another electrode - processed part. The speed of the arc at such treatment can achieve V=150 m/s, the current is regulated in the range of 100-1500.

When large surfaces are treated with fast-moving electric arc important characteristic is the overlap ratio K_n defined by the ratio $K_n = S/d_n$ (d_n - width of the arc spot). Strips can be imposed with an overlap (when K_n<1) and without overlap (when K_n> 1). In many instances, hardened strips are located at some distance from each other so that laser-treated area was 20...50% of the total area. This allows you to achieve the greatest durability.

The treatment of flat surfaces is carried according to two schemes the imposition of strips: linear (Figure 1, a) and cellular (Figure 1, b). After imposing each strip the detail or the electrode is periodically moved in the transverse direction at a distance equal to the step S. When using the cellular scheme by having strips is to cast at right angles to one another, and with the same pitch along both coordinates, there no difficulties in the case of strips imposing at a right angle and to harden when $S_x \neq S_y$. The treatment of cylindrical surfaces is performed according to the scheme of imposing strips along the generator of an axis with periodic rotation of the detail for displacement to step S (Figure 1, c). For the same scheme produced the hardening of internal cylindrical surfaces.

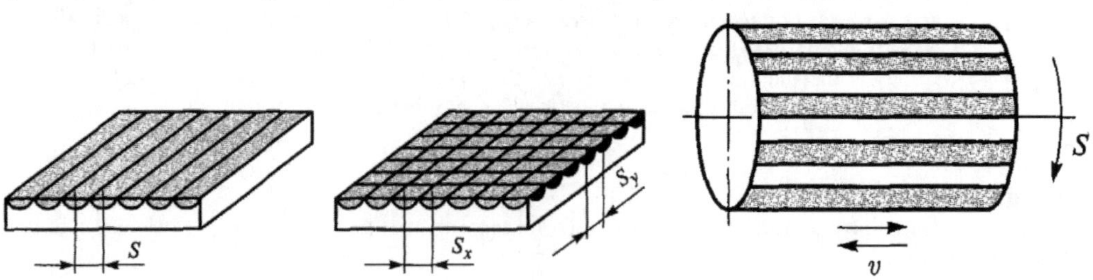

Figure 1.Schemes of imposing strips on different surfaces

The velocity of the arc is one of the main parameters which have a significant influence on the interaction of the arc to the electrodes and consequently on the process of hardening. Depending on the value of the arc speed is implemented one of the motion mechanisms: continuous or intermittent moving with different steps. Traces making by the moving arc on the surface of materials were conducted to determine the nature of the arc movement.

3. Results of experimental investigation

Experimental studies for steel electrodes showed, that depending on the polarity of the electrodes and their geometric sizes, the interelectrode gap and the arc combustion mode, the traces on the hardening detail essentially differ (Figure 2). For example, at a low speed of moving arc to 10 m/s traces are a continuous process (Figure 2, a). For relatively high currents (≥150A) and velocities (≥20m/s) (Figure 2, b) traces on the cathode represent a plurality of lines, approximately parallel to the direction of the arc movement. The number of lines are increasing when the current grow (Figure 2v). It is obvious that at the same time there are a large number of spots which have high mobility and move with the arc, their number are increasing when the speed and current increase. When moving many spots are running along the traces of other spots.

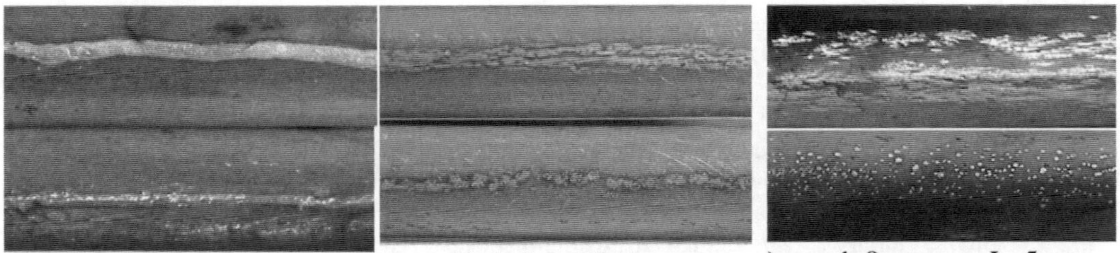

a) d=8mm, L=8mm, I=120A, U=46V, V=7,5m/s

b) d=8mm, L=4mm, I=230A, U=36V, V=25m/s;

c) d=8mm, L=5mm, I=1500A,U=70V,V=130m/s;

Figure 2.Pictures of the traces on the hardening surface of detail (at the top the detail is connected to "-" of power supply, from the bottom to ("+"), d – diameter of the rod-shaped electrode, L - the interelectrode gap.

It was investigated the thermal influence of moving arc on the internal structure of the material details, for this metallographic sections were prepared and the microstructure were investigated (Figure 3).

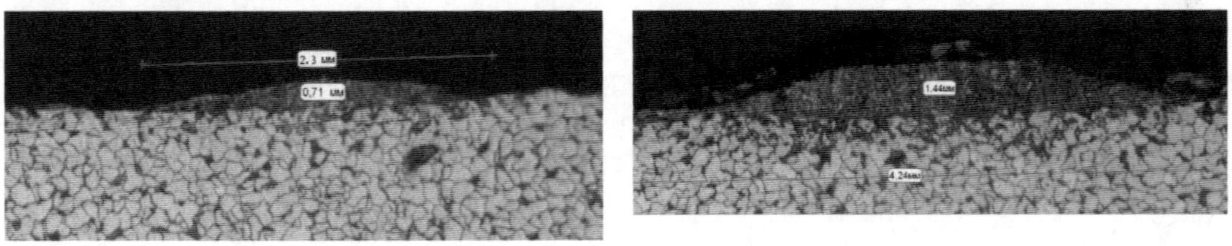

Figure 3.The microstructure of the hardened zone, material st20

Under the impact of moving arc the metal heat to various temperatures, so it has a stratified structure (the zone reflow, the zone hardening from the solid phase, the zone transition).

4. The conclusion

The results of the performed research show that the proposed method provides high performance and qualitative of surface hardening of detail with a depth of the zone hardening of 1.5 mm and widths of 5mm.

Thus, the plasma hardening of working surfaces parts of machines and tools is a progressive, accessible and efficient process of the local surface hardening, to reduce their deterioration, to increase the durability and overhaul life, and also to get real saving by increasing the efficiency and reducing the volume of purchases of new products.

References

[1] A.A. Berdnikov, G.V. Alisova, M.A. Filippov, Vas.V. Zapariy.*Plasma hardening steels - XXI engineering technology//* Materials XI All-Russian scientific conference "Bakuninskaya reading." Volume 2 - Yekaterinburg: UMC-UPI, 2013. - p.21-25.

[2] Kashapov N.F., Sharifullin S.N.2015 *IOP Conference Series: Materials Science and Engineering* **86** 012021

[3] R.M. Khisamutdinov,V.V. Zvezdin, Ruz.R. Saubanov, I.H. Israfilov, R.R. Rakhimov, A.A. Spirin.*Study of processes of steels surfaces modification with highly concentrated energy flows* / IOP Conf. Ser.: Mater. Sci. Eng., Volume 669, Issue 1, 2016, Article number 012024.

[4] Ruz.R. Saubanov, V.V. Zvezdin, I.H. Israfilov, I.I. Haybullin, R.R. Rakhimov. *Synthesis of oxidic powder in nonequilibrium low-temperature plasma with increase of indicators of*

quality of process // IOP Conf. Ser.: Mater. Sci. Eng., Volume 567, Issue 1, 2014, Articlenumber 012034.

[5] Az.T. Gabdrakhmanov, I.H. Israphilov, A.T. Galiakbarov, A.D. Samigullin, Al.T. Gabdrakhmanov.*Improving the efficiency of plasma heat treatment of metals*//Journal of Physics: Conference Series, Volume 669, 2016, Article number 012014

[6] Az.T. Gabdrakhmanov, I.H. Israphilov, A.T. Galiakbarov, D.A. Bashmakov, A.D. Samigullin *Pulse plasma surface thermostrengthening of machine parts*//IOP Conf. Ser.: Mater. Sci. Eng., Volume 69, Issue 1, 2014, Articlenumber 012037

[7] Gabdrakhmanov Az.T., Israfilov I.H., Galiakbarov A.T., Samigullin A.D.*The thermal characteristics of the process of pulsed plasma processing of steel* // News of the Tula State University. Vol. 6. Part 1. - Publishing House of the Tula State University, 2013. - S.253-260.

[8] Denisov D G, Kashapov N F and Kashapov R N 2015 *IOP Conference Series: Materials Science and Engineering* **86** 012005

[9] N.F. Kashapov, A.G. Luchkin.*The use of low-temperature plasma deposition of hardening coatings on plastics* // Proceedings of the higher educational institutions. Physics. 2014. T. 57. № 3-3. p. 160-163.

[10] Zaripov R G, Kashapov N F, Tkachenko L A and Shaydullin L R 2016 *J. Phys.: Conf. Ser.* **669** 012053

[11] Kashapov L N, Kashapov N F and Kashapov R N 2014 *J. Phys.: Conf. Ser.* **567** 012025

[12] N.F. Kashapov, R.N. Kashapov.*A study of plasma-electrolytic process for different ratios of the anode space to the cathode*// Proceedings of the higher educational institutions. Physics. 2014. T. 57. № 3-3. Pp 168-170.

Purification the surface of detail from biological contaminations

Az T Gabdrakhmanov, I H Israphilov, A T Galiakbarov, Al T Gabdrakhmanov

Department of Energy and Information, Kazan Federal University, Naberezhnye Chelny, Russian Federation

E-mail: ATGabdrahmanov@kpfu.ru

Abstract. More than 70% of biodegradation occur due to the corrosion processes. A biological corrosion causes the greatest damage to the oil and gas-production industry, the Navy and pipelines, constructions of water supply, means of communication. This paper proposes an effective method of purification various surfaces from biological contaminations by using of cold plasma.

1. Introduction

The biological corrosion - is the process of destruction of metals under the influence of living organisms (seaweed, bacteria, yeast, fungi) [1-4]. Microorganisms are located in the water environment and the soil may cause the serious corrosive damage.

The biological corrosion can exist as a separate independent type. But often the process of destruction occurs in parallel with soil, marine, atmospheric and other types of corrosion in water solutions and non-electrolytes. Pipelines, the subway, reservoirs, piles and underground pipelines and other constructions are most exposed to biological corrosion. Gas pipelines and pipelines of the oil industry are destroyed in seawater from biological corrosion.

Processes of biocorrosion begin with the appearance on the metal surface of small pits and irregularities, which are often filled by microorganisms and their metabolic products. Usually bacteria provoke the formation of ulcerous or pitting corrosion. By the mechanism action of the biological destruction is classified a number of types: electrochemical processes of destruction; chemical corrosion destructions; direct destruction under the influence microorganisms; - complex destructions (the influence of microorganisms, environmental conditions and other factors).

Microorganisms use the metal as a food medium or they secrete products, which are destroying it. In process life of microorganisms create the alkali, mineral and organic acids, peroxides, increasing the aggressiveness of the external corrosive environment. For example, some fungi, which are present in aviation fuel provoke of the process of biological corrosion of aluminum cisterns of air transport. Thiobacteria produce an acid, resulting in reduced the acidity of the soil, thereby changing of corrosion formation process.

By the nature of the biological factors the corrosion are divided into two types: bacterial; mycological.

Microorganisms are two types:

- aerobic. In the presence of oxygen perform their activities.

- anaerobic. Their existence extends by without the presence of oxygen.

Iron bacteria and sulfur bacteria, living in the soil, are the most dangerous for metals. And Aerobic and anaerobic bacteria have a common environment.

The environment of life microorganisms are: water; air; the soil; organic medium. Bacterial corrosion is more dangerous and more destructive. Bacteria are perfectly adapted to the changing existence conditions, multiply rapidly at temperatures between 5 to 40 degrees. The process of biodamage has a complicated mechanism. Live microorganisms can destroy the surface and also to stimulate the emergence and course of corrosion.

The biological corrosion for example, also includes the corrosion the taking of hands. On human skin are bacteria and sweat. Some metals have a low resistance to these factors. The biological corrosion causes the greatest damage to the oil and gas-production industry, the Navy and pipelines, constructions of water supply, means of communication [1, 2]. Over 70% of corrosion process becomes a biodegradation. The most frequently microorganisms provoke destructions on aluminum, steel, zinc, magnesium. Copper, lead, nickel are very resistant to biological degradation.

The main and effective method to protect of constructions from biological corrosion is a surface treatment product by antibacterial agents (chlorine, formalin, etc.) [1-4]. It is often bactericidal substances include in structure of paintwork materials and other coatings. But this method of protection is quite expensive and harmful for human and for the environment.

2. Experimental studies

The paper proposes a method of protection from biocorrosion by using cold plasma and device for its implementation. During the treatment by cold plasma creates a wide spectrum of environmentally sound particles (free radicals O and OH, ozone, nitrogen of oxides, ultraviolet radiation, etc.), which destroy the biological harmful contaminants – various microorganisms and chemical toxicants.

For purification various surfaces, a cold plasma generator was developed [5]. In this plasma generator occurs the ionization of the working gas and air at atmospheric pressure by using low-power pulsed arcs. The supply voltage of 3-10 kV with a frequency of 20kHz, the power consumption of up to 100 watts. The plasma temperature at the exit does not exceed $+45^0$C.

Test objects, infected with the bacterial suspension are subjected of the treatment. The bacterium Escherichia coli and Staphylococcus aureus were used as the samples. The results of treatment samples by cold plasma are presented in figures 1-2.

The results of these studies show that the cold plasma is an effective means for the inactivation of microorganisms and it destroys, depending on duration of influence (figures 1-2), to 99% of bacteria.

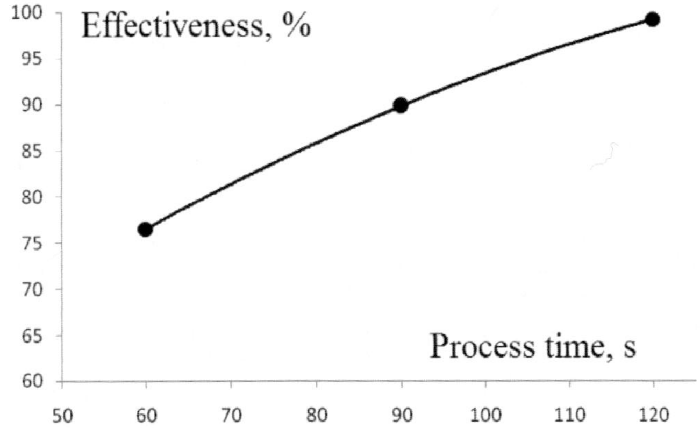

Figure 1. The diagram of relationship the efficiency of the treatment from duration of influence.

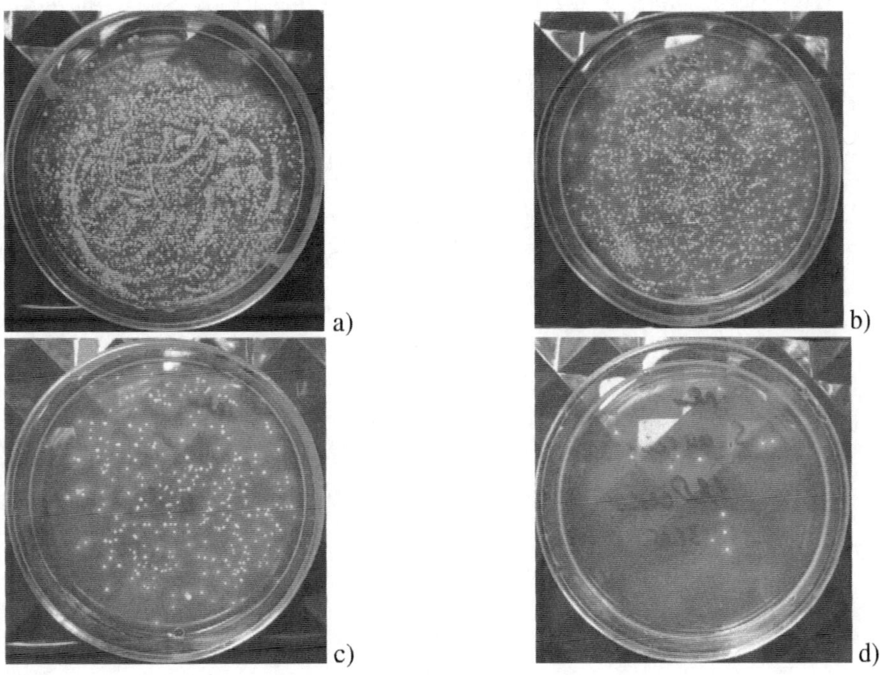

Figure 2. The samples after the plasma treatment: a) the control sample of bacteria Staphylococcus aureus without treatment; b) the sample after treatment for 60 seconds; c) the sample after treatment for 90 seconds; d) the sample after treatment for 120 seconds; bacteria - Staphylococcus aureus.

3. The conclusion

Despite the progress, achieved in the laboratory-scale trials [5-12], the treatment by cold plasma at atmospheric pressure is still not widely of the practical spread. From our point of view, new sources of cold plasma at atmospheric pressure can be widely used in medicine, in the protection of industrial materials from biodamage and bio-corrosion, the disinfection of food and food raw materials, the treatment of tissues, postal envelopes, polymer films, plastic cards and other fields.

References

[1] Samatov R.R. *Caution biocorrosion! Risks, myths and solutions* // Oil. Gas. Innovations. №10 (177) - Samara: «Oil. Gas. Innovations», 2013. - P. 51-57.

[2] Nanzatool J.V., Romankov N.V., Troshina M.V. Tsublova E.G. *Biocorrosion facilities of industrial enterprises and methods of protection from her* // Biosphere compatibility: people, region, technology. №4 (12) - Kursk: Southwest State University, 2015. - P. 79-87.

[3] Gorlenko V.M. *Microbial deterioration of industrial materials* // Microorganisms and lower plants - destroyers of materials and products. - Moscow: Nauka, 1979.

[4] Zlochevskaya I.V. *Biodegradation of stone materials by microorganisms and lower plants in atmospheric conditions* // Biodegradation in construction. - Moscow., 1984.

[5] Gabdrakhmanov Az.T., Galiakbarov A.T. Israphilov I.H. *Study generator of a cold plasma for sterilization* // Contemporary engineering sciences, Vol. 7, no. 20, pp. 973-978 (2014).

[6] Israfilov I.H., Saubanov R.R., Rakhimov R.R. *Prospective application of highly concentrated energy for the surface heat treatment products* // Socio-economic and technical systems: research, design, optimization. № 1, 2011, p. 25-30.

[7] Kashapov N.F., Sharifullin S.N. 2015 *IOP Conference Series: Materials Science and Engineering* **86** 012021

[8] Denisov D G, Kashapov N F and Kashapov R N 2015 *IOP Conference Series: Materials Science and Engineering* **86** 012005

[9] N.F. Kashapov, A.G. Luchkin. *The use of low-temperature plasma deposition of hardening coatings on plastics* // Proceedings of the higher educational institutions. Physics. 2014. T. 57. № 3-3. p. 160-163.

[10] Zaripov R G, Kashapov N F, Tkachenko L A and Shaydullin L R 2016 *J. Phys.: Conf. Ser.* **669** 012053

[11] N.F. Kashapov, R.N. Kashapov. *A study of plasma-electrolytic process for different ratios of the anode space to the cathode* // Proceedings of the higher educational institutions. Physics. 2014. T. 57. № 3-3. Pp 168-170.

[12] Kashapov L N, Kashapov N F and Kashapov R N 2013 *J. Phys.: Conf. Ser.* **479** 012011

Improving of stability of the volumetric glow discharge in the gas flow

I G Galeev and T Ya Asadullin

Kazan National Research Technical University named by A.N.Tupolev
420111 Russia Kazan, K.Marx str., 10

E-mail: galeev.physics@kstu-kai.ru

Abstract. The paper discusses a method of increasing the volumetric stability of the glow discharge in the flow of a working gas by reducing the normal current density at the cathode due to the injection of the inert additional gas in the near-electrode region of the discharge chamber. In this case a relatively thin near-cathode region of the discharge chamber is filled by additional gas but the main region contains only working gas. The additional gas is chosen by way that its value of the cathode potential fall is less one for the working gas but its value of thermal conductivity is more in contrast. Then the conditions for the occurrence of thermal-ionization instability are decreased and the stability of the volumetric glow discharge respectively increases.

As it is known there is a significant nonuniformity of discharge's characteristic distributions in the transverse glow both along and across the direction of the electric field. The electric field strength in the cathode region is much higher than one in the main region of the discharge. Due to the high cathode potential drop in the cathode region there is a significant energy release. The current region in the cathode region of the glow discharge tends to concentrate in a cathode spot with normal current density (in accordance with the Gel law). As a result of increased electric field strength near the cathode the possibility of development of instability in this region of the discharge is greater than in the rest region of the discharge.

In [1] the impact on the cathode discharge area is mechanically done by cowering a discontinuous dielectric coating on the surface of each cathode section of a partitioned cathode. Amid that an average current density in the cathode region decreases and volumetric stability of the glow discharge increases. In [2] a method of stabilizing combustion mode is presented with an adaptive control system of cathode sections potentials of the chamber for the glow discharge in the transverse gas flow. It is possible to implement algorithms of an advancing impact with either tracking type or the type of dynamic equilibrium to prevent compression of the cathode spot with the subsequent discharge contraction.

In [3,4] a significant effect is showed of a cross-section variability of the channel and an additional distributed injection of gas through the channel wall on the electrical and thermal characteristics of the discharge. The authors of the works [5,6] studied in detail the layered structure of the glow discharge taking into account nonlocality of the ionization coefficient in the cathode region. It is considered that the electric field greatly varies along the direction of the current and the coefficient of ionization is not a function only of the reduced electric field strength E/N.

According to the proposed method the additional gas is blown in the cathode region of the discharge with a velocity which normal component W_d to the direction of working gas flow satisfies the condition $\frac{l_c}{b} W_0 < W_{d\perp} < 0{,}1 W_0$, where l_c - length of the cathode layer, b is the distance along the main stream from the beginning of the injection of additional gas to the cathode unit and W_0 is the speed of the working gas. The additional gas must have lower values of the cathode fall of potential density and normal current density for a fixed cathode material and larger thermal conductivity in comparison with the corresponding values for the working gas.

While that organization of the discharge gas in the cathode region is different from the working gas. Both the area of the cathode spot is increased and the uniformity of discharge along the current also increased for a fixed total discharge current.

The cathode potential fall and the electric field strength are reduced in near-electrode region. That results in an increasing in time of instability in the cathode region and consequently increases the stability of the discharge. A higher conductivity of the additional gas provides efficient cooling of the cathode and the absorption of thermal nonuniformity which cause thermal-ionization instability.

The range of speed $W_{d\perp}$ for additional gas in the direction perpendicular to the flow of the working gas is determined due to the fact that eliminates the mixing of additional working gas in the main volume and provides the required properties of the working gas. When the values of $W_{d\perp}$ less than the lower limit the near-cathode area is not completely filled with additional gas. That results in an increasing of cathode potential drop and it not measurably increase the stability of the discharge. When the values $W_{d\perp}$ more than the upper limit there is a possibility of additional penetration of gas in the main volume of the discharge and it changes the specified properties of the working gas.

The method of organization of gas discharge in the stream was implemented with the help of the discharge chamber with the injection of additional gas in the cathode region, a schematic diagram of which is shown in Fig. 1. The horizontal arrows indicate the flow of the working gas and the vertical arrows - blowing additional gas in the cathode region of the partitioned cathode.

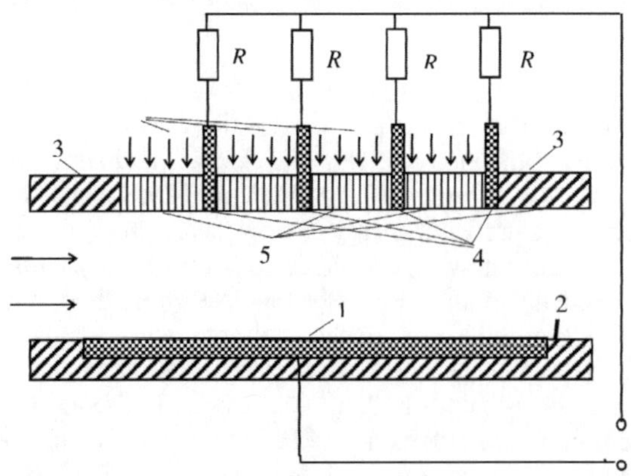

Fig.1

The anode 1 of the discharge chamber was made of one piece and is embedded flush in the dielectric wall 2. The cathode assembly is located between the dielectric plates 3 and performed partitioned. The cathodes 4 is embedded flush in the porous dielectric plate 5 and through the ballast resistance R is connected to the negative pole of the power source. Plate 5 allow the volumetric flow rate of the additional gas ~ 1% of the main gas flow. For instance air may be used as the working gas and argon as additional. On assessments layer of the additional gas was formed of a thickness more than 1 mm near the cathodes of the cathode block. This is approximately by order of magnitude more than the thickness of the cathode layer. For example for the combination of the cathode material iron is iron and the additional injected gas is argon, the length of the cathode region of l_c is determined by the ratio $p^* l_c = 0.33$ cm*tor, where p is the gas pressure [7]. Therefore we can assume that all the cathodic processes take place in an inert additional gas. In the discharge chamber (DC) with an electrode distance of 3 cm, a width of 5 cm at b = 1 cm, the gas flow of DC at pressure P = 50 tor amounted to G = 2,24 g/s. At this pressure the length of the cathode layer of the discharge in argon is less than 0.1 mm. The limit energy deposition in the area of the discharge has been increased approximately by 30%.

If helium is used as an additional gas and nitrogen – as the working gas, while using an iron cathode, the main discharge parameters have the values: U_c =131 V, j/p^2 = 2,2 μA/(cm*Tor)^2, $p^*\ l_c$ = 1.3 cm*tor, however if the main gas is nitrogen then similar parameters would be: U_c = 215 V, j/p^2=400 μA/(cm*Tor)^2, $p^*\ l_c$ = 0.42*cm Tor [7]. Then the density of the electric current at the cathode is estimated to be reduced by more than 180 times as a result of injection of additional gas in the cathode region of the discharge. The density of heat generation per unit length along the direction of the main flow decreases by more than 2 orders of magnitude, the thermal conductivity of the gas is increased more than 5 times, which leads to increased stability of the discharge.

References

[1] Galeev I G and Asadullin T Ya 2016 Enhancing of the glow discharge stability in chamber with cathode sections coated by a discontinuous dielectric coating J. of Phys. Conf. Ser. **669** 012015

[2] Galeyev I G, Asadullin T Ya and Timerkaev B A 2016 Way of stability increase of the glow discharge J. of Phys. Conf. Ser. **669** 012001

[3] Galeev I G, Tikhomirov I A and Asadullin T Ya 1995 Electrical characteristics of the positive trunk of a longitudinal glow discharge in a variable section axisymmetric channel Rus. Phys. J. **38** Is.10 1036-39

[4] Galeev I G and Timerkaev B A 1986 Effect of the distributed injection of atomic gas on the electric and thermal characteristics of the positive column of a low-pressure discharge J. of Eng. Phys. **51** Is. 3 1112-16

[5] Sayfutdinov A I and Timerkaev B A 2012 Drift model of glow discharge with account of nonlocal value of the electric field in the ionisation source J. of Eng. Phys. **85** N5 1104-09

[6] Sayfutdinov A I, Timerkaev B A and Petrova O A 2016 Self-organization of a layered structure of the normal glow discharge J. of Eng. Phys. **89** N2 477-82.

[7] Granovsky V L 1971 Electrical current in gas. The steady-state current (Moscow: Science)

Electroarc vapourtron for bitumen extraction and transportation

G R Ganieva, B A Timerkaev

General Physics Department, Kazan National Research Technical University – KAI, 420111, K. Marx St., 10, Kazan, Russia

E-mail: btimerkaev@gmail.com

Keywords: electroarc plasmatron, arcdischarge, electroarc vapourtron, natural bitumen

Abstract. In this article, we propose the unique submersible arc vapourtron designed to generate very large amount of highly overheated water vapour. The most prospective application of such vapourtron is producing and processing of heavy oil and natural bitumen. The vapourtron represents an electroarc plasmatron using water or water vapour as a working media. Firstly arriving from an external source water enters the cooling jackets of the cathode and the cylindrical anode, where it is preheated to high temperatures. In extreme cases. Water can be heated till the boiling point. Then, this water is supplied to the plasmatron where it is sprayed as the media for the plasma generation. A special twist of the water flow protects the anode walls from overheating.

1. Despite the increasing oil production in the world, we must remember that the oil reserves on Earth are limited. Soon these reserves will be exhausted. At the same time, subsoils contain enormous reserves of hardly extractable heavy oil and natural bitumen. Therefore, the search of new innovative technologies for the extraction, transportation and processing of heavy oil and natural bitumen becomes very topical. Recently the plasma-chemical methods of heavy hydrocarbons processing [1-10] become very popular due to its high efficiency and simplicity.

The Republic of Tatarstan is well-known for its large reserves of natural bitumen and high-viscosity oil, which are very promising sources for oil production. Located at depths of 50-400 m., heavy oil and bitumen have high viscosity and high coking ability. Itleads to problems during the extraction, transportation, and processing of raw materials. Nowadays, a lot of researchesaimed to find economically and technologically effective, as well as environmentally friendly ways for extraction and processing of natural bitumen.

One of the promisingextraction ways is heating of the layers of high viscosity oil and bitumen with overheated water vapor. However, water steam prepared on the surface causes thermal and power losses as a result of the environment heating and the transportation. Therefore, the economic efficiency of this method come into question.

In this work, we propose a uniquesubmersible arcvapourtron. It is designed to generate large amount of overheated to a high temperature water vapor close to the application area. The most prospective application of such vapourtron is producing and processing of heavy oil and natural bitumen. Vapourtron represents an electric-arcplasmatron using water or water vapour as a working media on the figure.

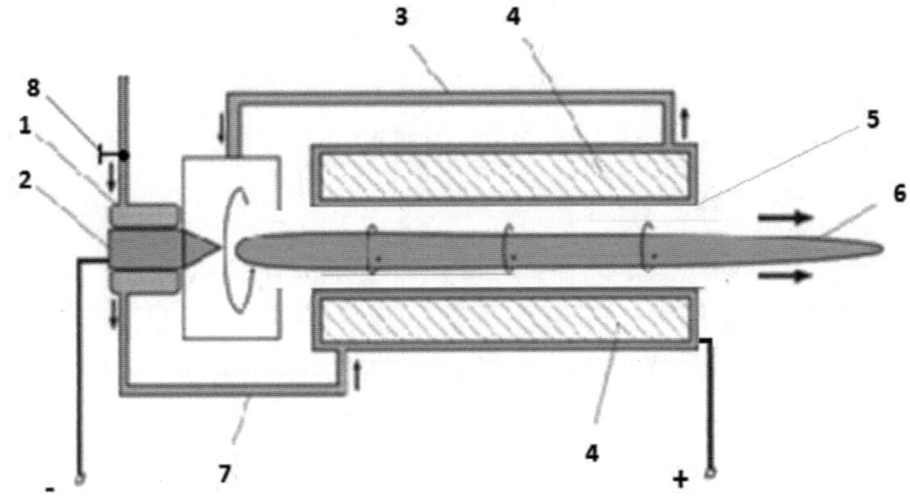

Figure 1. Scheme of the electric arc vapourtron. 1 – cooling jacket of the cathode; 2 – cathode; 3 –tube for the water supply to the discharge zone; 4 – jacket for anode cooling; 5 – cylindrical anode; 6 – vapour-gas plasma; 7 – tube drainage of water from the cathode to the anode; 8 – solenoid valve.

First, water from an external source is supplied to the cooling jacket of the cathode 1 and the cylindrical anode 5 through the electromagnetic valve 8. Second, it is heated to high temperatures. In extreme cases, water can even reach state of boiling. Then, this water is supplied to the plasmatron where it is sprayed as the media for the plasma generation. A special twist of the water flow protects the walls of the anode from overheating.

The experimental setup consists of an electric power supply, an arc ignition, a water cooling, a water supply, an electric arc vapourtron, a pump for water supplying, a pressure pipe, an output pipe, and a pump for raw materials evacuation. The body of the vapourtron is made of brass, which is corrosion resistant. The vapourtron consists of a water-cooled cylindrical anode 5, a tungsten (zirconium, graphite) cathode 2. The solenoid valve 8 is used to regulate the water supply.

Compact plasmatron should be placed on the bottom of the inlet water pipe, whereas the outer surface of the electrodes is covered by a dielectric material.

Water is pumped down by inlet water pipe. Isolated electric wires run down through the same pipe to the electrodes. An electric arc is generated by applying high voltage pulse. After the arc ignition the solenoid valve opens the water supply to vapourtron. Electric arc discharge is carried out in water vapour. Water vapor twisting stabilizes the arc and protects the inner walls of the anode from overheating.

By the action of superheated vapour, the layer gradually becomes heated, the viscosity of raw materials will decrease. Vapour temperature in the plasma core of vapourtron reaches several thousand degrees. Near the oily layer, the vapour temperature will reach several hundreds of degrees, leading therefore to a thermal cracking of bitumen. Light hydrocarbons will be formed from the excess hydrogen, while additional heat will be emitted within the layer due to the excess of free oxygen. Fortunately, both effects are desirable.

Preheated and pre-processed raw materials will be pumped up with the help of a special pump and piped in the tank.

Long-termprocessing significantly decreases the viscosity of bitumen and there fore facilitates its transportation.

We estimated the cost of 1 kg of water vapor, based on the heat balance equation:

$$IUt = c_1 m_1 \Delta T_1 + c_2 m_2 \Delta T_2 + \lambda m_2 + k\mu m_3 \qquad (1)$$

where: *I*, *U* – the total current and the discharge voltage, c_1, m_1, – heat capacity and mass of heated water, c_2, m_2–heat capacity and mass of the formed vapour, ΔT_1–the increment of water temperature, ΔT_2 – increment of the steam temperature; λ is the specific heat of vaporization; μ – the specific heat of dissociation of water molecules; m_3 - the mass of dissociated water molecules, and *k* – the degree of dissociation of water vapor.

Table 1 summarizes the experimental data taken to estimate the assessed results.

Table 1

Current*I*(A)	250
Voltage*U*(V)	200
Time of experiment*t*(s)	180
Temperature increment Δ*T* (K)	600

Let's assume the following constants:

water heat capacity – c_1.= $4187\frac{J}{kg \cdot K}$, water vapour heat capacity –c_2.= $2020\frac{J}{kg \cdot K}$, λ = $2{,}26\ 10^6\frac{J}{kg}$, $\mu = 52\ 10^6\frac{J}{kg}$.

Then the mass rate of water to steam conversion, without taking into account the dissociation of water molecules, can be presented in the form:

$$\frac{m}{t} = \frac{U \cdot J}{c_1\Delta T_1 + c_2\Delta T_2 + \lambda + \mu} = \frac{200 \cdot 250}{4187 \cdot 80 + 2020 \cdot 600 + 2260 \cdot 10^3} = 0{,}015 \cdot {}^{kg}/_{s} \quad (2)$$

that corresponds to about 50 kg/hour. On the basis of the received values, to produce 1 kilogram of steam at 600C,the experimental setup will consume 1 kW h of the electric energy.

It should be also considered, that about 10% of water molecules will disassociate to an atomic state (k = 0,1), then

$$\frac{m}{t} = \frac{U \cdot J}{c_1\Delta T_1 + c_2\Delta T_2 + \lambda + \mu} = \frac{200 \cdot 250}{4187 \cdot 80 + 2020 \cdot 300 + 2260 \cdot 10^3 + 0{,}1 * 520 \cdot 10^5} = 0{,}006 \cdot {}^{kg}/_{s} \quad (3)$$

Then, to produce 1 kilogram of steam at 600C the experimental setup will require about 2 kW h of the electric energy.

To conclude, the proposed experimental setup has heat use efficiency close to 100 percent as it can produce overheated water vapour directly inside the oily layer. Therefore, the vapourtron can undoubtly be used to generate vapour directly inside the oily layer.

References

[1] Ganieva G R Timerkaev B A Plasma-chemical decomposition of heavy hydrocarbons Petrochemicals Publishing(MIAK "Nauka / Interperiodica") 2016 vol no 9pp 1-5

[2] Slovetskii DI Plasma-chemical processing of hydrocarbons: current status and prospects Proceedings of The 3rd international Symposium on theoretical and applied plasma chemistry 2002 vol 1pp 55-58

[3] Ganieva G R Galeev I G Gismatullin N T Ziganshin I D Takhautdinov R S Timerkaev B A Decomposition of heavy hydrocarbons in a free electric arch Izvestiya Samara scientific center RAS Samara 2011 vol 13 No 4 pp 1156-1159

[4] Timerkaev B A GanievG RGaleevI GGismatullinN T Ziganshin D I The decomposition of heavy hydrocarbons in a recessed arc2012Vestnik KGTU im A N Tupolevvol 4 pp 184-188

[5] Peshnev B V Filimonov A S AbuD O Nikolaev A I Technology of production of carbon-based materials from the electrocrackingpaste. Moscow 2013 Proceedings of Conference p 35.

[6] Ganieva G R Timerkaev B A Plasma nanotechnology processing ofoil refinerywastesKazan IV international nanotechnology forum Kazan 2012 pp 119 – 122

[7] Ganieva G R Timerkaev B A Plasmachemical effect of recessed micro-arc discharge on liquid hydrocarbons / VI All-Russian scientific-technical conference "Low-temperature plasma in the processes of deposition of functional coatings" Kazan 2014 pp 196-199

[8] Ganieva G R Timerkaev B A Electrical microdischarges in liquids and prospects of their application in plasma chemistry May-June 2014 Journal of Engineering Physics and ThermophysicsMinsk vol 87 №3 pp 677-681

[9] Ganieva GR Timerkaev BAPlasmachemical processing of raw liquid hydrocarbons by sunkmicroarc- discharge IOP Conference Series: Materials Science and Engineering 012009 2015 vol 86

[10] Ganieva G R Timerkaev BA Decomposition of heavy hydrocarbons in argon arc with the sunken electrodes VII Conference on Low Temperature Plasma in the Processes of Functional Coating Preparation IOP Publishing JournalofPhysics: ConferenceSeries 669 2016 012061

IOP Conf. Series: Journal of Physics: Conf. Series **789** (2017) 012014 doi:10.1088/1742-6596/789/1/012014

Modification of carbon fabrics by radio-frequency capacitive discharge at low pressure to regulate mechanical properties of carbon fiber reinforced plastics based on it

A R Garifullin, I V Krasina, E A Skidchenko, M F Shaekhov and N V Tikhonova

PNTMC Department, Kazan National Research Technological University, Kazan, 420015, Russia

E-mail: darin-loko@yandex.ru

Abstract. To increase the values of mechanical properties of carbon fiber (CF) composite materials used in sports equipment production the method of radio-frequency capacitive (RFC) low-pressure plasma treatment in air was proposed. Previously it was found that this type of modification allows to effectively regulate the surface properties of fibers of different nature. This treatment method differs from the traditional ones by efficiency and environmental friendliness as it does not require the use of aggressive, environmentally hazardous chemicals. In this paper it was established that RFC low-pressure air plasma treatment of carbon fabrics enhances the interlaminar shear strength (ILSS) of carbon fiber reinforced plastic (CFRP). As a result of experimental studies of CF by Fourier Transform Infrared (FTIR) spectroscopy method it was proved that after radio-frequency capacitive plasma treatment at low pressure in air the oxygen-containing functional groups is grafted on the surface. These groups improve adhesion at the interface "matrix-fiber".

1. Introduction

Among a large number of carbon fiber modification technologies (e.g. acid treatment, electrochemical treatment, radiation grafting, application of nanoparticles on the surface) plasma modification method is one of the most environmentally friendly, universal and effective methods [1]. Using the unique RFC discharge at low pressure showed effectiveness in influencing on the figures of CF surface properties
[2, 3]. It was found that this method of plasma modification (RFC at low pressure) increases the amount of adhesive interactions at the interface of two phases (between fiber and matrix) in CFRP [2, 3].

2. Materials, methods and equipment

Carbon fabrics CC 245 woven from fiber Torayca T300 (Toray) was used as the object of study. Modification of fabrics was performed inside RFC low-pressure unit [4]. Treatment modes were selected according the following parameters: plasma gas – air, flow G = 0.04 g/s, pressure P = 50 Pa, anode voltage U_a = 3; 5 kV. Treatment time t = 20; 40 min.

Functional groups on the CF surface after RFC plasma treatment at low pressure in air were investigated by FTIR spectroscopy method carried out on FM-1201 instrument (Russia) using ATR method attachment and 50-fold accumulation in 1100-1900 cm^{-1} area.

The interlaminar shear strength study was conducted by short beam method on electromechanical universal testing machine Instron 5882 using a rig (figure 1).

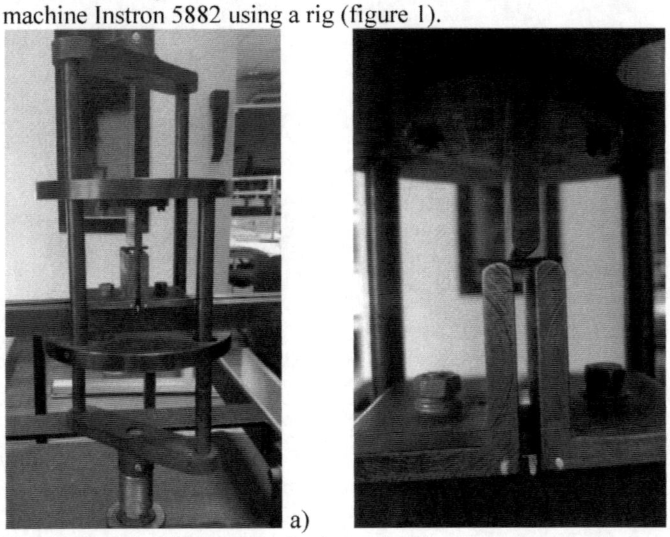

a) b)

Figure 1. Rig for the interlaminar strength determination by short beam method (a) and fixed sample (b)

Test conditions: temperature – 23-24 °C, humidity – 44-51 %, loading rate – 2 mm/min, sample pre-load force – 100 N.

3. Results

Table 1 shows the average results of the ILSS determination tests of CFRP samples (before and after modification) by short beam method.

Table 1. Interlaminar shear strength determination results

U_a (kV)	t (min)	Change of ILSS (%)
Control sample		100
3	20	116
5	20	116
3	40	119
5	40	123

It was proved that plasma treatment with the constant parameters: P = 50 Pa, G_{ait} = 0.04 g/s, t = 40 min and anode voltage U_a = 5kV provide ILSS increase by 23% in this test type.

To explain the rise of adhesion change in correlation with the structure samples were studied by FTIR spectroscopy method (figure 2).

The spectra analysis showed that oscillation band appears in the sample spectrum obtained by ATR method. It indicates about –C–O and C=O functional groups grafting due to oxidation process by plasma treatment in air that results in improvement of chemical adhesion with epoxy matrix in CFRP [5].

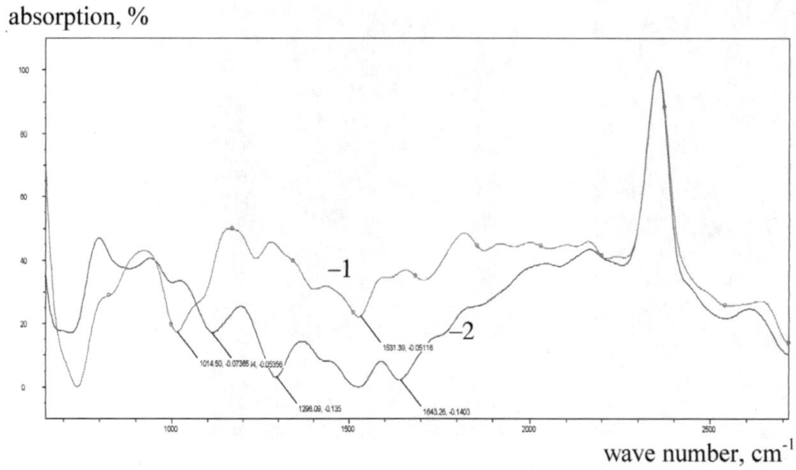

Figure 2. Spectrogram: CF before (1) and after (2) treatment

4. Conclusions

It was found that plasma RFC treatment at low pressure in air enhances adhesion interactions in CF-based composite material. This phenomenon was characterized by –C–O and –C=O functional groups grafting on the CF surface that improves mechanical properties of CFRP.

Acknowledgments
The study was carried out with the financial support of the Russian Ministry of Education, the project number 2196 from 01.02.2014.

References

[1] Garifullin A R and Abdullin I Sh 2014 The present state of the problem of surface treatment of carbon fibers for further use in polymer composites as reinforcing elements (*Vestnik Kazanskogo tekhnologicheskogo universiteta* vol 7) pp. 80-85 (In Russian)

[2] Garifullin A R and Abdullin I Sh 2014 A study of the properties of the carbon fibers were modify with radio-frequency capacitive discharge. (*Vestnik Kazanskogo tekhnologicheskogo universiteta* vol 18*)* pp. 32–33 (In Russian)

[3] Garifullin A R, Israfilov I Kh, Karimullin I I, Karnoukhov A E and Skidchenko E A 2016 Effect of low-temperature plasma treatment on shear strength of carbon microcomposites. (*Vestnik Kazanskogo tekhnologicheskogo universiteta* vol 8*)* pp. 81–82 (In Russian)

[4] Abdullin I Sh, Zheltukhin V S, Sagbiev I R and Shaekhov M F 2007 *Modifikatsiia nanosloev v vysokochastotnoi plazme ponizhennogo davleniia* [Modification of nanolayers in radio-frequency plasma of low pressure]. Kazan, (Izdatel'stvo Kazanskogo tekhnologicheskogo universiteta) p 356 (In Russian)

[5] Keming Ma, Ping Chen, Baichen Wang, Guiling Cui, Xinmeng Xu 2010 A study of the effect of oxygen plasma treatment on the interfacial properties of carbon fiber/epoxy composites Journal of Applied Polymer Science vol 118 pp. 1606–1614

The gap of electronic and gas temperatures at the border of the electric arc

Alexander Gerasimov, Alexander Kirpichnikov, Farida Sabirova, Igor Yakimov

Kazan National Research Technological University 420015, Karl Marx str., 68 Kazan, Russia
Kazan National Research Technological University 420015, Karl Marx str., 68 Kazan, Russia
Kazan National Research Technological University 420015, Karl Marx str., 68 Kazan, Russia
Kazan National Research Technical University named A. N. Tupolev 420111, Karl Marx str., 10 Kazan, Russia

E-mail: gerasimov@kstu.ru

Abstract. In this paper we consider fundamental, and one of the most interesting questions that arise in the theory and practice of two-temperature arc matter of breaking Te(R) and atomic-ion (gas) T(R) temperatures on the border of the electric arc. For the two-temperature calculations used channel model arc. Presented gap value calculated for different values of the current strength in the discharge. Calculations were made for an argon plasma at atmospheric pressure for two values of current intensity I = 78 A, and I = 200 A and various values of the wall temperature. Also shows the corresponding experimental results for the same conditions of other authors, we are in good qualitative agreement of the results of calculations with experimental data.

1. Introduction

To date, accumulated a large amount of material on the Te distribution in the arc plasma torches, including the Te value as a function of r at the border (in reality, of course, close to the border), stabilizing the discharge of a cylindrical tube, which plays the role of a thermostat for the atomic-ion gas [1-3].
In other words, the work discusses value jump temperature Te(R)-T(R) at the border of the primary heat sink zone (at r=R where R - radius of the bounding an arc of a cylindrical tube whose walls are maintained at a fixed low enough temperature T(R)).
The relevance of such a formulation is associated with the fact that current methods of direct measurement of the temperature of the atoms and ions are still insufficiently developed, while the measurement of the electron temperature or a temperature close thereto, spectral analysis techniques widely.
In this paper, this problem is solved in the framework of the known channel model M. Steenbeck [4] as the most suitable for the analysis of a wide class of problems of maintenance of the gas discharge plasma in the electric and magnetic fields [5, 6].

2. Method of solution

We use the built in [7] two-temperature channel model arc. Referring to the formulas obtained in this work dependencies electron and gas temperatures in the non-conductive zone arc from the current coordinate r

$$T_e(r)=T(R)+\frac{W}{2\pi\lambda}\ln\frac{R}{r}+\left(\frac{W}{2\pi\lambda}+\varepsilon\right)\cdot\frac{\lambda_e\Psi_0(R)+\lambda_{ai}\Psi_0(r)}{\lambda_{ai}\tilde{s}R\Psi_1(R)}; \quad (1)$$

$$T_e(r)=T(R)+\frac{W}{2\pi\lambda}\ln\frac{R}{r}+\left(\frac{W}{2\pi\lambda}+\varepsilon\right)\cdot\frac{\lambda_e\Psi_0(R)+\lambda_{ai}\Psi_0(r)}{\lambda_{ai}sR\Psi_1(R)}, \quad (2)$$

Where

$$\Psi_0(r)=I_0(sr)K_0(sr_\kappa)-I_0(sr_\kappa)K_0(sr);$$

$$\Psi_1(r)=I_1(sr)K_0(sr_\kappa)+I_0(sr_\kappa)K_1(sr)=\frac{1}{s}\frac{\partial\Psi_0(r)}{\partial r}.$$

$$s^2=\frac{3}{2}k\delta\nu n_e\left(\frac{1}{\lambda_e}+\frac{1}{\lambda_{ai}}\right)$$

In these formulas, r_c - the radius of the conductive channel, W specific (per unit column length) into the discharge power, λe=λe(T) and λai =λai(T) coefficients of thermal conductivity of the electronic and atomic-ion gas of heavy particles, k - is Boltzmann constant; δ - the proportion of the energy lost in the collision of electrons with heavy particles; ν - the frequency of these collisions; n_e - the concentration of electrons in the discharge. ε size in the formulas (1), (2) depends on discharge parameters and selected in each case based on some assumed physical model of interaction of the electron gas with the walls of the cylindrical tube. Case ε = 0 in this case corresponds to the adiabatic conditions for the adoption of the electron gas. Referring to the formula (2), putting in it, we obtain the following fundamental relation, the theory of two-temperature locking arc:

$$\Delta T(R)=T_e(R)-T(R)=\frac{W}{2\pi\lambda_{ai}}\frac{\Psi_0(R)}{sR\Psi_1(R)}. \quad (3)$$

Equation (3) can be simplified. We take into account that, in practice, we almost always $sR>>1$. In this case, simple calculations using the asymptotic representation [8] functions $I_{0,1}(x)$ and $K_{0,1}(x)$ for large values of the argument, you can get a simple relation

$$T_e(R)-T(R)\approx\frac{W}{2\pi\lambda_{ai}sR}$$

that is, to a first approximation by the decomposition of the formula (3) in a series in the small parameter $\frac{1}{sR}$ jump in the electron temperature near the boundary limiting the arc of a cylindrical tube with the same input to the discharge capacity of the W, it is inversely proportional to the radius of the tube:

$$\frac{T_e(R_1)-T(R_1)}{T_e(R_2)-T(R_2)}=\frac{R_2}{R_1}$$

At $\frac{W}{2\pi\lambda}>\varepsilon R_1,\varepsilon R_2$

$$\frac{T_e(R_1)-T(R_1)}{T_e(R_2)-T(R_2)}\approx 1$$

At $\frac{W}{2\pi\lambda}<\varepsilon R_1,\varepsilon R_2$

Although, of course, the fact that the gap with increasing R e and the gas temperature at the wall should decrease, that is, the condition must be

$\lim T_e(R)\to T(R)$ at $R\to\infty$

is a pre-obvious enough physically.

3. Results and conclusion

Tables 1-2 is represented by the gap value calculated for the different discharge formula amperage values (3). Calculations were made for an argon plasma at atmospheric pressure for two current intensity values I=78A and I=200A and various values of the wall temperature. It shows the corresponding experimental results in these tables, published in a series of papers [1-3]. As you can see, there is good qualitative agreement of the results of calculations with experimental data, which increases with decreasing temperature of the wall.

Table 1. Values $\Delta T(R)$ for I=78 A

	$\Delta T(R)$ for $T(R)$=1440 K	$\Delta T(R)$ for $T(R)$=4000 K
Calculation	7200 K	4700 K
Experimental data [3]	7860 K	5300 K

Table 2. Values $\Delta T(R)$ for I=200 A

	$\Delta T(R)$ for $T(R)$=4000 K
Calculation	4800 K
Experimental data [3]	6000 K

IOP Conf. Series: Journal of Physics: Conf. Series **789** (2017) 012015 doi:10.1088/1742-6596/789/1/012015

In conclusion, we note that the results obtained in this study can be used as a simple model for the assessment fairly complex physical processes, and in some engineering calculations, including those relating to the use of low-temperature plasma arcs in the functional coating processes.

References

[1] Asinovskii E. I., Pakhomov E. P., Yartsev Y. M. Issledovanie kharakteristik laminarnogo potoka plazmy argona v elektricheskoi duge // Khimicheskie reakcii v nizkotemperaturnoi plasme. (Chemical reactions in low-temperature plasma) Moscow: INKHS (Institute of petrochemical synthesis). 1977. P.83-103.

[2] Asinovskii E. I., Pakhomov E. P., Yartsev Y. M. Opredelenie vyazkosti plazmy argona s pomosh'yu stabilizirovannoi elektricheskoi dugi // Teplofizika Vysokikh Temperatur (High Temperature) 1978. V. 16. № 1. P.28-36.

[3] Asinovskii E. I., Pakhomov E. P. Analiz temperaturnogo polya v cilindricheski simmetrichnom stolbe electricheskoi dugi // Teplofizika Vysokikh Temperatur (High Temperature) 1968. V. 6. № 2. P.333-336.

[4] Engel, A., Steenbeck, M., Physics and Engineering of a Discharge of Electric ity in Gases,Vol. 2, Properties of Gas Discharges. Engineering Application (in German), Julius Springer Verlag, Berlin, 1934

[5] Raizer Yu. P. Osnovy sovremennoi fiziki gasorazryadnykh pocessov. (Basis of a modern physics of gas-discharge processes) Moscow: Nauka. 1980. 416 p.

[6] Raizer Yu. P. Fizika gasovogo razryada. (Physics of Gas Discharge) Moscow: Nauka. 1987.

[7] Gerasimov A.V., Kirpichnikov A.P. Quasi-equilibrium channel model of an arc of a constant current //Thermal Science. 2003 No 1. P. 101-108.

[8] Dwight G.B. Tables of Integrals and other mathematical data, The Macmillan Company, New York 1957 176 p

Transmission and reflection of sound wave from a layer of liquid with gas bubbles

D A Gubaidullin, D D Gubaidullina, Yu V Fedorov

Institute of Mechanics and Engineering, Kazan Science Center, Russian Academy of Sciences, 2/31 Lobachevsky str., Kazan 420111, Russia
Kazan (Volga Region) Federal University, 18 Kremlyovskaya str., Kazan 420008, Russia

E-mail: kopperfildd@yandex.ru

Abstract. The problem of reflection and transmission of sound wave through the two-layer medium containing a layer of bubble liquid is considered. A comparison of the reflection and transmission of wave coefficients with known experimental data is presented.

1. Introduction

Waves propagation in complex layered media is an active subject of research. A layered medium containing bubble liquid is of particular interest [1] - [10]. This is due both to natural processes, such as bottom layers containing gas bubbles in the seas, lakes or near the surface [1, 2], and technological processes [3, 4]. For example, in [4] the description of construction and characteristics of the combined panel consisting of two glass or polycarbonate, the space between them is filled with bubble liquid, to protect information both on acoustic, optical, electromagnetic, and technical channels is shown.

There are several experimental studies. In [7] the two-layer medium water - water with bubbles - air is considered and reflection coefficient when the acoustic wave falls normally on the water side is experimentally determined, and in [8] the two-layer medium water - water with bubbles - water is considered and transmittance through bubble screen is found experimentally.

Problems of acoustics of bubble liquids in a particular setting are considered in [11] - [22].

In this paper on the basis of the methodics described in the book [25] and dependence of the wave number on the frequency of the disturbances for the bubble liquid, obtained in [13] reflection and transmission of sound wave through the bubble liquid are investigated.

2. Mathematical model.

To determine the transmission and reflection from two-layer medium coefficients, a diagram of which is shown in fig. 1, the following method of calculations is used. According to [21], the result of the reflection and transmission of a plane monochromatic wave $\exp i(\omega t - Kx)$ from multi-layer media will be plane waves $R \exp i(\omega t + Kx)$ and $T \exp i(\omega t - Kx)$, where R and T – reflection and transmission of the waves coefficients, which are defined through layers impedances Z_j and input impedances of layers borders Z_j^{in}. For a two-layer medium reflection and transmission coefficients are of the form

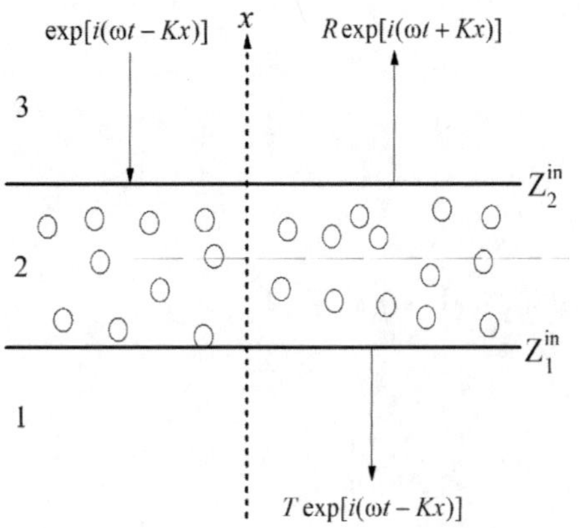

Figure 1. Scheme of the media.

$$R = \frac{Z_2^{in} - Z_3}{Z_2^{in} + Z_3}, T = \prod_{j=1}^{2} \left(\frac{Z_j^{in} + Z_j}{Z_j^{in} + Z_{j+1}} \exp iK_j d_j \right) \tag{1}$$

$$Z_j = \rho_j \frac{\omega}{K_j}, Z_2^{in} = \frac{Z_1 - iZ_2 \tan(K_2 d_2)}{Z_2 - iZ_1 \tan(K_2 d_2)} Z_2, i^2 = -1, j = 1, 2, 3$$

Here d_2 – the thickness of the bubble layer, K – wave number, ω – frequency of disturbances, ρ – density of the layer. For a homogeneous layer the wave number is defined as $K_j = \omega / C_j$, where C_j – the speed of sound in the j -th layer. Since in the second layer bubble liquid will be considered, i.e., non- homogeneous and absorbing medium, to determine, in general, a complex wave number $K_2 = K_*$ the relation obtained in [13] will be used. This dependence of a complex wave number on the disturbances frequency is given by

$$\left(\frac{K_*}{\omega} \right)^2 = \frac{1}{C_f^2} + \frac{3\alpha_{20}\alpha_{10}\rho_{10} \langle Q \rangle}{3\gamma_{20}p_{10} - \langle QS \rangle} \tag{2}$$

$$Q = 1 + \phi, \phi = 3(\gamma_{20} - 1)\frac{y \coth y - 1}{y^2}, y = \left(-\frac{i\omega a^2}{\kappa_{20}} \right)^{1/2}$$

$$S = \frac{i\omega a^2 h \rho_{10}}{1 + ht}, h = \frac{4\nu_{10}}{a^2} - i\omega, t = \frac{a}{C_1 \alpha_{20}^{1/6}}$$

$$\langle h \rangle = \frac{1}{\rho_2} \int_{\Delta a} N_0(a) g_0(a) h da, \rho_2 = \int_{\Delta a} N_0(a) g_0(a) da, g_0(a) = \frac{4}{3}\pi a^3 \rho_{20}, \Delta a = [a_{\min}, a_{\max}]$$

Here, the subscript 1 refers to the parameters of the liquid phase, index 2 – to the parameters of the bubbles, 0 – the initial undisturbed state. The parameters α determine the volume content, ρ_{10} – liquid density, ρ_{20} – density of gas in bubbles, p_{10} – pressure, C_f – frozen velocity, is almost equal to the speed of sound C_1 in pure liquid, γ – the adiabatic index, κ – coefficient of the gas thermal conductivity, ν_{10} – the kinematic viscosity of the fluid, $N_0(a)$ – function of bubble size distribution.

3. The results of calculations.

Let us consider the acoustic signal passing through the two-layer medium: water - water with bubbles of air - water. Let the thickness of the bubble layer is 4 mm, the radius of air bubbles

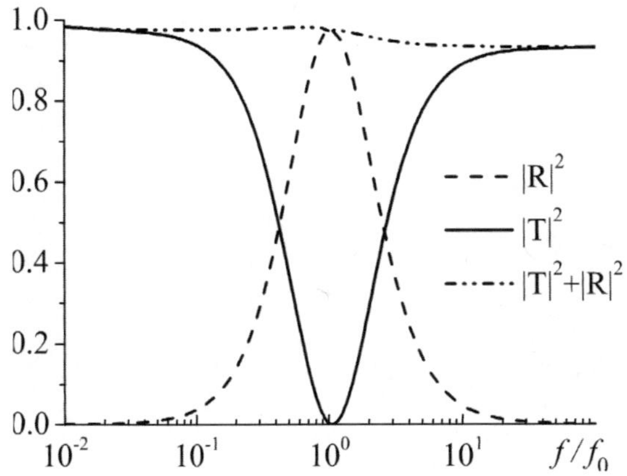

Figure 2. The coefficients of reflection and transmission of sound through the bubble screen: $\alpha_{20} = 0.01$, $a_0 = 2$ mm, $d_2 = 4$ mm, $f_0 = 1630$ Hz.

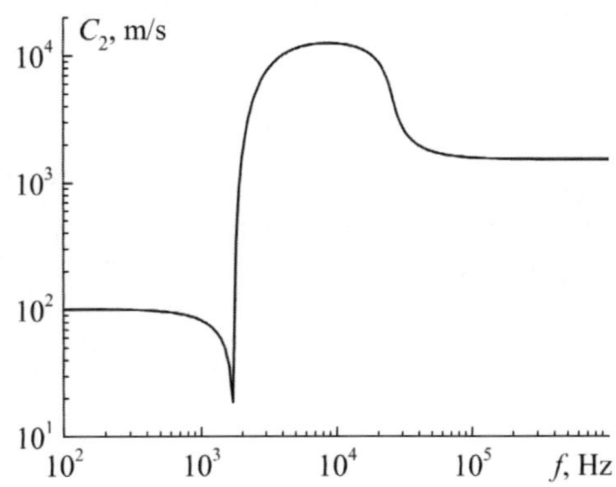

Figure 3. Dependence of phase velocity on frequency of disturbances in bubble layer: $\alpha_{20} = 0.01$, $a_0 = 2$ mm.

is 2 mm, the volume content of 1%. The pressure of a mixture is 0.1 MPa, temperature is 288 K. The calculations we will make using the formulas (1) and the equations (2). Fig. 2 shows the dependences of reflection and transmission of the waves coefficients on the dimensionless disturbances frequency through the bubble layer. As can be seen from this figure, in the resonant frequency area of bubbles the minimum of transmission coefficient and the maximum of reflection coefficient are observed, i.e., at a given frequency bubble layer almost completely reflects incident acoustic wave. If you add up the values of these coefficients, the result is always less than one, because the wave energy is partially absorbed by the bubbles at oscillation due to viscosity and process of thermal conductivity.

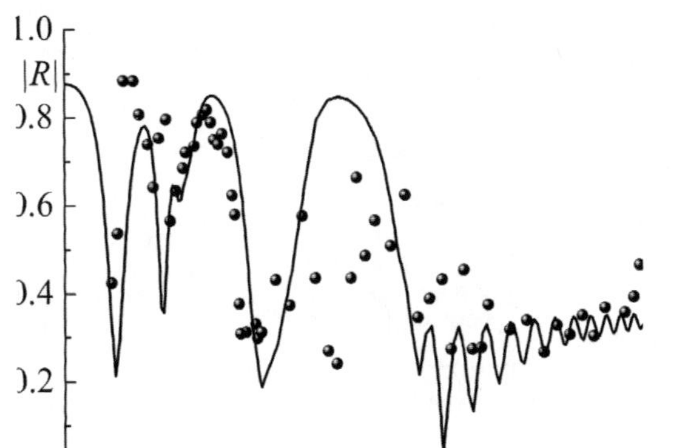

Figure 4. Comparison of reflection coefficient with experimental data [7] for medium: water – bubbly liquid – air; $\alpha_{20} = 0.032\%$, $d_2 = 10$ cm.

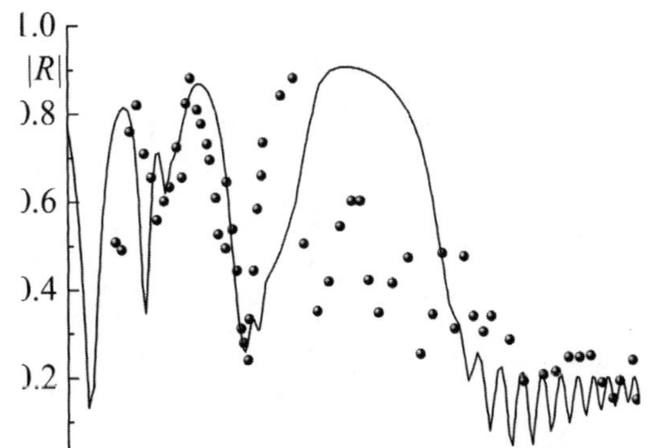

Figure 5. Comparison of reflection coefficient with experimental data [7] for medium: water – bubbly liquid – air; $\alpha_{20} = 0.052\%$, $d_2 = 10$ cm.

Figures 4 – 6 show a satisfactory agreement between the theoretical curves of reflection and transmission coefficients with known experimental data.

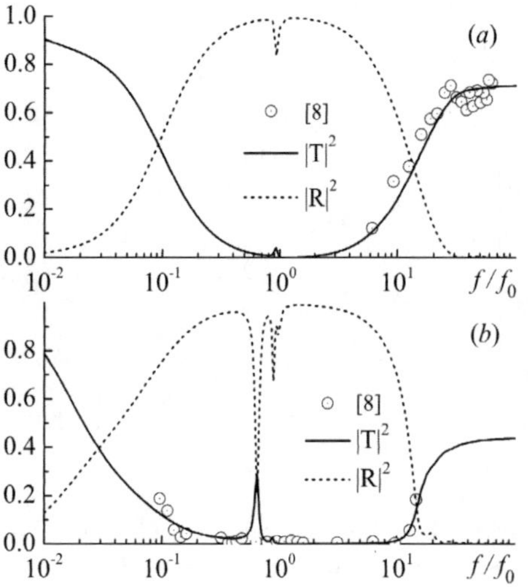

Figure 6. Comparison of reflection and transmission coefficients through the bubble screen with experimental data [8]; a – $\alpha_{20} = 0.015$, $a_0 = 2$ mm, $d_2 = 1.5$ cm, $f_0 = 1630$ Hz, b – $\alpha_{20} = 0.01$, $a_0 = 0.5$ mm, $d_2 = 1.2$ cm, $f_0 = 6523$ Hz.

Acknowledgment
This work was performed with the assistance of Council for grants of the leading scientific school (NSH-6987.2016.1).

References
[1] Shagapov V Sh and Sarapulova V V 2015 *Acoustical Physics* **61** 37
[2] Gusev V A and Rudenko O V 2015 *Acoustical Physics* **61** 152
[3] Leroy V, Strybulevich A, Page J H and Scanlon M G 2008 *J. Acoust. Soc. Am.* **123** 1931
[4] Zelmanskiy O B, Petrov S N and Kazeka A A 2013 *Doklady BGUIR* **78** 30
[5] Nigmatulin R I, Gubaidullin D A and Nikiforov A A 2014 *Doklady Physics* **59** 286
[6] Baranowska A 2012 *Archives of Acoustics* **37** 287
[7] Lee K, Choi B K and Yoon S W 2002 *Journal of the Korean Physical Society* **40** 256
[8] Tien T M 2001 *Sound propagation through a bubble screen at finite gas-volume fraction* (Taiwan: National Cheng Kung University)
[9] Gubaidullin D A and Fedorov Yu V 2016 *Doklady Physics* **61** 499
[10] Marr K W 1981 *On the design of an acoustically isolating bubble screen for the Carr inlet acoustic range* (California: Naval Postgraduate School)
[11] Commander K W and Prosperetti A 1989 *J. Acoust. Soc. Am.* **85** 732
[12] Gubaidullin D A, Nikiforov A A and Gafiyatov R N 2012 *High Temperature* **50** 250
[13] Gubaidullin D A and Fedorov Yu V 2013 *Journal of Applied Mathematics and Mechanics* **77** 532
[14] Gubaidullin D A, Gubaidullina D D and Fedorov Yu V 2013 *Fluid Dynamics* **48** 781
[15] Nigmatulin R I, Gubaidullin D A and Fedorov Yu V 2013 *Doklady Physics* **58** 261
[16] Fuster D and Montel F 2015 *J. Fluid Mech.* **779** 598
[17] Gubaidullin D A and Gafiyatov R N 2014 *J. Phys.: Conf. Ser.* **567** 012020
[18] Gafiyatov R N, Gubaidullin D A and Nikiforov A A 2013 *Fluid Dynamics* **48** 366
[19] Gubaidullin D A and Fedorov Yu V 2015 *Fluid Dynamics* **50** 61
[20] Gubaidullin D A, Gubaidullina D D and Fedorov Yu V 2014 *J. Phys.: Conf. Ser.* **567** 012011
[21] Gubaidullin D A, Nikiforov A A and Gafiyatov R N 2016 *J. Phys.: Conf. Ser.* **669** 012019
[22] Gubaidullin D A, Gubaidullina D D and Fedorov Yu V 2016 *J. Phys.: Conf. Ser.* **669** 012011
[23] Gubaidullin D A, Nikiforov A A and Gafiyatov R N 2015 *High Temperature* **53** 240
[24] Gubaidullin D A and Fedorov Yu V 2016 *Acoustical Physics* **62** 179
[25] Brekhovskikh L M and Godin O A 1989 *Acoustics of Layered Media* (Moscow: Nauka)

Experimental studies for flow velocities of oscillations gas when transiting through resonance in the area of an open end of pipe

D A Gubaidullin[1], N F Kashapov[2], R G Zaripov[1], L A Tkachenko[1, 2] and L R Shaydullin[2]

[1] Institute of Mechanics and Engineering, Kazan Science Center, Russian Academy of Sciences, 2/31 Lobachevsky str., Kazan 420111, Russia
[2] Kazan (Volga Region) Federal University, 18 Kremlyovskaya str., Kazan 420111, Russia

E-mail: LiRShajdullin@kpfu.ru

Abstract. Mean velocities of gas in the area on an open end of pipe while gas oscillated in a non-linear way near its first eigenfrequency were researched experimentally. A non-monotonic character of the change of the mean velocity in the radial direction was identified.

Introduction

Researches of features of non-linear oscillations of gas in pipes near resonance points are important today both to improve efficiency of existing applications and to develop new prospective ones: in the vibration engineering, pipeline systems, vibration-based cleaning of reservoirs. In case of such oscillations, shockwaves, middle streams, eddy structures, thermo-acoustic effects occur, etc. [1-4]. Lack of consideration of such effects when developing technical devices may result in a dramatic decrease of their service life or to occurrence of emergency situations. In the course of research of resonance oscillations of gas in an open pipe, in terms of theory [2], significant difficulties arise caused with various effects of the non-linear wave field at the outflow section. In research papers [5, 6], the field of instantaneous velocities in the axial and the radial directions were researched in detail in the area of the open end of a pipe. Motions of a flat particle in the external wave field were also studied [7]. It seems necessary to continue the works [5, 6, 8] for experimental studies of the field of medium velocities of oscillating gas near its first eigenfrequency in the area of the open end of a pipe.

Experimental conditions

The research was performed in an installation based on a vibration generator 1 of TIRAvib S 5220/LS make with a power amplifier 8 model BAA 1000-ET firm of the TIRA (fig. 1). Into the table of the vibration generator, a flat piston 3 with the diameter of d = 100 mm was screwed. A rod was oscillating inside the piston linked with a nozzle with a glass pipe 3 glued with epoxy resin Dimensions of the pipe were as follows: the length of L = 918 mm, the internal diameter d_0 = 10,05 mm, and the wall thickness of 2 mm.

The vibration installation was controlled and monitored with a computer 11 by means of a piezoelectric IEPE accelerometer 2 of 4513 model by Bruel & Kjaer firm and a controller 8 of VR9500 type by Vibration Research Corporation firm via the specialized software VibrationVIEW.

The positioning controller 9 of TMC-2 model was used to adjust the table to the vibration generator. A cooling fan 14 was applied to prevent overheating of the vibration coil and the field excitation coil.

Gas pressure was measured with a piezoelectric gauge 4 of 8530C-15 model by Bruel & Kjaer firm, and the signal from the gauge was transferred to a three-channel voltage bridge amplifier 13 of ENDEVCO brand of 136 model by Bruel & Kjaer firm to be further registered with a digital oscilloscope 12 of DSO 3062A model by Agilent Technologies firm. The oscilloscope was connected to the computer via the RS-232 interface to process the data. The pressure gauge was screwed in the lower nozzle of the pipe.

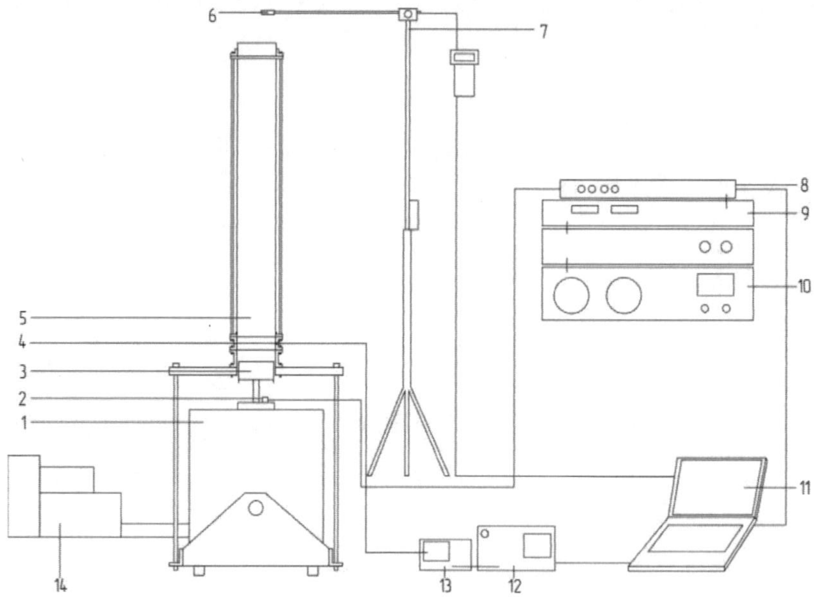

Figure 1. Basic diagram of the experimental installation.
1 – vibration generator, 2 – piezoelectric IEPE accelerometer, 3 – flat piston, 4 – pressure sensor, 5 – glass pipe,6 – thermal mass flow meter, 7 – traversing probe, 8 – controller, 9 – positioning controller TMS-2, 10 – power amplifier, 11 – control computer, 12 – oscilloscope, 13 – three-channel bridge voltage amplifier, 14 – cooling fan.

Mean velocities of gas v were measured with a thermal mass flow meter 6 of ATT-1004 model by AKTAKOM firm in the radial and the axial directions using the traversing probe 7 by DICA firm that measured the distance in mm in the non-linear wave field near the open end of the pipe. The range of measurements for the mean gas velocity varied from 0.1 m/s to 25.0 m/s. The definition was 0.01 m/s, the allowable measurement error was ±5% of the measured value or ±1% of the range. The traversing probe was fixed in a stand placed next to the vibration installation which enabled to shift it vertically for a set distance (the value of $x = 0$ corresponded to the edge of the pipe).

Experimental results

The experiments resulted in oscilloscope charts of gas pressure for different excitation frequencies when transiting through resonance for the piston displacement amplitude equal to $l = 0.75$ mm.

The experimental data showed that oscillations of gas at the resonance are of non-linear character as the steepness of the leading edge is greater than that of the training edge. However, the shape of the wave remains continuous. At the same time, the range of oscillation of the pressure at the resonance reaches 0.04 bar.

Figure 2 shows dependencies of the mean velocity of gas in the area of the open end of the pipe in the axial x and the radial r directions ($r = 0$ corresponds to the axis of the pipe) at the resonance frequency of ν = 90.3 Hz at the amplitude of the piston displacement of $l = 1.5$ mm for the respective distances to the open end: $x = 5$ mm; 60 mm; 100 mm; 200 mm. At the distance of 5 mm, 60 mm from the open end of the pipe, structure of the gas stream contains an area of constant value of the gas velocity near the symmetry axis – the core of the stream corresponding to the velocity of 15 m/s, 7.5 m/s, further in the directions of edges of the pipe, there are areas of increase and then of an abrupt fall in the flow velocity with the corresponding velocities of 22.5 m/s, 15 m/s.

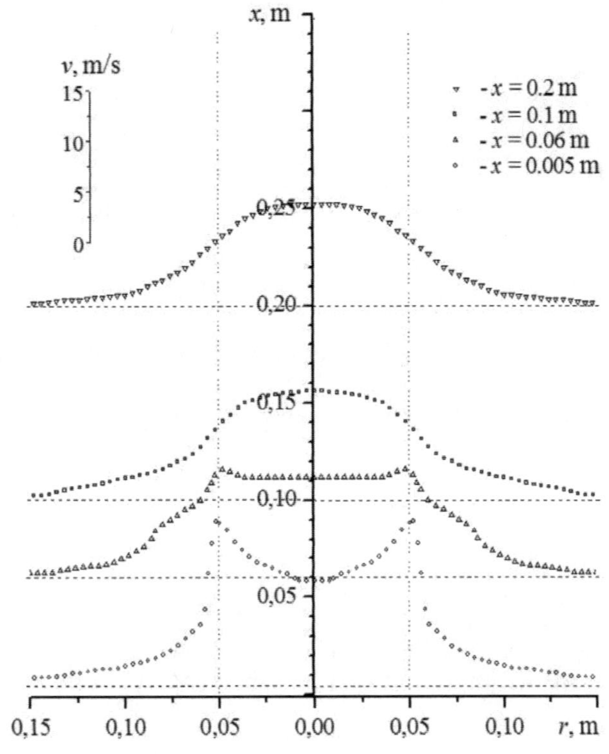

Figure 2. Distribution of the mean velocity of gas in the radial direction in different sections of the external wave field of the pipe at the resonance frequency.

At the same time, the maximum value of the velocity is near the edges of the pipe. Along with the recession from the end of the pipe to the external wave field, the difference of the velocities decreases, and the maximum value of the velocity is placed near the symmetry axis and equals to 14.5 m/s, while at the edges, an area of monotonous decrease of the velocity is observed. The occurrence of the velocity maximum near the edges is explained with a formation of an eddy motion when gas flows out of the pipe. A similar phenomenon was observed for the field of instantaneous gas velocities [5, 6]. At the distance of 200 mm, the mean velocity at the axis of the pipe equals to 13 m/s, which is less than the velocity near the open end at $x = 5$ mm. This is due to the expansion of the cylindrical stream along with recession from the open end.

Figure 3 shows values of the gas velocity depending on the excitation frequency at the axis of the pipe for $x = 5$ mm at $l = 1.5$ mm.

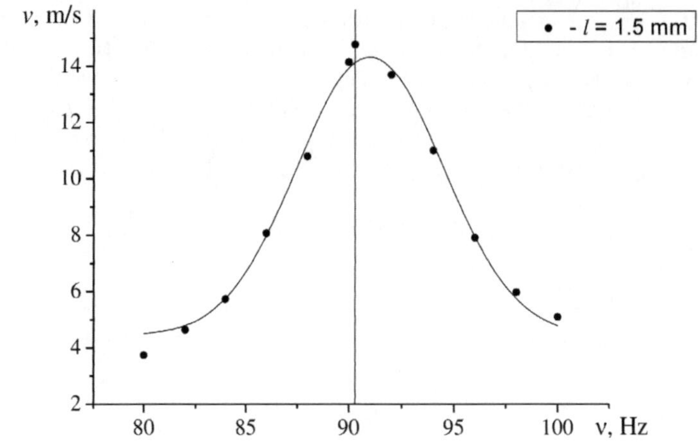

Figure 3. Values of the gas velocity depending on the excitation frequency at the axis of the pipe for $x = 5$ mm at $l = 1.5$ mm. Dots mark the experimental results, and the solid line is Gauss approximation.

Along with an increase of the frequency from a pre-resonance oscillation mode, a rapid increase of the mean velocity in observed with the maximum value equal to 14.5 m/s at the resonance. After the resonance, along with a further increase of the frequency, the values of the velocity are decreased just as rapidly.

Conclusion

Mean velocities of gas in the area of the open end of the pipe were studied experimentally with non-linear oscillations of gas near its first eigenfrequency. The non-monotone character was identified for distribution of velocities related to the vortex motions at the outflow of oscillating gas from the pipe. Away from the open end, the structure of the flow similar to that of the stationary outflow from the pipe is observed.

References

[1] Nakoryakov V E, Burdukov A P, Boldyrev A M, Terleev P N 1970 *Heat and mass transfer in an acoustic field* (Novosibirsk: Nauka. Sibir. Otd-e) p 253

[2] Ilgamov M A, Zaripov R G, Galiullin R G, Repin V B 1996 *Appl. Mechanics Reviews* **49** 137

[3] Zaripov R G, Galiullin R G and Galiullina E R 2001 Actual problems of continuum mechanics. On the 10th anniversary of IME KazSC RAS 19-35 (Kazan: IME KazSC RAS)

[4] Galiullin R G, Tkachenko L A 2007 *Nonlinear Oscillations of a Gas in the Tubes* (Kazan: Izd. Kazan. Gos. Univ.) p 116

[5] Vasil'ev L S, Zaripov R G, Magsumova A T, Sal'yanov O R 1991, *Inzhenerno-Fiz. Zh.*, **61** (5) p 714

[6] Zaripov, R G, Davydov, R I, Sonin, N V 2001, *Russ. Aeronaut.* **8** (2) pp 1-3

[7] Gubaidullin D A, Zaripov R G, Tkachenko L A and Shaidullin L R 2015 *J. of Engin. Physics and Thermophysics* **88** (4) pp 843-847

[8] Zaripov R G, Kashapov N F, Tkachenko L A, Shaydullin L R 2016 *J. Phys.: Conf. Series* **669** 012053

Propagation of acoustic waves in multifractional polydisperse gas suspension

D A Gubaidullin[1], **E A Teregulova** [1]

[1] Institute of Mechanics and Engineering, Kazan Science Center, Russian Academy of Sciences, 2/31 Lobachevsky str., Kazan 420111, Russia

E-Mail: teregulova@inbox.ru

Abstract. The propagation of acoustic waves in multifractional polydisperse gas suspension is studied. A mathematical model is presented, the dispersion equation is obtained, dispersion curves are calculated. The influence of the particle size and the parameters of the dispersed phase for multifractional gas mixture with ice particles, aluminum and sand on dissipation and dispersion of sound waves is analyzed.

1. Introduction

The acoustics and wave dynamics of multiphase media is of great theoretical and practical interest, due to the wide spread of such media in nature and use them in practice. The study of non-stationary wave processes is one of the urgent problems of mechanics of multiphase media. Basic models of wave dynamics of dispersed media and a number of results in this area are presented in [1]. Work [2] is devoted to the problems of the investigation of two-phase flows with solid particles, droplets and bubbles. In monography [3] a brief overview of the results on the study of acoustic disturbances in monodisperse gas mixtures without phase transformations is given. Influence of polydisperse composition of the gas mixture on the distribution of monochromatic disturbances in one-component mixtures of gas and particles, or vapor with droplets is studie
d in [4]. Peculiarities of the propagation of monochromatic waves in two-component polydisperse mixtures of gas and vapor and liquid droplets are investigated in [5]. Propagation of spherical and cylindrical waves of small amplitude in polydispersed fogs with phase transformations is considered in [6]. A general dispersion relation of the wave number to the oscillation frequency and thermal properties of the phases is obtained. The anomalous effect of nonmonotonic dependence of dissipation of weak harmonic and pulse disturbances on the mass concentration of droplets in monodisperse aerosols with heat and mass transfer is studied in [7]. Fairly complete presentation of the propagation linear theory of flat disturbances in mono-and polydisperse two-phase gas-vapor-liquid droplet mixtures is given in [8]. The propagation of acoustic waves of different geometry in two- fraction gas mixtures with particles of different sizes and materials, without taking phase transformations into account is studied in [9]. Peculiarities of plane, cylindrical and spherical waves of small amplitude in the gas-vapor-droplet mixtures with solid particles are analysed in [10], [11] and [12]. In this paper the

[1] To whom any correspondence should be addressed.

propagation of plane, cylindrical, and spherical waves of small amplitude in multifractional mixtures of gas with an arbitrary number of solid particles of different sizes and substances with significantly different thermal properties is studied.

2. Basic equations and dispersion equation.

Similarly to [8], the linearized system of equations of disturbed motion of multifractional polydisperse gas suspension, in a coordinate system relative to which the undisturbed medium is at rest, can be written as

$$\frac{\partial \rho_1'}{\partial t}+\rho_{10}\left(\frac{\partial v_1'}{\partial r}+\theta\frac{v_1'}{r}\right)=0,\quad \frac{\partial \rho_j^{i\prime}}{\partial t}+\rho_{j0}^{i}\left(\frac{\partial v_j^{i\prime}}{\partial r}+\theta\frac{v_j^{i\prime}}{r}\right)=0,\quad \left(j=\overline{2,N},i=\overline{1,M_j}\right)$$

$$\rho_{10}\frac{\partial v_1'}{\partial t}+\frac{\partial p_1'}{\partial r}+\sum_{j=2}^{N}\sum_{i=1}^{M_j}n_{0j}^{i}\left(6\pi a_j^i\mu_1\left(v_1'-v_j^{i\prime}\right)+6\left(a_j^i\right)^2\sqrt{\pi\mu_1\rho_1^{\circ}}\int_{-\infty}^{t}\frac{\partial}{\partial\tau}\left(v_1'-v_j^{i\prime}\right)\frac{d\tau}{\sqrt{t-\tau}}\right)=0,$$

$$\rho_{j0}^{i}\frac{\partial v_j'}{\partial t}=n_{0j}^{i}\left(6\pi a_j^i\mu_1\left(v_1'-v_j^{i\prime}\right)+6\left(a_j^i\right)^2\sqrt{\pi\mu_1\rho_1^{\circ}}\int_{-\infty}^{t}\frac{\partial}{\partial\tau}\left(v_1'-v_j^{i\prime}\right)\frac{d\tau}{\sqrt{t-\tau}}\right),\quad \left(j=\overline{2,N},i=\overline{1,M_j}\right)$$

$$\rho_{10}c_{p1}\frac{\partial T_1'}{\partial t}=\alpha_{10}\frac{\partial p_1'}{\partial t}-\sum_{j=2}^{N}\sum_{i=1}^{M_j}2\pi n_{0j}^{i}a_j^i\lambda_1 Nu_{1j}^{i}\left(T_1'-T_{\Sigma j}^{i\prime}\right),\tag{1}$$

$$\rho_{j0}c_j\frac{\partial T_j'}{\partial t}=-2\pi n_{0j}^{i}a_j^i\lambda_j^i Nu_{1j}^{i}\left(T_j^{i\prime}-T_{\Sigma j}^{i\prime}\right),\quad \left(j=\overline{2,N},i=\overline{1,M_j}\right)$$

$$\lambda_1 Nu_{1j}^{i}\left(T_1'-T_{\Sigma j}^{i\prime}\right)+\lambda_j^i Nu_{1j}^{i}\left(T_j^{i\prime}-T_{\Sigma j}^{i\prime}\right)=0,\quad \left(j=\overline{2,N},i=\overline{1,M_j}\right)$$

$$p_1'=\frac{C_1^2}{\gamma_1\alpha_{10}}\rho_1'+\frac{p_0}{T_0}T_1'.$$

The system of equations (1) for the parameter $\theta=0$ describes plane waves in a Cartesian coordinate system, when $\theta=1$ - cylindrical waves in a cylindrical coordinate system, when $\theta=2$ - spherical waves in a spherical coordinate system. Here and further, variables with subscript 1 refer to the carrier phase, with indexes $j=\overline{2,N}, i=\overline{1,M_j}$ to the particles of the dispersed phase of the radius a_j^i. Index 0 corresponds to the initial undisturbed state. Dashes above are used to denote disturbed parameters. Here and further ρ- reduced density, ρ° - true density , v - speed , α- volume content , p - pressure , μ_1- coefficient of the dynamic viscosity of the carrier medium, T_1- the temperature of the carrier phase, $T_{\Sigma j}^{i}$ - the temperature in the surface Σ- layer of a particle of the radius a_j^i, T_j^i- temperature of solid particles of the radius a_j^i, Nu_{1j}^{i} - the dimensionless (Nusselt number) coefficient of the carrier phase heat transfer with the border of the interface of particles of the radius a_j^i, Nu_j^i - the dimensionless (Nusselt number) coefficient of the heat transfer of particles of the radius a_j^i with the border of the interface, λ - coefficient of thermal conductivity.

The system of equations (1) is closed and can be used to investigate the propagation of acoustic disturbances in multifractional polydisperse gas suspension in flat, spherical and cylindrical cases.

Let us introduce into the system of equations (1) the phase velocity potentials and explore the solutions of the resulting system of equations in the form of progressive waves for disturbances

$$\eta'=A_\eta\psi\tag{2}$$

where

$\psi=\exp\left[i\left(K_* r-\omega t\right)\right]$ - for flat disturbances

$\psi = H_0^{(1)}(K_* r)\exp(-i\omega t)$ - for cylindrical disturbances

$\psi = \frac{1}{r}\exp\left[i(K_* r - \omega t)\right]$ - for spherical disturbances

$K_* = K + iK_{**}$, $C_p = \frac{\omega}{K}$, $\sigma = 2\pi\frac{K_{**}}{K}$.

Here A_η - the oscillation amplitude of parameters disturbances , K_* - the complex wave number , i - the imaginary unit, C_p - the phase velocity, K_{**} - the linear attenuation coefficient, σ - the attenuation decrement on the wave length, $H_0^{(1)}(z)$ - Hankel function , which is a combination of Bessel functions of the first and second kind of order zero $J_0(z)$ and $Y_0(z)$ $\left(H_0^{(1)}(z) = J_0(z) + iY_0(z)\right)$.

Substituting the velocity potentials and solutions of the form (2) in the system of equations (1), we obtain the following system of linear algebraic equations and from conditions of the existence of nontrivial solution of a system of linear algebraic equations the following dispersion equation is obtained

$$\left(\frac{C_1 K_*}{\omega}\right)^2 = V(\omega)D(\omega), \qquad (3)$$

where

$$V(\omega) = 1 + \sum_{i=2}^{M}\sum_{j=1}^{N_i}\frac{m_j^i}{1 - i\omega\tau_{vj}^{i*}},$$

$$D(\omega) = 1 + (\gamma_1 - 1)\frac{\sum_{i=2}^{M}\frac{c_i}{c_{p1}}\sum_{j=1}^{N_i}\frac{m_j^i}{1 - i\omega\overline{\tau}_{Tj}^{i*}}}{1 + \sum_{i=2}^{M}\frac{c_i}{c_{p1}}\sum_{j=1}^{N_i}\frac{m_j^i}{1 - i\omega\overline{\tau}_{Tj}^{i*}}}, \quad \overline{\tau}_{Tj}^{i*} = \tau_{Tj}^{i*} + \frac{m_j^i c_j}{c_{p1}}\tau_{T1j}^{i*}.$$

3. Results.

As an example the mixture of air and particles of aluminum of the radius $r_a = 10^{-6}$ m, sand of the radius $r_s = 8\cdot10^{-6}$ m and ice of the radius $r_l = 10^{-5}$ m. Fig. 1,2 illustrate the effect of inclusions mass content on the dependences of the relative speed of sound and attenuation decrement on the wavelength on the dimensionless oscillation frequency $\omega\tau_{vs}$ respectively. The calculated curves are built using the dispersion relation (10). The graphs show that with increasing mass content of particles the sound speed dispersion and dissipation of the waves increase. Accounting of the threefraction composition and difference of thermal parameters of fractions lead to the appearance of characteristic inflexions for the dependence of the relative speed of sound in the frequency field inversely proportional to the characteristic relaxation times of the phase velocities τ_{va}, τ_{vb} and τ_{vs} (Fig. 3). As shown in Fig. 4 , the difference between the sizes of inclusions and thermal parameters of particles fractions leads to appearance of three maxima for the dependence of attenuation decrement on the wavelength at the characteristic values , $\omega\tau_{va}$, $\omega\tau_{vl}$ and $\omega\tau_{vs} = 1$.

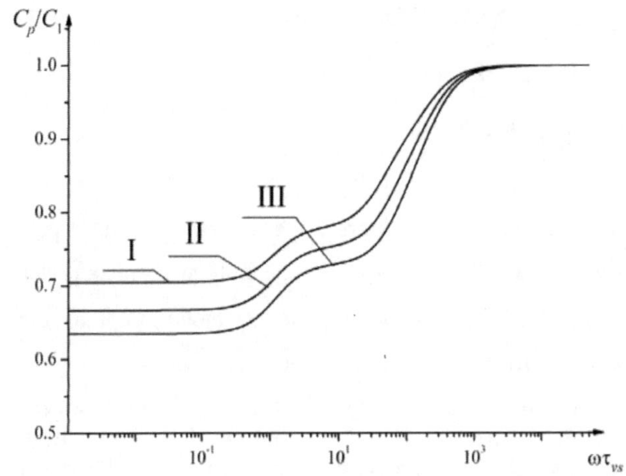

Figure 1. Dependence of the relative speed of sound on the dimensionless oscillation frequency for a mixture of air with particles of ice, aluminum and sand with different mass content of dispersed phase (I: m = 0.2, II: m = 0.4, III: m = 0.6)

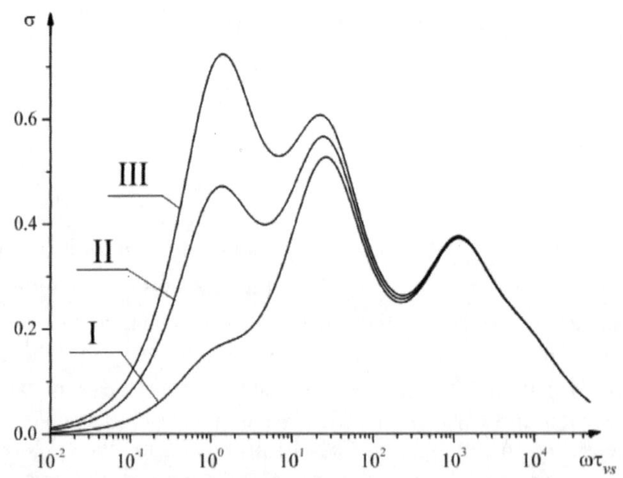

Figure 2 .The attenuation decrement dependence on the wavelength on the dimensionless oscillation frequency for a mixture of air with particles of ice, aluminum and sand with different mass content of dispersed phase (I: m = 0.2, II: m = 0.4, III: m = 0.6).

4. Conclusion.

This paper presents a closed system of linear differential equations of motion for multifractional polydisperse gas suspension. A dispersion equation, determining the propagation of plane, spherical and cylindrical disturbances of small amplitude derived. The dispersion curves are calculated. The influence of the parameters of the dispersed phase for three-fraction gas mixture with aluminum particles, carbon black, and ice on the dissipation and dispersion of sound waves is analysed. It is found that the difference of thermal properties and inclusions sizes significantly affects on the dynamics of weak waves in multifractional polydisperse gas suspension and it must be considered in the development of methods of acoustic diagnostics of considered environments.

References

[1] Nigmatulin R I 1991 Dynamics of Multiphase Media vol 1,2 (Washington: Hemisphere)
[2] Varaksin A Yu 2013 Fluid dynamics and thermal physics of two-phase flows: Problems and achievements *High Temperature* **51(3)** 421-455
[3] Temkin S 2005 *Suspension acoustics: An introduction to the physics of suspension* (Cambridge: Cambridge University Press) p 398
[4] Gumerov N A and Ivandaev A I 1988 Sound propagation in polydispersed gas suspensions *Journal of Applied Mechanics and Technical Physics* **5** 115-124.
[5] Gubaidullin D A and Ivandaev A I 1993 Effect of polydispersion on sound propagation in gas mixtures with vapor and liquid drops *Journal of Applied Mechanics and Technical Physics* **4** 75-83
[6] Gubaidullin D A 2003 Spherical and Cylindrical Low-Amplitude Waves in Polydisperse Fogs with Phase Transitions *Fluid Dynamics* **5** 85-94
[7] Nigmatulin R I, Ivandaev A I and Gubaidullin D A 1991 Effect of the non-monotonic dependence of the dissipation of sound on the concentration of drops in a suspension in gas *Dokl. USSR Academy of Sciences* **316(3)** 601-605
[8] Gubaidulln D A 1998 *Dynamics of twophase gas-vapor-droplet media* (Kazan: Kazan mathematical society) p 153
[9] Gubaidullin D A, Nikiforov A A and Utkina E A 2009 Propagation of acoustic waves to two-fractional gas mixtures with particles of different materials and sizes *Izvestiya vysshih uchebnykh zaavedenii. Problemi Energetiki* **1-2** 25-33.
[10] Gubaidullin D A, Nikiforov A A and Utkina E A 2011 Effect of the phase transformations on propagation of acoustic waves in two-fraction mixtures of gas with vapor, droplets and solid particles of different materials and sizes *High Temperature* **49(6)** 942-947
[11] Gubaidullin D A, Nikiforov A A and Utkina E A 2011 Acoustic waves in two-fraction mixtures of gas with vapor, droplets and solid particles of different materials and sizes in the presence of phase transitions *Fluid Dynamics* **1** 95-103
[12] Nikiforov A A, Utkina E A and Gafiyatov R N 2011 Acoustic disturbances in vapor-gas-liquid systems *Vestnik of Lobachevsky State University of Nizhni Novgorod* **4(3)** 1017-1018

IOP Conf. Series: Journal of Physics: Conf. Series **789** (2017) 012019 doi:10.1088/1742-6596/789/1/012019

Sound waves in multifractional liquids with bubbles

D A Gubaidullin[1] and R N Gafiyatov[1]

[1] Institute of Mechanics and Engineering, Kazan Science Center, Russian Academy of Sciences, 2/31 Lobachevsky str., Kazan, 420111, Russia

E-mail: gafiyatov@mail.ru

Abstract. The propagation of sound waves in multifractional mixtures of liquid with vapor–gas and gas bubbles of different sizes and different compositions with phase transitions is studied. The dispersed phase consists of $N+M$ fractions having various gases in bubbles and different in the bubbles radii. Phase transitions accounted for N fractions. The total bubble volume concentration is small (less than 1%). The dispersion relation is derived and dispersion curves is built. The evolution of the weak pulsed perturbations of the pressure in this mixture was calculated numerically.

1. Introduction

A propagation of acoustic waves in liquid with gas bubbles is studied intensively both theoretically and experimentally. Nigmatulin [1], Nakoryakov et al. [2], Temkin [3] considered various basic aspects of acoustics of multiphase flows. Hao and Prosperetti [4] obtained the results of study of the vapor bubble dynamics in liquid. Commander and Prosperetti [5] presented a model of the propagation of plane waves with small amplitude in a mixture of liquid and of gas bubbles. This model is adequate for the frequencies far from resonance at the volume content of a disperse phase of 1-2%. Nigmatulin et al. [6] showed the need of allowing for the effect of liquid compressibility for problems of acoustics of bubbly liquids. Shagapov [7] considered a problem of propagation of small disturbances in liquids with polydisperse bubbles. Nigmatulin et al. [8] studied the small radial oscillations of vapor-gas bubbles in liquid under the influence of an acoustic field. They showed that capillary effects and phase transitions lead to the new resonant frequency of small bubbles. This frequency is different from the Minnaert frequency. The propagation of acoustic waves in a two-fractional mixture of liquid with bubbles is investigated in [9-11]. The propagation of sound waves in two-fractional liquids with polydisperse bubbles is investigated in [12-15]. A mathematical model of the propagation of sound waves in the two-fractional mixture of liquid with polydisperse vapor-gas and gas bubbles with phase transitions is shown in [16-18]. The problem of reflection of an acoustic wave from a two-layer medium containing a layer of bubble liquid is considered [19]. The propagation of acoustic waves in multifractional mixtures of liquid with gas and vapor-gas bubbles of different sizes and different compositions is studied in [20-21].

In the present paper, a propagation of sound waves in multifractional liquids with bubbles is studied.

2. Dispersion relation

Linearized equations for one-dimensional disturbances in a multifractional mixture of liquid with bubbles are obtained from the general equations of motion for bubbly mixtures [1]. Solving this system this dispersion relation is obtained

$$\left(\frac{K_*}{\omega}\right)^2=\frac{1}{C_f^2}+\frac{\rho_{10}}{p_0}\frac{\left(\sum_{i=1}^{M}H_{3i}+\sum_{j=1}^{N}H_{3j}\right)\left(\sum_{i=1}^{M}H_{1i}+\sum_{j=1}^{N}H_{1j}\right)}{\sum_{i=1}^{M}\left(\frac{m_i}{\tau_{T1i}}+H_{2i}\right)+\sum_{j=1}^{N}\left(\frac{m_j}{\tau_{T1j}}+H_{2j}\right)-i\omega}+$$
$$+\frac{\rho_{10}}{p_0}\sum_{i=1}^{M}\left(\frac{\alpha_{20i}}{N_{Ri}}\left(1-\frac{M_{4i}}{M_{3i}}\right)\right)+\frac{\rho_{10}}{p_0}\sum_{j=1}^{N}\left(\frac{\alpha_{20j}}{N_{Rj}}\left(1-\frac{M_{4j}}{M_{3j}}\right)\right) \quad (1)$$

Dispersion relation (1) (i.e. the function of the complex wave number K_* on the frequency ω) determines a propagation of sound waves in multifractional mixtures of liquid with vapor-gas and gas bubbles (different initial radii, different initial volume contents and different thermal properties of fractions) with the interphase diffusion mass transfer.

3. Results

Figure 1 shows the phase velocity and attenuation coefficient in three-fraction mixture of water with bubbles, for the radius of vapor-air bubbles of $a_{0N1}=1.5\cdot10^{-3}$ m, carbon dioxide with vapor bubbles – $a_{0N2}=10^{-3}$ m, helium bubbles – $a_{0M1}=2\cdot10^{-3}$ m, calculated by dispersion relation (1). In unperturbed state a pressure of a three-fractional mixture is $p_0=0.1$ MPa. Curves *1* are obtained for volume concentration of disperse phase $\alpha_2=0.01$, curves *2* – $\alpha_2=0.015$, curves *3* – $\alpha_2=0.021$.

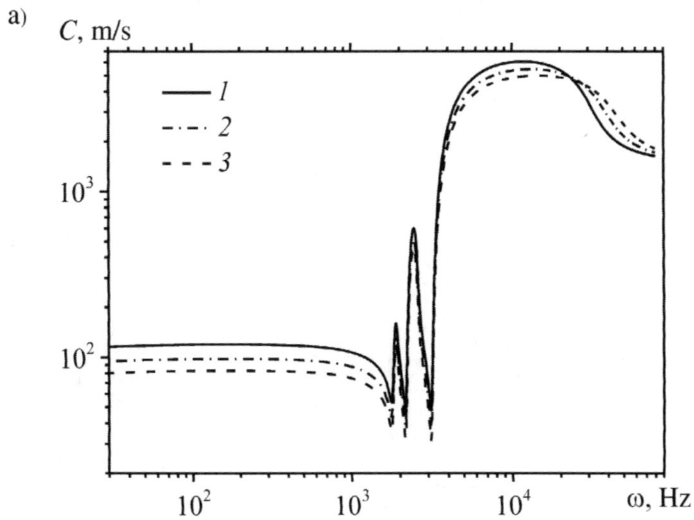

Figure 1. Frequency dependences of the phase velocity (a) and of the attenuation coefficient (b) for the three-fraction mixture of water for three different volume concentration of fractions: α_2=0.01 (solid line), α_2=0.015 (dash-dotted line), α_2=0.021 (dashed line).

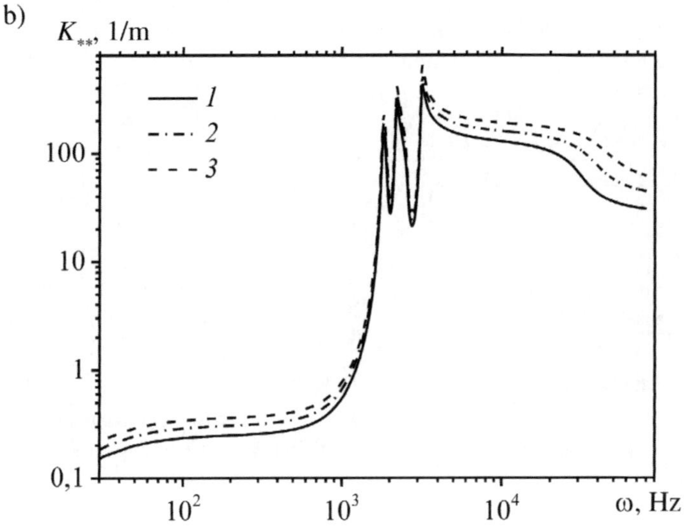

Figure 1 shows that increasing the volume concentration of bubbles leads to decreasing the value of the phase velocity and increasing the value of the attenuation coefficient.

Let us consider now the evolution of the pressure pulses of the type of the Gaussian curve, produced on the boundary of the bubble medium, when the initial shape of pulse is described by the following function: $p(0,t)=\exp\left[-((t-t_*)/N)^2\right]$, where t_* is half of the pulse duration and N is the parameter determining the width of the pulse peak. The calculations were performed by means of the dispersion relation (1) using the subroutines of the fast Fourier transformation.

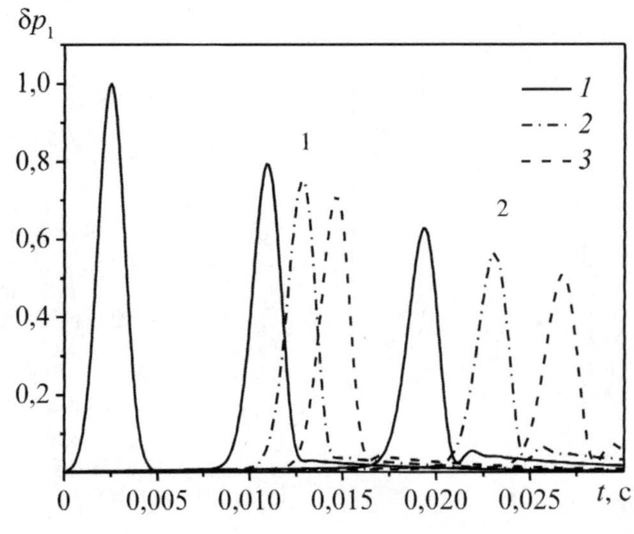

Figure 2. Evolution of pulsed pressure perturbation in the three-fraction mixture of water for three different volume concentration of fractions: α_2=0.01 (solid line), α_2=0.015 (dash-dotted line), α_2=0.021 (dashed line).

Calculated profiles built with the same parameters as in Figure 1. The character of pulse propagation affect the basic low-frequency harmonics, the propagation velocity is close to the equilibrium speed of sound. Therefore, increasing the volume concentration of bubbles leads to a significant decrease in the propagation velocity of the pulse disturbance.

4. Conclusions

The propagation of sound waves in multifractional mixture of liquid with vapor-gas and gas bubbles taking into account the phase transitions is studied. The influence of the volume concentration of bubbles on the dispersion curves and the propagation of the pressure pulses is shown. Changing the volume concentration of bubbles has an effect on the rate of decay of long-wave impulse disturbances.

References

[1] Nigmatulin R I 1991 *Dynamics of Multiphase Media* (New York: Hemisphere)
[2] Nakoryakov V E, Pokusaev B G and Shreiber I R 1993 *Wave Propagation in Gas-Liquid Media* (New York: CRC Press)
[3] Temkin S 2005 *Suspension Acoustics: An Introduction to the Physics of Suspensions* (New York: Cambridge University Press)
[4] Hao Y and Prosperetti A 1999 *Physics of Fluids* **11** 2008
[5] Commander K W and Prosperetti A 1989 *Journal of the Acoustical Society of America* **85** 732
[6] Nigmatulin R I, Shagapov V Sh and Vakhitova N K 1989 *USSR Academy of Sciences Doklady* **304** 1077
[7] Shagapov V Sh 1977 *Journal of Applied Mechanics and Technical Physics* **18** 77
[8] Nigmatulin R I, Khabeev N S and Nagiev F B 1981 *International Journal Heat and Mass Transfer* **24** 1033
[9] Gubaidullin D A, Nikifirov A A and Gafiyatov R N 2012 *High Temperature* **50** 250
[10] Gafiyatov R N, Gubaidullin D A and Nikiforov A A 2013 *Fluid Dynamics* **48** 366
[11] Gubaidullin D A and Gafiyatov R N 2014 *J. Phys.: Conf. Ser.* **567** 012020
[12] Nigmatulin R I, Gubaidullin D A and Fedorov Yu V 2012 *Doklady Physics* **57** 451
[13] Nigmatulin R I, Gubaidullin D A and Fedorov Yu V 2013 *Doklady Physics* **58** 261
[14] Gubaidullin D A and Fedorov Yu V 2013 *Journal of Applied Mathematics and Mechanics* **77** 532
[15] Gubaidullin D A, Gubaidullina D D and Fedorov Yu V 2014 *J. Phys.: Conf. Ser.* **567** 012011
[16] Gubaidullin D A and Fedorov Yu V 2015 *Fluid Dynamics* **50** 61
[17] Gubaidullin D A and Fedorov Yu V 2016 Acoustical Physics **62** 179
[18] Gubaidullin D A, Gubaidullina D D and Fedorov Yu V 2016 *J. Phys.: Conf. Ser.* **669** 012011
[19] Gubaidullin D A and Fedorov Yu V 2016 Doklady Physics **61** 499
[20] Gubaidullin D A, Nikiforov A A and Gafiyatov R N 2015 *High Temperature* **53** 240
[21] Gubaidullin D A, Nikiforov A A and Gafiyatov R N 2016 *J. Phys.: Conf. Ser.* **669** 012019

Technical procedure features of nickel dry-powder developer manufacturing by method of plasma-electrolytic dispergating

A Z Gumerov, R T Nasibullin, L R Sarimov

Branch of Kazan Federal University in Naberezhnye Chelny, 68/19 Mira Street, Naberezhnye Chelny, 423810, Russia

E-mail: Nasibullin.ramil@mail.ru

Abstract. The following work describes means of nickel dry-powder developer manufacturing by plasma-electrolytic dispergating. For that purpose, a plasma-electrothermic device was developed that allows to manufacture dry-powders with wide range of parameters. Technical conditions for nickel dry-powder developer manufacturing by method of plasma-electrolytic dispergating were defined. By factorial design method regression equations were obtained, that make it possible to plan the process of dry-powders manufacturing with set average particle size and set manufacturing process rate. The described means allows to manufacture nicol dry-powder developer with a particle size of aprox. 100 nm.

1. Introduction

Dry-powder developers based on ferrous-ferric oxide, maghemite and other metal ferrites are widely used in mechanical engineering and aerospace industry through their unique properties. Magnetic features of the dry-powder developers make them effective in high-frequency magnetic conductors units and SHF cannels manufacturing; moreover, their relatively high conductivity and outstanding resistance to anodic dissolution make them valuable in the manufacturing of nonexpendable electrodes for electrochemical industry as well as of induced current cathodic protection and thermal protection, corrosion-resistant, radar-absorbent coating.

The modern technology extension offers a challenge of constant improving of dry-powder developers performance and technical features. Main disadvantage of dry-powder developers is heterogeneity of the chemical composition as a result of unconverted feedstock incorporation appearance. Solving this problem by manufacturing dry-powder developers through nicol plasma-electrolytic dispergating might be advantageous as the previously mentioned process shows simplicity of implementation, low cost and availability of basic materials, broad range of possibilities for automation, which promotes homogeneous structure of dry-powder developers.

There are various means and devices for electrical discharge generation under atmospheric pressure between metal and electrolytic electrodes [1, 2]. Complexity of such devices is a disadvantage. To manufacture nickel dry-powder developer through plasma-electrolytic dispergating a plasma-electrothermic device was developed; one of the features of this device is simplicity of its design.

In order to regulate the process of nickel dry-powder developers manufacturing, the device is provided with control system. This particular way or regulation is of a great interest, as detecting element is simple. A current sensor can be used as a detecting element. Current sensor might be installed on the out-port of discharge power supply unit, before the discharge chamber. To process the signals from this sensor a special control system should be developed, that would transform the current sensor signals into drive engine control signals. By combinations of control actions on technic parameters it is possible to set various process modes to manufacture dry-powder developers and to provide their quality and necessary optimum properties. Such kind of control system implements inverse link between input parameter (current) and output parameter (electrode position).

2. Research methods

The device (Figures 1) works in a wide range of parameters of electrode gap $l = 1 \div 100$ mm, electrode diameter $d = 3 \div 15$ mm, discharge current $I = 0{,}01 \div 10$ A, discharge voltage $U = 0{,}1 \div 4$ kV, volume electrolyte velocity $G = 15 \div 34$ g/s, electrolyte pump velocity $\upsilon = 0{,}02 \div 0{,}05$ m/s.

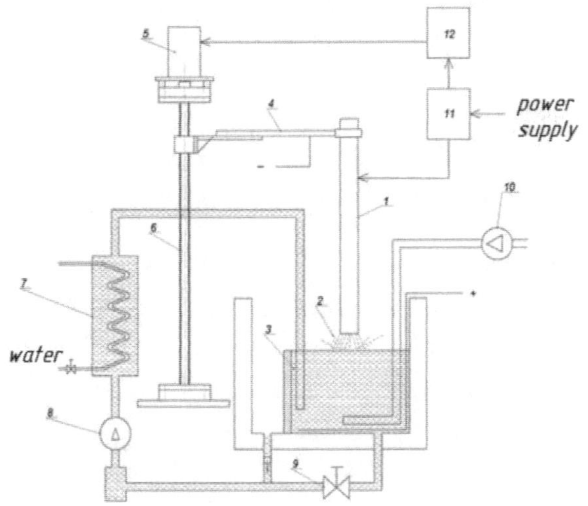

Figure 1. Flow chart of plasma device (PD) for nickel dry-powder developer manufacturing under atmospheric pressure
1 - powdered metal electrode; 2 - manufactured powder; 3 - continuous electrolyte flow bath; 4 – electrode holder; 5 - step motor; 6 - screw-type shaft; 7 - tank with heat exchange unit; 8 - eletrolyte transfer pump; 9 - isolation valves; 10 - air compressor; 11 - current sensor; 12 - PD control system.

With suggested means the nickel dry-powder developer manufacturing was achieved through discharge allowance between flow electrolytic anode from utility water and nickel electrode Mk N-0 (H-0 ru); setting the current within $I = 0{,}6 - 1{,}3$ A, voltage $U = 400 - 800$ V under atmospheric pressure, electrode gap $l = 1 - 30$ mm. Dry-powder developer manufactured through method of plasma-electrolytic dispergating was collected in container and purged with ethanol. Ethanol overflow was removed, the powder was air-cured up to its constant weight. After that the powder weight was measured with microchemical balance Ohaus Adventurer pro AV2264.

Under atmospheric pressure, $jc = 1$ A/cm^2, $U = 600$ V, powder production rate is 14 g/h. Decrease of electrode diameter leads to increase of dry-powder developer production rate. Microscopic examination with light microscope Mk Micromed MET, as well as electron microscope Zeiss EVO-40 showed that nickel powder particles have spherical shape with diameter within 0,1 to 1,5 μm. Consequently, depending on the operating mode and conditions it is possible to manufacture nickel dry-powder developer of particular degree of dispersion; in comparison to another know method [3] the outcome is as much as 6 times more [4].

3. Results and discussions

By factorial design method coefficients of regression equation were calculated and checked, where linear effects and interaction effects were taken into account. Based on preliminary investigations were chosen main factors of the process that affect manufacturing rate of dry-powder developer as well as degree of dispersion the most [5, 6]: j – current intensity on anode, A/cm^2; P – pressure, Pa; L – electrode gap, mm.

As output parameters the following were picked: P – dry-powder production rate, g/h; D – dry-powder particles average diameter, μm.

According to the factorial design method equations of regression were calculated for average size of particles and for production rate using nickel N-O as metal cathode.

Table 1. Results of the factorial design method equations

Set of tests	Average particles size D_{av}, μm	Dispersion assessment S_D^2	Production rate P_{av}, g/h	Dispersion assessment S_p^2
1	0.1000	0.0011	8.6333	0.2022
2	0.3000	0.0066	7.5000	0.1666
3	0.2000	0.0066	14.0000	2.0000
4	0.4000	0.0066	8.3333	1.5555
5	0.5666	0.0022	13.0000	0.6666
6	0.6333	0.0022	10.6666	0.2222
7	0.5333	0.0022	8.0000	0.6666
8	0.4333	0.0155	8.9000	0.4866

Coefficients calculations for equations of regression:

$$b_0 = \frac{1}{N}\sum_{j=1}^{N} y_j \tag{1}$$

$$b_l = \frac{1}{N}\sum_{j=1}^{N} x_{jl} \cdot y_j \tag{2}$$

$$b_{lm} = \frac{1}{N}\sum_{j=1}^{N} x_{jl} \cdot x_{jm} \cdot y_j \tag{3}$$

$$b_{lmn} = \frac{1}{N}\sum_{j=1}^{N} x_{jl} \cdot x_{jm} \cdot x_{jn} \cdot y_j \tag{4}$$

Conforming to it (1) – (4), obtained,

for average size of particles: b_{D0} = 0.3958; b_{D1} = 0.1458; b_{D2} = - 0.0041; b_{D3} = 0.0458; b_{D12} = - 0.0541; b_{D13} = - 0.0541; b_{D23} = - 0.0208; b_{D123} = - 0.0208.

for dry-powder manufacturing production rate: b_{P0} = 9.8792; b_{P1} = 0.2625; b_{P2} = - 0.0708; b_{P3} = - 1.0291; b_{P12} = - 1.6208; b_{P13} = 0.6708; b_{P23} = - 0.1625; b_{P123} = 0.9708.

Considering orthogonality of multifactor experiment, evaluation of dispersion degree is the same for each coefficient calculation.

Subsequently:

$$S_{(b_D)} = \frac{S_{(D)}}{\sqrt{8}} = \frac{0,0254}{\sqrt{8}} = 0,0090 \tag{5}$$

$$S_{(b_P)} = \frac{S_{(P)}}{\sqrt{8}} = \frac{0,2887}{\sqrt{8}} = 0,1020 \tag{6}$$

Test of significance of equation of regression coefficients was conducted through Student's coefficient, eq. 2,12.

Equation coefficient is to be considered significant if the following condition works:

$$|b| \geq T \cdot S_{bi}$$

Conforming to it and (5) – (6):

$$|b_D| \geq 2{,}12 \cdot 0{,}0090 = 0{,}0190$$

$$|b_P| \geq 2{,}12 \cdot 0{,}1020 = 0{,}2163$$

Therefore, coefficients b_{D2} are considered as insignificant for equation of dependence regression of average particles size. Coefficients b_{P2} and b_{P23} are considered insignificant for equation of dependence regression of dry-powder manufacturing production rate.

In such case equations of regression are written as:

For average particles size:

$$D_{Nic} = 0{,}395 + 0{,}145 \cdot x_1 + 0{,}045 \cdot x_3 - 0{,}054 \cdot x_1 \cdot x_2 - 0{,}054 \cdot x_1 \cdot x_3 - 0{,}02 \cdot x_2 \cdot x_3 - 0{,}02 \cdot x_1 \cdot x_2 \cdot x_3$$

For dry-powder manufacturing production rate:

$$P_{Nic} = 9{,}879 + 0{,}262 \cdot x_1 - 1{,}029 \cdot x_3 - 1{,}62 \cdot x_1 \cdot x_2 + 0{,}67 \cdot x_1 \cdot x_3 + 0{,}97 \cdot x_1 \cdot x_2 \cdot x_3$$

4. Conclusion

With help of multifactor experiment method equations of regression were deduced. The equations allow to pick the necessary parameters of processing for nickel dry-powder developer manufacturing with required particles size and appropriate production rate. The deduced regression equations can be used to calculate the operation modes for plasma devices with both flow-type and non-flow-type electrodes in a wide range of current and discharge voltage, pressure and electrode gap.

References

[1] Gaisin F M 1992 Physical Processes in Gaseous Discharges with Solid, Liquid and Plasma Electrodes (*Dissertation for the degree of Ph.D*) (Moskow)

[2] Gaisin F M, Gaziatullina F A and Dautov G Y 1983 Device for glow discharge generation under atmospheric pressure *License* 1088086

[3] Gumerov A Z 2006 Properties of Spray Multichannel Discharge Between Electrolyte Anode (flow-type, non-flow-type) and Metal Cathode under Atmospheric Pressure (*Dissertation for the degree of Ph.D in Engineering*) (Kazan)

[4] Gaisin A F, Nasibullin R T, Sarimov L R and Fakhrutdinova I T 2012 Means of manufacturing of metallic powder *License* 2486032

[5] Gaisin A F and Sarimov L R 2011 *XXXVIII International (Zvenigorod) Plasma Physics and Controlled Fusion Conference* p 287

[6] Gaisin A F and Nasibullin R T 2011 *Plasma Physics Reports.* **37(10)** 896

High Frequency High Power Full Bridge Converter for Plasma Torches

A Hakki[1], K Sadikov[1], N Kashapov[1]
[1]Physical Institute,
Kazan (Volga-Region) Federal University,
420008, Kremlyovskaya St, bdg.18, Kazan, Russia

E-mail: hakki.adnan@gmail.com

Abstract. This paper presents a full bridge converter with a high speed rectifier and filter for plasma torches. Full bridge circuit with high frequency pulse width modulation control is used and high frequency transformer to insulate is adopted. For Argon (Ar) gas Plasma Torches, and by increasing the pulse width the average power consumption has modified from 1KW to 5.6KW (positive load). For Nitrogen (N2) gas Plasma Torches, and by increasing the pulse width the average power consumption has reduced from 7.7KW down to 3KW (negative load). For the Plasma Torches load the frequency of converter was steady at 50KHz.

Key words. Full Bridge Converters, Power Pulse, Mosfet, Filters, Plasma Torches

1. Introduction

Full bridge converters are recommended to use for high power applications. There are many applications for pulse power supplies. This paper provides a new method to improve the performance and flexibility for pulse power supplies .This design includes using proper adjustment and full bridge converter to obtain the output voltage with rapid raising time. The output voltage in this method depends on the width of pulse control signals. The control circuit for this full bridge converters is very simple and cheap, high frequency full bridge converters using ferrite material core for transformers and very fast power Mosfet transistors as switches, so these kind of power supplies are small size, compatible, and the efficiency factor is very high [1,2,3].

2. Experimental Setup

Fig. 1. Shows the schematic of plasma torches power supply, when Switch Q1 and Q3 turns on, current flows from V_{IN} through Q1, primary of transformer T_1, and Q3 to ground. At the same time the current flows in the secondary and gets rectified and filtered. When Q2 and Q4 turns on current flows from V_{IN} through Q2 primary of transformer T_1, and Q4 and to ground, the voltage is reflected to the secondary proportional to the turns ratio. The rectified voltage is filtered. The current though the transformer primary reverses every alternate half cycle. When a short circuit happens at the output load (in this case the feedback voltage should be more than 4V) the current transformer T_4 will sensor this short and the control unit will automatically stops generate the pulses (U_1, U_2), so this circuit is protected from short circuit. For the transformers T2, T3 turns ratio $n = 4$, the maximum amplitude of signals driver is $V_{gs} = 15V$. To start plasma process it need a one high voltage pulse, the control unit generates the pulse U_3 , and then the ignition circuit will give a high voltage pulse, the amplitude of this pulse is about 6KV , the width is about $2\mu sec$, the amplitude and the width of ignition pulse depend on the parameters of plasma chamber [4,5,6].

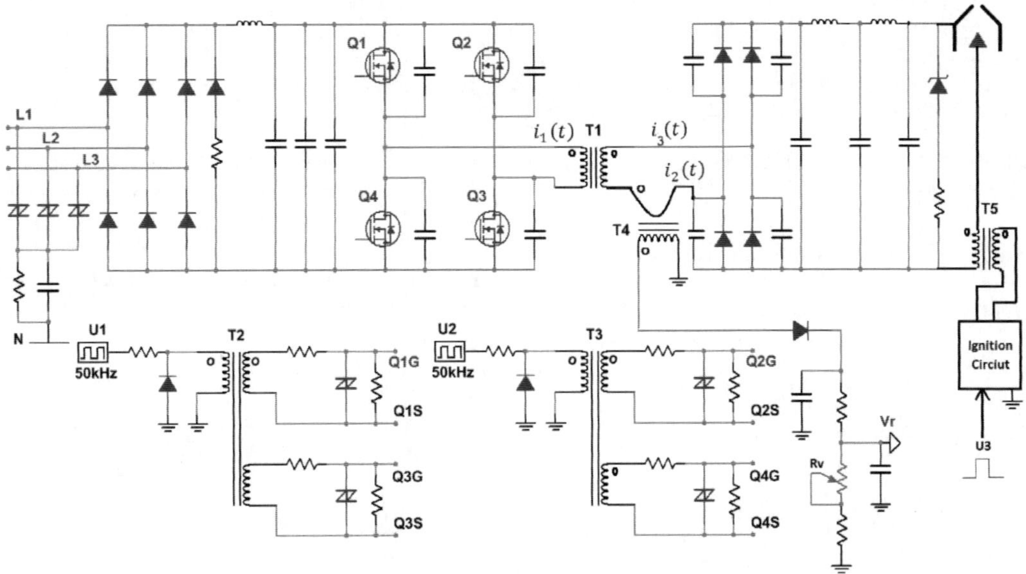

Figure 1. Schematic of Plasma Torches Power Supply

3. Results and discussion

Fig. 2. shows the currents of primary and secondary of transformer T_1 [7,8].

$$I_{1,RMS} = \sqrt{\frac{1}{2T}\int_0^{2T} i_1{}^2(t)dt} = \frac{N_2}{N_1} I_{o,max}\sqrt{\delta}$$

$$I_{2,RMS} = I_{3,RMS} = \sqrt{\frac{1}{2T}\int_0^{2T} i_2{}^2(t)\, dt}$$

$$I_{2,RMS} = I_{3,RMS}\frac{1}{2} I_{o,max}\sqrt{\delta + 1}$$

Where $Duty\ Ratio = \delta = \frac{T_{on}}{T}$, $T = \frac{1}{f}$

The output voltage V_o is given by

$$V_o = \frac{1}{T}\int_0^T V_{oi}(t)\, dt$$

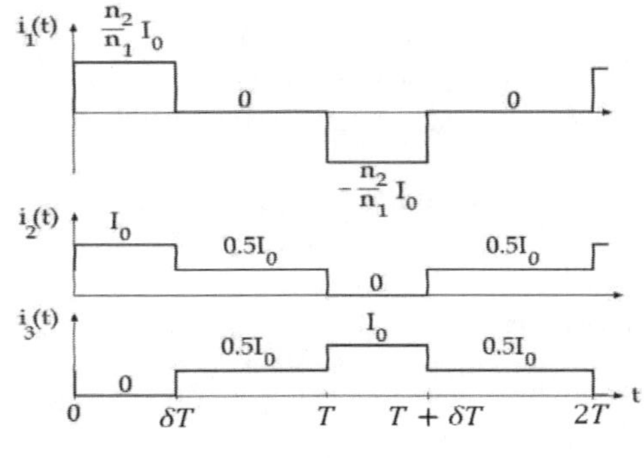

Figure 2. RMS currents

$$V_o = \frac{1}{T}\left(2\int_0^{T_{on}} \frac{n_2}{n_1} V_d\, dt + \int_{T_{on}}^{T} 0\, dt\right)$$

$$V_o = 2\frac{n_2}{n_1}\frac{1}{T} V_{IN}\, \delta\, T$$

$$\frac{V_o}{V_{IN}} = 2\frac{n_2}{n_1}\delta = 2\, n\, \delta$$

Where T_{on} is the width of pulse U_1, U_2 as shown in Fig. 4.

Where n is the turns ratio of the transformer T_1.
In this case $n = 1$

$$V_o = 2. V_{IN}. \delta$$

Working frequency:

$$f \approx 50 \text{ KHZ} \Rightarrow T = \frac{1}{f} = 20\mu sec$$

V_{IN} Input DC voltage

$$V_{IN} \approx 1.41 V_{p-p} \approx 535\text{V}$$

Where $V_{p-p} = 380V$ phase to phase voltage.
by changing the period T_{on} (the width of pulses U_1, U_2) regulate the output voltage value.

$$K = \mu_0 An * \frac{R_{out}}{L_c},$$

Where μ_0 is the magnetic constant, A area of one loop, n number of turns, R_{out} the output resistance.
Thus, the load current is determined by the formula:

$$i = \frac{1}{k} * V_{out}$$

Finally the output current is given by:

$$i_{out} = \frac{1}{k} * V_{out} = \frac{1}{0.5} * V_{out} = 2 * V_{out}$$

$$0 \le \delta \le 0.5$$

In our case $\delta_{Max} = 0.45$
Delay time (death time) $\text{T.D} = \frac{T}{4} - \frac{T_{on}}{2}$
Note that during delay time T. D, all transistors are in cut off situation.

$$\text{T.}D_{min} = \frac{T}{4} - \frac{T_{onmax}}{2} = 0.5\mu sec$$

$T_{onmax} = T.\delta_{Max} = 9\mu sec$
Target efficiency of full bridge converter $\eta = 0.9$

Input power Assuming 90% Efficiency the input power is:

$$P_{IN} = \frac{P_{out}}{0.9} = 8.5KW$$

Maximum average Input current:

$$I_{IN} = \frac{P_{IN}}{V_{INmin}} = 17A$$

Parameters of the HF transformer

the core of the transformer T_1 must be ferrite material.

Parameters	The value
V_{IN}	535 V
V_{INmax}	570 V
V_{INmin}	500 V
V_{outmax}	550 V
I_{IN}	17 A
F	50 KHz
η	90%
D_{max}	0.45

4. Conclusion

In short, the working region of the plasma torches can be in the forward conduction region (positive resistor), or in the reverse region (negative resistor), for Argon gas (Ar) plasma, and by increasing the pulse width the average power consumption has modified from 1KW to 5.6KW (positive load) [9-16]. For Nitrogen gas (N2) plasma, and by increasing the pulse width the average power consumption has reduced from 7.7KW down to 3KW (negative load). for the both plasma load (Ar,N2) the frequency of full bridge converter was steady at 50KHz. In other words, by varying the value of resister (Rv) showing in fig.1 the load current can be adjusted.

References

[1] Schonknecht A., De Doncker R., Novel Topology for Parallel Connection of Soft-Switching High-Power High-Frequency Inverters, IEEE Transactions on Industry Applications, Vol. 39, (2003), 550 – 555
[2] D. Aggeler, J. Biela, and J. W. Kolar, “A compact, high voltage 25 kW, 50 kHz DC-DC converter based on sic jfets,” in Proc. Twenty-Third Annual IEEE Applied Power Electronics Conference and Exposition APEC 2008, 24–28 Feb. 2008, pp. 801–807.
[3] K. A. Alberto Guerra and S. Fimiani, “Ultra-fast recovery diodes meet todays requirements for high frequency operation and power ratings in smps applications,” International Rectifier, Tech. Rep., 2000.
[4] R. Severns (2005). Design of Snubbers for Power Circuits. Available from: http://www.cde.com/tech/design.pdf
[5] F. Wang, A. Kuthi, and M. A. Gundersen, “Compact high repetition rate pseudo spark pulse generator,” vol. 33, no. 4, pp. 1177–1181, Aug. 2005.
[6] Magnetic Core. Wikipedia, Available from: http://en.wikipedia.org/wiki/Magnetic_core, (2010).
[7] Design of Planar Power Transformers. Ferroxcube, Available from: http://www.ferroxcube.com/appl/info/plandesi.pdf
[8] K. Venkatachalam, C. R. Sullivan, T. Abdallah, and H. Tacca, “Accurate prediction of ferrite core loss ith nonsinusoidal waveforms using only steinmetz parameters,” in Proc. IEEE Workshop on Computers in Power Electronics, 3–4 June 2002, pp. 36–41.
[9] Gavrilova V A, Fazlyyyakhmatov M G and Kashapov N F 2013 *J. Phys.: Conf. Ser.* **479** 012010
[10] Fayrushin I, Kashapov N and Dautov I 2014 *J. Phys.: Conf. Ser.* **567** 012009
[11] Kashapov L N, Kashapov N F and Kashapov R N 2013 *J. Phys.: Conf. Ser.* **479** 012011
[12] Kashapov L N, Kashapov N F and Kashapov R N 2014 *J. Phys.: Conf. Ser.* **567** 012025
[13] Zaripov R G, Kashapov N F, Tkachenko L A and Shaydullin L R 2016 *J. Phys.: Conf. Ser.* **669** 012053
[14] Denisov D G, Kashapov N F and Kashapov R N 2015 *IOP Conference Series: Materials Science and Engineering* **86** 012005
[15] Saifutdinov A I, Fairushin I I and Kashapov N F 2016 *JETP Lett.* **104** 180–185
[16] Sadikov K, Fayrushin I, Shamsutdinov A, Kashapov N 2016 *J. Phys.: Conf. Ser.* **669** 012048

Investigation of mechanical properties of thermal coatings obtained during plasma spraying of powder zirconium dioxide

A R Ibragimov[1], T A Ilinkova[2], L N Shafigullin[1], A I Saifutdinov[1]

[1]Kazan Federal University
[2]Kazan National Research Technical University AN Tupolev - KAI

E-mail: ibragimov.a.r@mail.ru

Abstract. Thermal coatings of zirconia partially stabilized with yttrium, deposited by low-temperature plasma, are the basis for the thermal protection of aircraft engine. At the same time there is an actual problem of selection of coating systems "ceramic layer - underlayer" of great thickness, providing better thermal protection, but having low strength characteristics due to the accumulation of internal stresses. To determine the optimal thickness of the test method used in the 4-point bending to allow the surface coating to explore in the elastic-plastic behavior of the field of coatings and strength. Based on the experimental results established the role of underlayer in the formation of the complex mechanical properties of thermal barrier coatings. With a well formed underlayer (PVNH16U6) system becomes sensitive to a change in thickness of the coating, to optimize the response on the strength and deformation criteria. According to the results the optimum ratio of the thickness of the ceramic layer and the underlayer should be regarded as the ratio of 3-5 for which the highest strength values were obtained for all the test coating systems.

1. Introduction

In modern aircraft engine application of thermal barrier coatings (TBC) based on zirconium dioxide detail on combustor of a gas turbine engine, can significantly reduce the temperature at the work piece surface, to eliminate local overheating and thus, ensure the service life of the combustion chamber. Designers put a problem applying TBC to the thickness of the heat of the ceramic layer at 300 microns or more. However, with increasing thickness of the coating that is applied using a low temperature plasma dramatically increases the number of structural defects as pores, cracks, and the level of residual stresses, which drastically reduces the durability of coatings [1-11].

Ceramics which are sprayed, as well as metallic sub layer has a fine-grained structure. The test zirconium oxide has a polymorphism. The low-temperature phase monoclinic (m-ZrO2) is stable to 1170 ° C and above this temperature it is transformed into a tetragonal (t-ZrO2) and at 2370 ° C - the cubic (c-ZrO2). At 2680 ° C zirconium dioxide begins to melt. Phase changes occurring both during heating and cooling associated with significant changes of volume, which can be eliminated zirconium stabilized yttrium oxide, cerium (Y2O3, CeO2). Spraying with low-temperature plasma creates a special condition [12], zirconium crystallizes martensitic mechanism. T'- metastable phase is formed in the coating. However, depending on the type of the stabilizer and its concentration in the coating composition, except t'- metastable tetragonal phase may be present, cubic and monoclinic phases. These phase transformations lead to the appearance of internal stresses. Their knowledge is of great practical importance, since it allows improved multivariable process with low-temperature plasma deposition and the establishment of reasonable criteria for coating systems. For testing coated plates we use the four-point bending method. In contrast to the three-point at which the pure bending, in this case, provided the load distribution and lateral shifts (Fig.1).

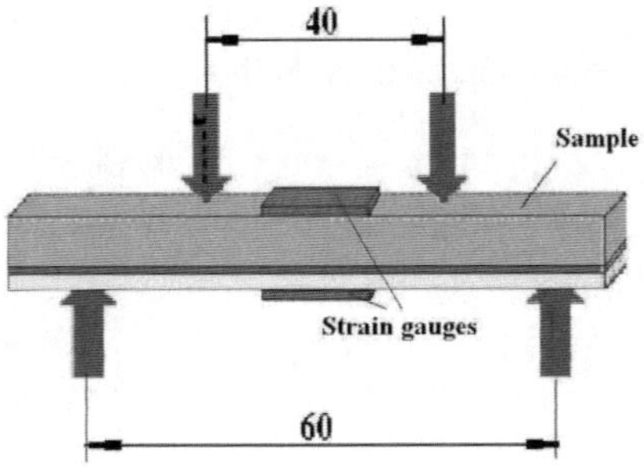

a) Scheme

б) Installation complete

Fig. 1. Schematic illustrating 4-point bending

This mainly shows the properties of coating systems «base metal - underlayer - ceramic layer».

2. Methods

For the investigations, we used plate dimensions which are equal to 2*10*80 mm. For spraying, we used plasma installation UPU-8M. The superalloy VH-4A (EP648) was chosen as a base metal used for the manufacture of aircraft parts in the form of a sheet 2 mm thick. VH-4A alloy has a high content of chromium, has high heat resistance and corrosion resistance in hostile environments at temperatures up to 1100 ° C. As the underlayer materials three types of powders were chosen: two nickel-based (PVNH16U6It; PNH20K20U13) and one based on cobalt powder AMDRY 995C. For spraying the external ceramic layer were used three types of zirconia powder stabilized with 8.7% yttrium oxide: SIO-7-10-50, Z7Y-10-90 и Metco 204 NS. The loading device as tensile testing machine was used FPZ 100/1, ensuring smooth loading small loads not exceeding 10 kN with a minimum loading rate (0.02 ... 0.84 mm / min).

Measurement of the small size of deformation performed using strain gauges type 2PKB that stuck on the sample from the coating and substrate and connected to the hardware and software system for measuring deformations. The loading of the samples was carried out smoothly, so that the coating was subjected to tensile loads. When the maximum load carried smooth unloading sample. As a result of the measurement diagram is constructed in coordinates: «Load, P (N) - displacement (absolute deformation) - L, um».

3. Results

Investigations were carried out in the area of the elastic-plastic deformation. Upon reaching the load of 760 N in the ceramic coating layer is developed elastic-plastic deformation, as occurs with the full deformation hysteresis unloading samples. Calculation of the difference in the absolute values of deformation (displacement) at fixed loads (200, 400, 600N) during loading and unloading allowed to allocate a full share of the plastic deformation. It is found that the ceramic coating layer SIO-7-10-50 (curve 1) is the most as compared with plastic Z7Y-10-90 (curve 2) and Metco 204-NS (curves 3 and 4) (Fig.2).

Dual heat treatment coating Metco 204-NS compared with single (vacuum diffusion annealing) promotes an increase of the elastic properties (curves 3 and 4).

From the curves of deformation was determined angle curves, tangent which can be identified with the rigidity of the system "coating-base" (Fig. 3a, 3b). It is found that the curves are grouped depending on the underlayer composition: more rigid system - with underlayer PVNH16U6 less severe - with PNH20K20UЮ13 underlayer. System SIO-7-10-50 with PVNH16U6 has the most narrow range of movements under load to 400N, which indicates the stability of the properties of this system in the elastic region. System covering the Metco 204-NS+ AMDRY 995C color has the greatest stiffness comparable values of the thickness of the ceramic layer with the rest of studied coatings systems.

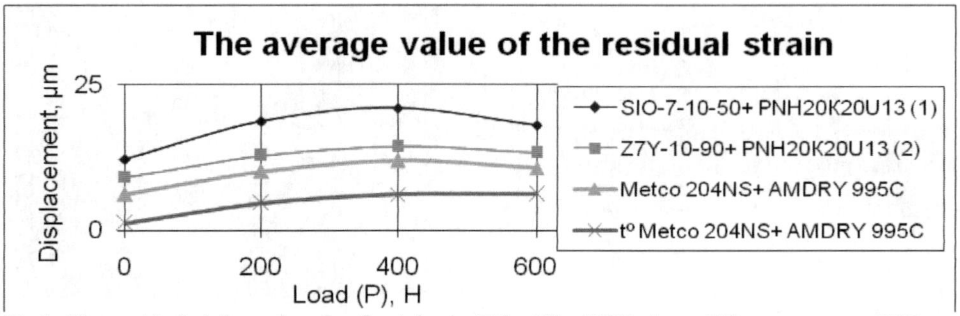

Fig.2. The residual deformation for fixed loads 200, 400, 600N three different systems TZP ceramic composition layer and subsequent heat treatment (SIO-7-10-50+ PNH20K20U13, Z7Y-10-90+ PNH20K20U13 - diffusion annealing of 1050 ° C - 4 hours, oxidative annealing of 850 ° C - 20 hours; Metco 204NS+ AMDRY 995C - diffusion annealing 1050 ° C - 4 hours; Metco 204NS+ AMDRY 995C - diffusion annealing 1050 ° C - 4 hours, oxidative annealing of 850 ° C - 2 hours).

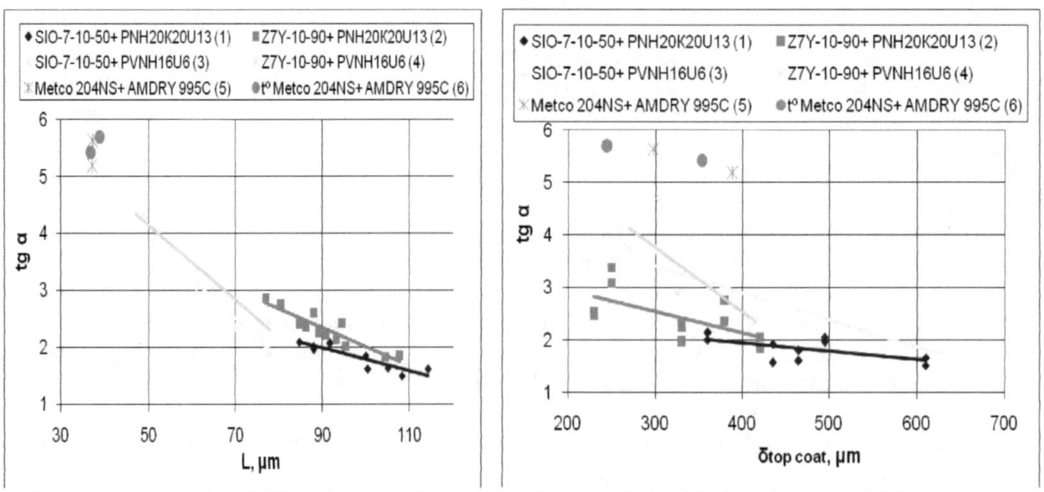

Fig. 3. Interconnection rigidity of system "Base - a two-layer coating" with absolute value of deformation at P = 400N a); and ceramic layer thickness - b)

The samples were examined for strength. All coating systems showed high strength characteristics in the range of the ratio of ceramic layer thickness - underlayer - 3-5.

4. Discussion

The porous, cracked ceramics has no such strong effect on the stiffness of the system, as a more dense metal underlayer. However, analyzing the influence of the thickness of the ceramic layer to change the rigidity of systems with well-formed underlayer PVNH16U6 (Fig. 3b), can make a difference in composition systems and ceramics. Thus, accumulation of structural defects in the ceramic layer with an increase its thickness more negatively affects the stiffness of the system Z7Y-10-90+ PVNH16U6: with increasing thickness of the ceramic is only 1.5 times from 260 to 400mkm slope decreases 1,72 times - from 4.3 to 2.5. System SIO-7-10-50+PVNH16U6 in a wide range of thicknesses of 300 ... 670 mm shows a decrease in stiffness of about 1.8 times. In systems with loose ceramic sub layer

(PNH20K20U13) effect of layer thickness are less effect on reducing the rigidity. These systems are deformed in a wide displacement range independently of the thicknesses of both layers ТЗП and thus exhibit high volatility properties. These dependencies allow select powder materials system which must be sensitive to the geometry of the coating. A preferred system is SIO-7-10-50+ PVNH16U6, since it is a smooth decrease of rigidity with an increase ceramic layer thickness over a wide range.

5. Conclusions

Determined the role of underlayer in the formation of the complex of mechanical properties thermal barrier coatings. With a well formed underlayer (PVNH16U6) system becomes sensitive to a change in thickness of the coating, which allows to optimize the response on the strength and deformation criteria. According to the results the optimum ratio of the thickness of the ceramic layer and underlayer should be regarded as the ratio of 3-5, for which the highest strength values were obtained for all the test coating systems.

A.I.S grateful to the support by grant #16-38-60187_mol_a_dk of the Russian Fund of Basic Research (RFBR)

References

[1] Budovskikh E.A., Gromov V.E., Romanov D.A. The Formation Mechanism Providing High Adhesion Properties of an Electric-Explosive Coating on a Metal Basis // Doklady Physics – 2013. – Vol. 58. – № 3. – P. 82–84

[2] M.A. Popova, V.P. Kuznetsov, V.P. Lesnikov, N.A. Popov, I.P. Konakova. The structure and mechanical properties of single-crystal nickel alloys with Re and Ru after high-temperature holds // Materials Science & Engineering A 642 (2015) P. 304–308

[3] Fayrushin I, Kashapov N and Dautov I 2014 *J. Phys.: Conf. Ser.* **567** 012009

[4] Kashapov L N, Kashapov N F and Kashapov R N 2013 *J. Phys.: Conf. Ser.* **479** 012011

[5] Kashapov L N, Kashapov N F and Kashapov R N 2014 *J. Phys.: Conf. Ser.* **567** 012025

[6] Zaripov R G, Kashapov N F, Tkachenko L A and Shaydullin L R 2016 *J. Phys.: Conf. Ser.* **669** 012053

[7] Denisov D G, Kashapov N F and Kashapov R N 2015 *IOP Conference Series: Materials Science and Engineering* **86** 012005

[8] Galyautdinov R T, Kasparov N F and Luchkin G S 2002 *Inzhenerno-Fizicheskii Zhurnal* **75** 170–173

[9] Saifutdinov A I, Fairushin I I and Kashapov N F 2016 *JETP Lett.* **104** 180–185

[10] Galyautdinov R T and Kashapov N F 2003 *Svarochnoe Proizvodstvo* **(3)** 27–31

[11] Kornienko E E, Lapushkina E J, Kuzmin, V I, Vaschenko S P, Gulyaev I P, Kartaev E V, Sergachev D S, Kashapov N, Sharifullin S, Fayrushin I 2014 *Journal of Physics: Conf. Series* **567** 012010

[12] Gabdrakhmanov AT, Israfilov IH, Galiakbarov AT, AD Samigullin The thermal characteristics of the process of pulsed plasma processing of steel / Proceedings of the Tula State University, vol. 6, Part 1: Publishing House of the Tula State University, 2013. - S.253-260.

Gas dynamic characteristics of glow discharge chamber for functional coating deposition.

D.I.Israphilov

Kazan Federal University

diisraphilov@yandex.ru

Abstract. In article shown conditions of glow discharge existence at low pressures. Described possibility of organizing glow discharge at very low pressures due to the supersonic gas flow in a limited area of the discharge chamber. Procedure of supersonic flow regime calculating in vacuum chamber for coating deposition shown.

1. Relevance

Field of glow discharge application for plasma processing of details surfaces continuously expanding. Ion-plasma sputtering of the material or modification of surfaces is increasingly used in optic industry, electronics and other fields. One of new area is additive technologies.

In ion-plasma technological process is necessary to organize smooth achievement of the particles to processing surface. To achieve this condition free path length λ of particles should be several times larger than the distance L from the area of movement beginning to processing surface.

One of conditions for glow discharge existence is presence of all near-electrode zones. Length of each zone is determined by the number of collisions of an electron with neutral particles. With pressure decreasing length of cathode zones is increasing and certainly there comes a time that the electrode gap is not enough space for all near-electrode zones. Therefore, at very low pressures, the discharge can not exist. To find the critical pressure, at which it is still possible glow discharge, we assume condition of cold discharge. We assume that inter-electrode distance of 10 cm and number of collisions 10, so mean of free path λ is 1 cm. Calculate critical pressure for exisring of glow discharge [1]:

$$P_{kr} = \frac{k_b \cdot T}{\sqrt{2} \cdot \lambda \cdot \sigma} = 4.2Pa = 3.15 \cdot 10^{-2} Torr \quad (1)$$

Thus, existence of glow discharge at pressures below 10^{-2} Torr hampered. Therefore, there is an insoluble problem: on one side discharge existence pressure above 10^{-2} Torr is necessary and from another glow discharge application in technological processes of coating deposition required pressure below 10^{-2} Torr.

One solution to this problem is magnetron sputtering application [1, 2, 3]. Magnetic field may increase the probability of collision of electrons with neutral atoms to order more times at low pressures by the joint action of electric and magnetic fields on moving charges. All this gave opportunity to magnetron discharge application at a pressure of about 10^{-3} Torr. That led to a widespread of magnetron systems in installations for ion-plasma treatment.

In works [4-7] is proposed a fundamentally new solution to problem. Namely, they proposed the creation of zones of different concentrations of neutral particles in electrode gap. For example, in cathode area to create a high vacuum with a minimum concentration of neutral particles, in anode area to provide a relatively high concentration of neutral particles. Electrons in region with high concentration of neutral particles can experience dozens of collisions with neutral particles, while the processing of the material will take place at the maximum possible low pressures. For this condition, it is necessary to create a stream with particles of the working gas in the interelectrode space and at the same time to maintain the vacuum in the discharge chamber below 10^{-2} Torr. To achieve such a separation of zones in the electrode gap may allow the supersonic gas flow. The density of pumped gas through the working area should provide more than a dozen collisions of electrons with neutral particles, which will allow maintaining the glow discharge, keeping chamber pressure below 10^{-2} Torr.

To maintain different particle concentrations in chamber device shown (Fig.1), providing the organization with a continuous supersonic flow and its pumping from the vacuum chamber using a supersonic nozzle and mixing chamber. In mixing chamber, flow restore and removed after by vacuum pump. At certain conditions device can operate as a jet pump that will not only keep the initial chamber pressure, but also lowering it.

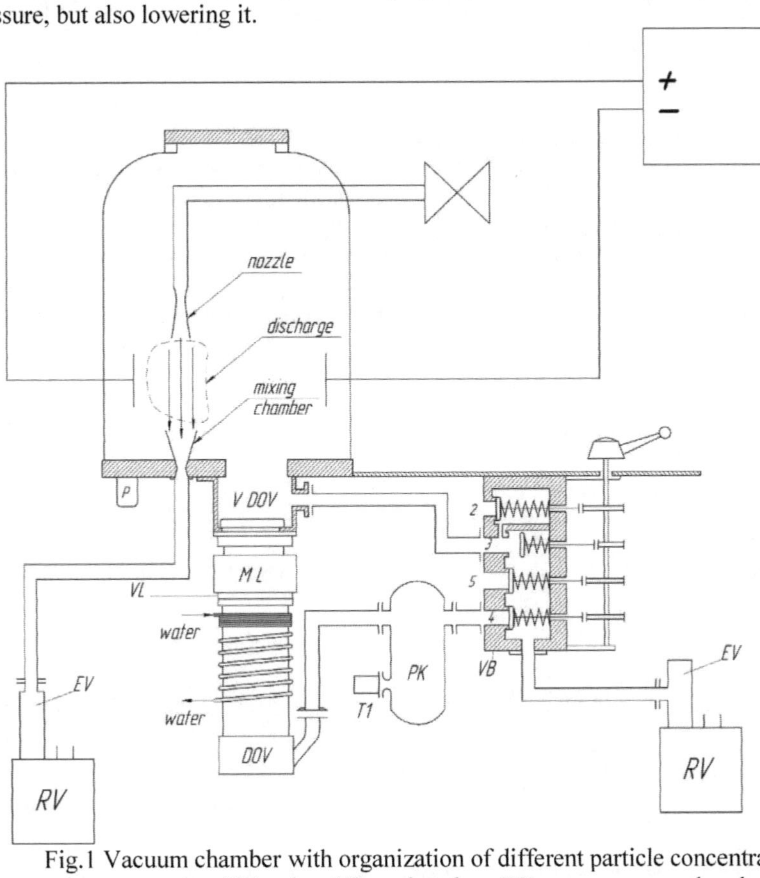

Fig.1 Vacuum chamber with organization of different particle concentration.
RV- rotary vacuum pump, EV-valve, VB- valves box, PK- prev vacuum chamber, DOV - diffusion vacuum pump, ML - oil trap, VL - nitrogen trap, R – vacuum chamber, T1, T2 - thermocouple vacuum gauges, P - high vacuum gauge.

The aim of this work is to determine required conditions of supersonic flow and calculate system diffuser-confusor for supersonic gas flow regime creation in nozzle and mixing chamber.

Nozzle and mixing chamber consists of a diffuser and confusor parts and passage of critical cross-section flow velocity reaches critical speed, with the further expansion of flow occurs the growth rate.

In calculation of nozzle, main dependencies deduced from analysis of equations of flow continuity and the first two laws of thermodynamics [7].

2. Calculating and results

Calculating mixing chamber geometric parameters and operating gas flow regimes in supply system, field of continuous flow and pumping will allow carrying out pilot studies to obtain a new coating method in vacuum.

The calculations carried out in MathCAD.

Calculations [8] need to start with gas flow regime determining by comparing the differential pressure inlet and outlet of the nozzle. To start we define input pressures P_{in} and P_{out} of the output of the nozzle. It is necessary to receive the supersonic flow with diameter of 5 mm, while for 10 collisions, requires mean free path of electrons λ_e more than 0.5 mm, using (1) we obtain P_{out}:

$$P_{out} > \frac{k \cdot T}{\sqrt{2} \cdot \lambda_e \cdot \sigma} = 85Pa = 64 \cdot 10^{-2} Torr$$

Determine P_{cam}=0.5Pa, as a pressure of magnetron coating deposition. To determine P_{in} use the formula for determining differential pressure $\beta_{out} = P_{out}/P_{in}$, and to obtain supersonic flow is necessary that the differential pressure was lower critical $\beta_{out} < \beta_{cr}$, which is calculated:

$$\beta_{kr} = \frac{2}{k+1}^{\frac{k}{k-1}} = 0.53 \quad (2)$$

From (2) we see that critical pressure ratio is only the physical properties of gas function. In this case, operating gas air, for which the adiabatic index k = 1.4, gas constant R = 208 J / (kg * K), from where P_{in} > 170 Pa., assign P_{in} = 100 000 Pa, approximately equal to atmospheric pressure.

For a given pressure differential obtain $\beta_{out} = P_{out}/P_{in} = 0.00085$, a condition $\beta_{out} < \beta_{cr}$ satisfied. Given that the pressure Pi along axis of nozzle will be reduced from P_{in} to P_{out}, we obtain:

$$\beta_i(P_i) = P_i/P_{in} \quad (3)$$

Initial temperature T_0 take as a normal 293 K and considering that operating gas temperature along the length of of the nozzle is reduced, define T_i at every point along axis of nozzle:

$$T_i(P_i) = T_0 \cdot \beta_i(P_i)^{\frac{k-1}{k}} \quad (4)$$

Specific volume is calculated using the ideal gas law (5) and density reciprocal of the specific volume (6):

$$v_i(P_i) = \frac{R \cdot T_i(P_i)}{P_i} \quad (5)$$

$$\rho_i(P_i) = \frac{1}{v_i(P_i)} \quad (6)$$

Define flow velocity along axis of the nozzle:

$$\omega_i(P_i) = \sqrt{\frac{2 \cdot k}{k-1} \cdot R \cdot (T_0 - T_i(P_i))} \quad (7)$$

Local sound velocity (8) and Mach number M_i - ratio of flow rate to local sound velocity in it:

$$a_i(P_i) = \sqrt{k \cdot R \cdot T_i(P_i))} \quad (8)$$

$$M_i = w_i / a_i = 1.97;$$

knowing M <1 - subsonic flow; M = 1 - sound flow; M> 1 - supersonic flow, we define supersonic flow.

Through cross-sectional area of the nozzle f_i define diameter of the nozzle d_i:

$$f_i(P_i) = \frac{G}{\rho_i(P_i) \cdot \omega_i(P_i)} \quad (9)$$

Gas consumption defined through characteristic of the used vacuum pump 2NVR-5DM with ultimate total pressure $P_n = 5*10^{-3}$ Torr, pumping speed G = 5.5 l/s. By use coefficient 0.95 at 1 Torr pressure get G = 4.9 l/s , knowing density of operating gas at the outlet of the nozzle $P_{out} = 85\ Pa$ translate consumption in $3.81*10^{-5}$kg/s.

To calculate length of confusor l_{ci} and diffusor l_{ci} of nozzle find full critical pressure P_{cr} and the critical diameter d_{cr}, through the critical cross-sectional area f_{cr}:

$P_{cr} = \beta_{cr} * P_{in}$ =5.28*10^4 Pa

We obtain d_{cr} = 0.453 mm. Confusor angle of nozzle take α_{in} =45°, diffusor angle α_{out} =30°:

$$l_{ci} = \frac{d_i - d_{kr}}{2 \cdot \tan(\frac{\alpha_{in}}{2})} = 3.06mm$$

$$l_{di} = \frac{d_i - d_{kr}}{2 \cdot \tan(\frac{\alpha_{out}}{2})} = 4.72mm$$

Full length of nozzle is $l = l_{di} + l_{ci}$=7.78 mm.

To calculate mixing chamber parameters it is necessary to know pressure at its output. To carry out this characteristics of gas pumping system from the vacuum chamber and operating gas calculated. Pumping system, consisting of piping system and pump, should have enough bandwidth to provide supersonic flow rate in a vacuum chamber with pressure lower than 10^{-2} Torr.

To determine mixing chamber parameters it is necessary know provided by pump inlet pressure P_t in pumping system, considering flow resistance and flow rate G.

The pumping system consists of three pipeline diameter Dn = 25 mm, length m L_{t1} = 0,5, L_{t2}= 1 m, L_{t2}= 0,5 m along the length of the system from the pump up to chamber has two rotatings 90 ° and valve. After calculations [18] determine total conductivity of the pumping system U_t = 154.7 L / s:

$$S_{eff} = \frac{S_n \cdot U_t}{S_n + U_t} = 4.84 L/s$$

$$P_t = \frac{G}{S_{eff}} + P_n = 138.3Pa$$

So pump can provide the maximum flow rate of the mixed gas 4.84 L / s at an outlet pressure of mixing chamber 138 Pa. To define parameters of mixing chamber, pressure difference across the inlet and outlet calculated:

$\beta_c = P_t / P_{out} = 1.65$ *(10)*

From (10) comparing differential and critical pressure, define $\beta_c > \beta_{cr}$, therefore, there is a subsonic flow regime. Rate of the mixed flow velocity ω_c at the outlet of confusor is (11):

$$\omega_c = \sqrt{2 \cdot \frac{k}{k-1} \cdot R \cdot T_i(P_t) \cdot (\beta_c^{\frac{k-1}{k}} - 1)} = 109.34 \frac{m}{s} \quad (11)$$

Rate of critical mixed flow velocity ω_{ccr} (12) and Specific critical volume υ_{ccr} are (13):

$$\omega_{ccr} = \sqrt{2 \cdot \frac{k}{k+1} \cdot P_{out} \cdot v_i(P_t)} = 113.28 \frac{m}{s} \quad (12)$$

$$v_{ccr} = \frac{v_i(P_{out})}{\beta_c^{\frac{1}{k}}} = 91.9 \frac{m^3}{kg} \quad (13)$$

Minimum cross sectional area f_{ccr} and diameter d_{ccr} of mixing chamber are:

$$f_{c\min} = \frac{G \cdot v_{ccr}}{\omega_{ccr}} = 27.82mm^2$$

Area of inlet cross-section adopt as a f_{inc} = 2 * f_{ccr} = 55.4 mm2, diameter of inlet d_{inc} = 8.4 mm.

3. Conclusion

After calculations, nozzle and confusor geometric parameters and gas flow regimes in supply and pumping region of the continuous supersonic flow are determined. Defined steady pressure in the gas stream for confusor and cathode region. Specific parameters for this case were as follows: pressures inlet and outlet of nozzle (P_{in}, P_{out}) and (P_{out}, P_t) of mixing chamber, also their geometric characteristics.

Area of inlet cross-section adopt as a f_{inc} = 2 * f_{ccr} = 55.4 mm2, diameter of inlet d_{inc} = 8.4 mm.

References

[1] Rayzer J.P. Physics of gas discharge. 2009. - 736 c.
[2] S.B. Nesterov et.all. Methods of vacuum systems calculation. — M.:,2004. — 220 p.
[3] E.V. Berlin, L.A. Seydman. Ion-plasma processes in thin-film technology. Moscow: 2010.-528p. ISBN: 978-5-94836-222-9
[4] B.A.Timerkaev, B.R. Zalyaliev. Glow Discharge in a Transverse Supersonic Gas Flow at Low Pressures. High Temperature, 2014, Vol. 52, No. 4, pp. 471–474.
[5] B.A.Timerkaev, B.R. Zalyaliev, B.R.Karimov, D I Israfilov. Behavior of a glow discharge in plasma deposition installations in distributed supersonic gas flow / Vestnik KGTU. 2013. №4.P. 198-291.
[6] B A Timerkaev, B R Zalyaliev, A I Saifutdinov. Glow Discharge Characteristics in Transverse Supersonic Air Flow. Journal of Physics: Conference Series 567 (2014) 012032.
[7] A.I.Saifutdinov, B.A.Timerkaev, and B R Zalyaliev. Control of glow discharge parameters using transverse supersonic gas flow - numerical experiment Journal of Physics: Conference Series 567 (2014) 0120313.
[8] E.J. Sokolov, N.M. Zinger. Stream appararus.— M: 1989.— 352 p. ISBN 5-283-00079-6.
[9] Saifutdinov A.I. Drift model of glow discharge with nonlocal value of the electric field in the ionization source / A.I. Saifutdinov, B.A.Timerkaev - Journal of Engineering Physics, T. 85, №5, 2012 S.1104-1109. (A.I. Saifutdinov, B.A. Timerkaev - Journal of Engineering Physics and Thermophysics, Vol. 85, No. 5, 2012, P. 1202-1207).
[10] B A Timerkaev, M M Ahmetov, B R Zalyaliev, O A Petrova and D I Israfilov. Longitudinal distribution of electrical parameters in normal glow discharge. Journal of Physics: Conference Series 567 (2014) 012036.
[11] G.J. Dautov, B A Timerkaev. Nonequilibrium gas-discharge plasma generators. 1996, 200p.
[12] I.G.Galeev, V.E. Goncharov, B A Timerkaev. Glow discharge features in a supersonic gas flow. Teplofizika Vysokikh Temperatur, 1990. V.28, №5, p 843-846.
[13] I.G.Galeev, V.E. Goncharov, B A Timerkaev. Characteristics of gas discharge in micronozzles. Teplofizika Vysokikh Temperatur, 1992. V.30, p 439-444.
[14] Alferov, V.I. Bushmin, A.S. Dmitriev, L.M. Nozzle array and ballast resistance for producing a glow discharge in a gas flow. (1985) Instruments and experimental techniques New York
[15] Longitudinal distribution of electrical parameters in normal glow discharge
[16] Timerkaev, B.A., Ahmetov, M.M., Zalyaliev, B.R., Petrova, O.A., Israfilov, D.I. 2014 Journal of Physics: Conference Series 567 (1), 012036
[17] Timerkaev, B.A., Ahmetov, M.M., Zalyaliev, B.R., Petrova, O.A., Israfilov, D.I. Longitudinal distribution of electrical parameters in normal glow discharge. 2014 Journal of Physics: Conference Series 567 (1), 012036
[18] Vacuum equipment: Reference book /E.S. Frolov et. All.. – M.:, 1992. – 480 p. ISBN 5-217-0140-1

Determination of regression materials mircohardness, processed by low-temperature plasma dependence on process conditions

A A Khafizov[1], Yu I Shakirov, R A Valiev, R I Valiev

Kazan Federal University Naberezhnye Chelny Institute, 68/19 Mira Street, Naberezhnye Chelny, 423810, Russian Federation

E-mail: Almazok75@yandex.ru

Abstract. The influence of conditions of plasma surface treatment on hardening of carbon steel technological process was analyzed. Hardening was carried out in plasma electrothermal line with an electrolytic cathode. When processing, steel crystal grains are crushed and the structure is changed from ferrite-pearlite to bainite- troostite and martensite, depending on the processing conditions. In this case the surface microhardness increase in 2 - 3 times. The dependence of the carbon steel surface microhardness on the discharge current (2 - 10 A), the distance between the heat source and the surface, the plasma gas flow rate and treatment duration was found. On the basis of multifactor experiment planning methods and the method of least squares, the formula that describes this relationship was found. This allowed to conduct a targeted search of optimal conditions of processes of hardening steel and improve the efficiency and quality of research.

1. Introduction

Looking for ways to modify the properties of materials - this is one of the main ways of materials development. Thus methods for surface treatment are preferred, since in most cases, the characteristics of surface determine the level of material properties overall. In addition, surface modification is more attractive in relation to methods of changing the bulk characteristics of the materials and products from an economic point of view. All this has led to the rapid development of research and development aimed at improving the existing and creation of new methods of impact on the surface in order to give it the required characteristics of the operation conditions.

Methods for surface modification are divided into two groups: the modification and coating deposition with a change of state, structure, surface properties [1-6]. This approach allows the effective use of the combination of the properties of the base material and a modified surface layer, a surface modification processes can be quite easily integrated into the process of production or repair products.

All the more prominent place in modern technology is occupied by the methods of hardening materials by highly concentrated radiation sources using plasma torches. The advantages of these techniques are: the high technological flexibility, independence from the mechanical strength of the processed materials, the possibility of an effective mechanization and automation of technological processes.

[1] To whom any correspondence should be addressed.

2. Statement of the problem

One of the tasks that you want to optimize the plasma technology process is the selection of the most favorable ratios of the numerous factors that influence the final result, such as the energy potential of the plasma jet, the distance from the source to the surface to be treated, plasma gas flow, processing time, etc. The systematic study of the influence of each individual factor leads to unnecessarily high losses of time and money.

The objective of this work is to determine the regression dependence of the strength characteristics of the materials processed by the process conditions.

3. Research method

To harden the surface of metals the plasma electrothermal line [7] with an electrolytic cathode (Figure 1) was used. Obtaining plasma using pairs of "metal-electrolyte" [8-10] is much more favorable than pairs of "metal-metal" because electrolytes are cheap and based on their manufacturing processes are environmentally friendly, and surfaces made from them have a very good quality [11]. This method of construction hardening steels allows to obtain a layer with high physical-mechanical characteristics, therefore, significantly increase the service life of parts.

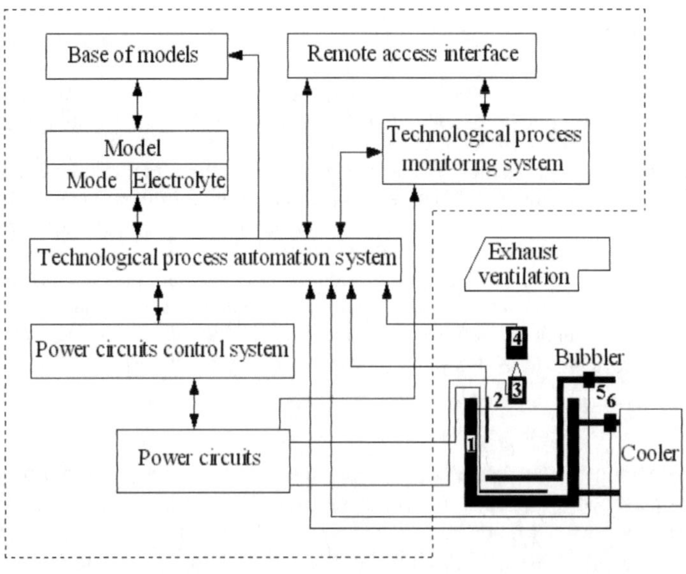

Figure 1. A generalized block diagram of a plasma processing electrothermal line (1 – electrolytic cell; 2 – electrolyte temperature sensor; 3 – head of plasma electrothermal line; 4 – workpiece; 5 – air flow sensor; 6 – water pressure sensor in the electrolytic cell cooling system).

In order to optimize plasma hardening process we need to determine product regression dependence of the hardened layer hardness on some of the input parameters [12]. To do this, we use multifactor experiment planning method. Application of multifactor experiment planning methods allows to minimize experimental efforts, improves the efficiency and quality of research, allows a targeted search for optimal flow conditions of metal hardening processes. Since all of the input parameters considered in the study is not possible to be selected, setting and hardening process were selected as the most significant parameters: discharge current (I), the distance from the heat source to the metal surface (h), plasma gas flow (Q), and the processing time (τ). As the output value (y) the parameter optimization HV (metal hardness) was selected. The lower and upper levels of the factors, as well as the range of variation of their values are given in Table 1.

Table 1. Levels and intervals of varying factors.

Factors	τ, с	*Q*, м^3/с·10^{-5}	*h*, м	*I*, A
Main level	3.5	31.5	0.0725	6
Interval of varying	1.5	12.2	0.0325	4
Upper level	5.0	43.6	0.1050	10
Lower level	2.0	19.3	0.0400	2
Code	x_1	x_2	x_3	x_4

4. Results and discussion

The mathematical processing of the experimental results by the method of least squares [13, 14] and MathCad software package yielded a regression equation for steel grade 40 , as follows:

$$y=387{,}4+25{,}4\,x_1+46{,}5x_2+58x_3+38{,}6\,x_4+47{,}3\,x_1x_2+35{,}1x_1x_3-10{,}3\,x_2x_3$$

Samples were analyzed for the distribution of microhardness and microstructure. As a result, it found that treatment significantly alters the microstructure of the steel. Thus, when a combination of factors x1-x4 to the surface treatment of parts composed of ferrite and pearlite in a grid along the grain boundaries of pearlite, and thereafter, recrystallization occurs on the surface layer to a depth of 0.6 mm. Grid ferrite was broken and the metal start to be consisted of bainite and troostite. As a result the initial sample microhardness was 240HV, after treatment of 520HV.

In another combination the hardened layer with the structure of martensite and troostomartensita of different dispersion was formed on material surface. In this case the crystal grains are milled, microhardness as compared with the original structure of the sample to a depth of 0.5-1.5 mm on average increased by 3 times. It was getting larger as getting closer to the center of the plasma jet, longer processing time and discharge capacity (Figure 2).

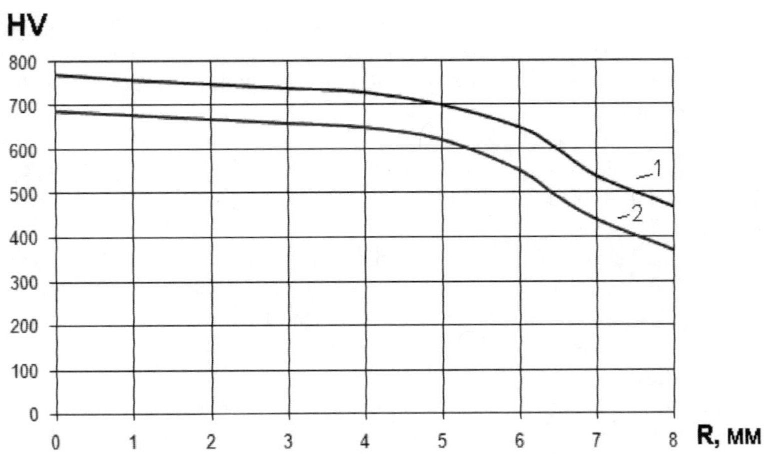

Figure 5. Dependence of the microhardness of steel 40 at a distance R from the center of the plasma jet on the processing time, the distance from the surface of the product to the device and discharge power (1 – t = 10 c, l = 5 mm, N = 10 kW; 2 – t = 8 c, l = 7 mm, N = 8 kW).

The adequacy of the model was tested using a F-test [15]. The error mean square

$$\sigma^2(y) = \frac{1}{N(n-1)}\left[\sum_{i=1}^{N}\sum_{j=1}^{n}(y_{ij} - \bar{y}_i)^2\right]$$

and adequacy dispersion were determined

$$\sigma_{ag}^2 = \frac{1}{F_{ag}}\left[\sum_{i=1}^{N} n(\bar{y}_i - y_{im})^2\right]$$

where n = 3 is the number of repeat experiments at each point of the space factor; N = 8 is the number of different experiments; $y_i = \frac{1}{n}\sum_{j=1}^{N} y_{ij}$ - the average real value of the response in the i-th point of the factor space; y_{ij} - response value in a single experiment; y_{im} - response value predicted in the i-th experiment on the received model; F_{ag} - number of degrees of freedom.

It has been proposed that the ratio $F = \sigma_{ag}^2 / \sigma^2(y)$ does not exceed the tabulated value of F-test at a significance level α = 0,05. Thus, the adopted model can be considered adequate.

Testing of the significance of individual regression coefficients was performed by Student's t-distribution: parameter was calculated

$$t_k = \frac{a_k}{\sigma^2(a_k)}$$

$$(\text{где } \sigma^2(a_k) = \frac{\sigma^2(y)}{N}),$$

which equalized with a table at a significance level α = 0,05. The coefficients for which $t_k > t_{table}$, were considered as minor.

Taking into account the procedures described the regression equation took the following form:

$$y=387,4+25,4x_1+46,5x_2+58x_3+38,6x_4+47,3x_1x_2+35,1x_1x_3$$

5. Conclusion

The formula that describes the impact of the discharge current, the distance between the heat source and the surface to be treated, the plasma gas flow rate and duration of exposure to the plasma on the hardness of the surface of materials treated with plasma electrolytic cathode was found. The dependence found can be used to determine the optimal process conditions of plasma treatment.

References

[1] Dautov G Y, Nail K F, Viktor L, Renat G Z, Raphae G T, Fayrushin I, Kashapov R N 2013 Low Temperature plasma in the processes of functional coating preparation. *J. of Phys.: Conf. Ser.* **479 (1)**

[2] Kashapov N F, Sharifullin S N 2015 Hardening of the surface plasma jet high-frequency induction discharge of low pressure. *IOP Conference Series: Materials Science and Engineering* "International Scientific and Technical Conference "Innovative Mechanical Engineering Technologies, Equipment and Materials-2014" 12021.

[3] Khafizov A A, Valiev R I, Shakirov Yu I, Valiev R A 2014 Steel surface modification with plasma spraying electrothermal installation using a liquid electrode *J. of Phys.: Conf. Ser.* **567** 012026

[4] Khafizov A A, Shakirov Yu I, Valiev R A, Valiev R I, Khafizova G M 2016 Study of thermal and electrical parameters of workpieces during spray coating by electrolytic plasma jet. *J. of Phys.: Conf. Ser.* **669** 012030

[5] Guzel R. Gibadullina, Almaz H. Tazmeev, Ramilya N. Tazmeeva 2015 The creation of high temperature steam flow for plasma chemical gasification of polymer waste *International Journal of Applied Engineering Research* 10(24) p 45015-21

[6] Zvezdin V V, Pesoshin V A, Israfilov I N, Saubanov R R, Mirgajazov R R 2015 Issledovanie processov plazmennoj termoobrabotki stali 45 s vysokimi jekspluatacionnymi svojstvami. *Nizkotemperaturnaja plazma v processah nanesenija funkcional'nyh pokrytij* **1(6)** p 191
[7] Khafizov A A, Shakirov B Yu, Shakirov Yu I 2014 Surface hardening of components by automatic plasma electrothermal installation with molten cathode *IOP Conf. Series: Materials Science and Engineering* **69** 12019
[8] Tazmeev Kh K., Tazmeev A Kh 2014 Gas discharge with liquid electrolyte cathode in the mode of occurrence of the constricted channels *J. of Phys.: Conf. Ser.* **567** 012035
[9] Tazmeev A K, Tazmeeva R N, Sarvarov F S 2016 The features of high-current gas discharge in a narrow gap between the liquid electrolyte and solid electrode *J. of Phys.: Conf. Ser.* **669** 012056.
[10] Gajsin F M, Son Je E 1990 *Himija plazmy* **16** p 120.
[11] Shakirov Ju I, Valiev R I, Hafizov A A, Shakirova G Ju 2011 Mnogokanal'naja plazmennaja ustanovka s jelektroliticheskim katodom *Nauchno - tehnicheskij zhurnal «Avtomobil'naja promyshlennost'* **2** p 36
[12] Ionov Ju G, Klubnikin V S, Kenhi R, Tumin Ja A 1977 *Teorija i praktika plazmennogo napylenija* **1** p 68
[13] Linnik Ju V 1962 *Metod naimen'shih kvadratov i osnovy matematiko-statisticheskoj teorii obrabotki nabljudenij* p 238
[14] Adler Ju P, Markova E V, Gramovskij Ju V 1976 *Planirovanie jeksperimenta pri poiske optimal'nyh uslovij* p 279
[15] Kendall M, St'juart A 1973 *Statisticheskie vyvody i svjazi* p 899

Hybrid Calcium Phosphate Coatings for Titanium Implants

E. Kharapudchenko[1], V. Ignatov[2], V. Ivanov [3], S. Tverdokhlebov[1]

[1]The Department of Experimental physics, National Research Tomsk Polytechnic University, 30 Lenin av., Tomsk, 634050, Russian Federation
[2]The Department of Technology of Silicates and Nanomaterials, National Research Tomsk Polytechnic University, 30 Lenin av., Tomsk, 634050, Russian Federation
[3]Siberian State Medical University, 2 Moskovsky trakt, Tomsk, 634055, Russian Federation

e-mail: kleine_harey92@mail.ru; tverd@tpu.ru

Abstract. Hybrid multilayer coatings were obtained on titanium substrates by the combination of two methods: the micro-arc oxidation in phosphoric acid solution with the addition of calcium compounds to high supersaturated state and RF magnetron sputtering of the target made of synthetic hydroxyapatite. 16 different groups of coatings were formed on titanium substrates and in vitro studies were conducted in accordance with ISO 23317 in the solution simulating body fluid. The studies using SEM, XRD of the coatings of the samples before and after exposure to SBF were performed. The features of morphology, chemical and phase composition of the studied coatings are shown.

1. Introduction

Implants must be compatible with living tissues. Modern medical materials science aims at improving the implant survival. It can be achieved by forming biocompatible and bioactive coatings. The implants made on the base of Titanium and its alloys capable of surviving under mechanical loads in the recovery process find a wide application in traumatology. To reduce the duration of treatment different biocoatings are formed on the titanium surface. These coatings in particular calcium phosphate coatings on the base of hydroxyapatite must combine bioactivity and high mechanical strength. A number of technologies have been applied in industry. The calcium phosphate coatings on titanium implant materials formed using the techniques developed at Tomsk Polytechnic University are shown in Fig.1.

The electrochemical method is the simplest, widely used and effective method of forming coatings. MAO allows us to form the coatings with the thickness to 30 μm and high adhesion with substrate surface. However, this method has some limitations: MAO coating can be formed only on the valve group materials, their chemical composition of are poorly controlled, and they have low elasticity.

One of the promising methods for the formation of calcium phosphate coating is a method RFMS of target. Using this method the dense coatings with high adhesion can be formed on complex shapes surfaces maintaining stoichiometric composition of a sputtered target and controlled chemical composition. Method RFMS of target has the following features: low rate of coating growth and low porosity of coatings.

The combination of MAO and RFMS methods allows us to use the advantages of each method, masking their shortcomings.

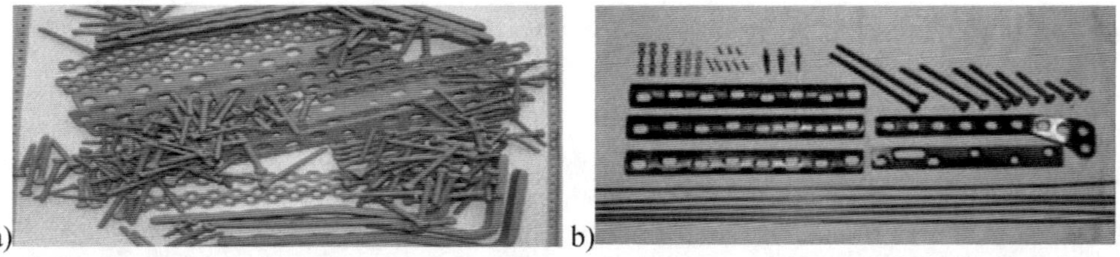

Fig.1. Implants with calcium phosphate coatings. a) micro-arc oxidation (MAO), b) high-frequency magnetron sputtering (RFMS) of target

2. Experimental part

For the study the samples were made of titanium plate (BT1-0) of the following size:10×10×2,8 mm^3. Preparation of samples before forming coatings included sandblasting and chemical cleaning in an ultrasonic bath with distilled water.

Formation of calcium phosphate coatings by the micro-arc oxidation method was performed in a saturated solution of CaO in 10% H3PO4 with the addition of phase dispersion of hydroxyapatite with particle size of 70 microns.

The following technological conditions of formation of MAO coatings were used: dense coatings were formed at constant voltage of about 150 V for 15 minutes, the porous - 170 V, 30 minutes. The pore size of the dense coatings is 1 - 2 µm, porous coatings of 5 - 7 µm.

The formation of calcium phosphate coatings by RF magnetron sputtering target of hydroxyapatite was produced using the setup developed at Tomsk Polytechnic University (see Fig. 2.), under the following technological conditions: preliminary pressure of chamber is $7 \cdot 10^{-3}$ Pa, operating pressure of Ar is $5 \cdot 10^{-1}$ Pa, RF power is 2 kW, deposition duration is 8 hours, distance between the target and samples is 35 mm.

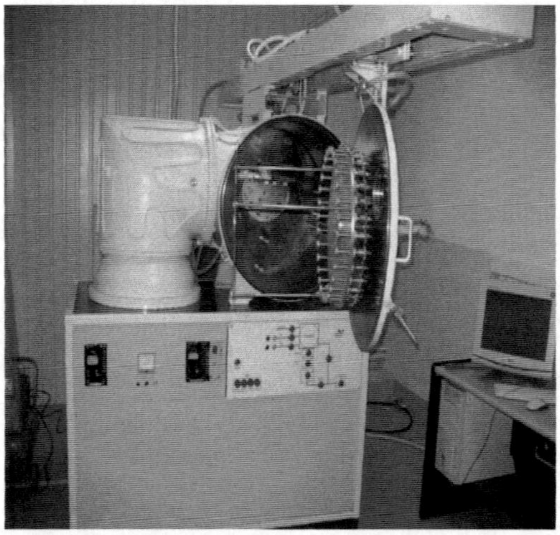

Fig.2 Setup of RFMS developed in Tomsk Polytechnic University

Using the combination of MAO and RFMS methods, 16 different types of coatings were formed.

Crystallization of calcium phosphate coatings was carried out at temperature 730-770° C.

Roughness of the coatings was determined by the profilometer "Talysurf 5-120". The morphology of the coating surface and elemental analysis were examined by scanning electron microscopy "JEOL-6000".

In vitro evaluation for apatite-forming ability of implant materials was carried out according to ISO 23317 [1].

3. Results and discussion

Using the data of evaluation for apatite-forming ability of implant materials the best coating from 16 kinds was chosen for studies in vivo. For obtaining this coating it is necessary to carry out the following sequential technological operations: MAO-annealing-RFMS of target- annealing. [2]
According to the energy dispersive analysis of MAO coatings, the coatings have calcium-deficient composition. In contrast, RFMS coating has a calcium-rich composition. Thus, using a combination of these methods it is possible to obtain balanced chemical composition of the coating.
Fig. 3 shows SEM images of surface of the hybrid coating made after exposure to SBF solution. The analysis of images shows that apatite crystallites form on the titanium surface. The apatite crystallites are supposed to contribute to osseointegration of implant.

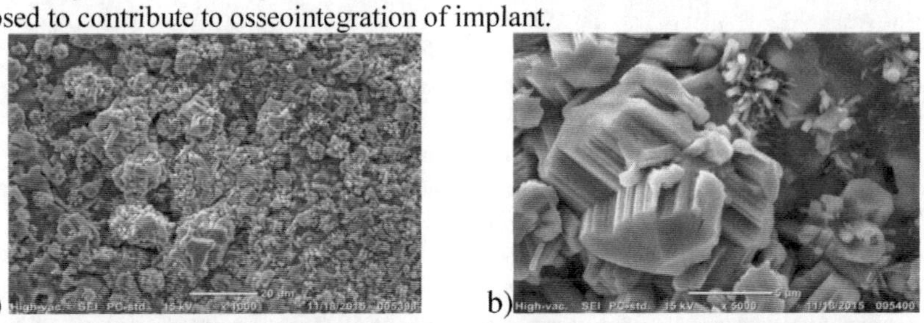

a) b)

Fig.3 SEM of hybrid calcium phosphate coating after exposure of samples to SBF solution with magnifying a -1000; b -5000

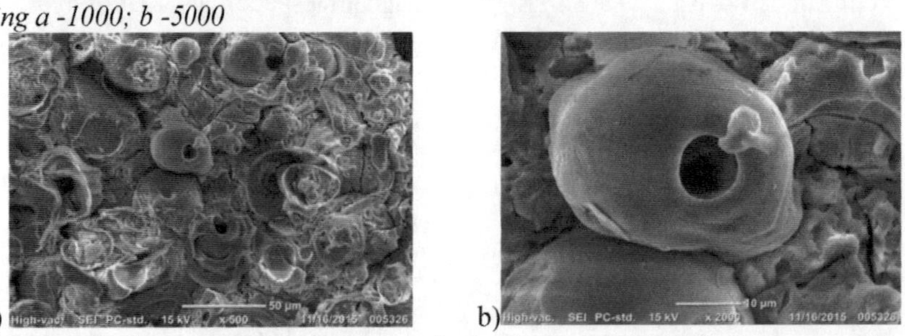

a) b)

Fig.4 SEM of calcium phosphate coating obtained using MAO after exposure of samples to SBF solution with magnifying a -500; b -2000

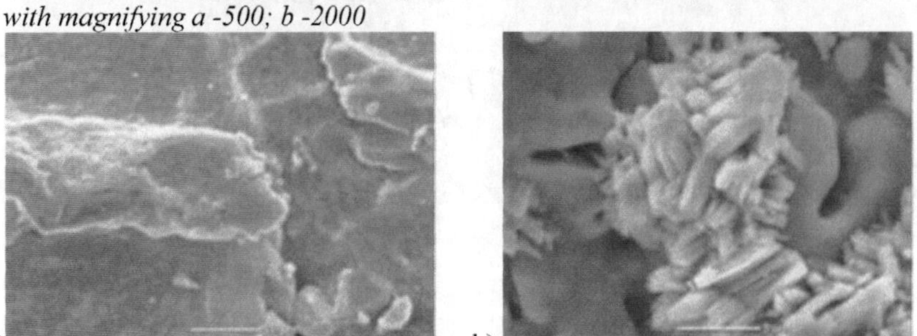

a) b)

Fig.5 *SEM of calcium phosphate coating obtained using RFMS after exposure of samples to SBF solution with magnifying a -1000; b -5000*

This conclusion was confirmed by the results of the energy dispersion analysis of surfaces of titanium samples with hybrid coatings samples before and after exposure to SBF, (Table 1). After exposure to SBF the content of P, O increase and content of Ca, Ti decrease. Thus, exposure of samples in SBF allows us to bring the value of calcium to phosphorus ratio to normal in vivo.

Table 1. Chemical content of hybrid CaP coating after exposure of samples to SBF according energy dispersion analysis

	Ca, at.%	P,at.%	Ca/P	O, at.%	Ti, at.%
Before SBF	32.43	5.52	5.87	36.89	25.16
After SBF	25.81	16.61	1.55	42.10	15.49

Table 2. Chemical content of CaP coating obtained using MAO after exposure of samples to SBF according energy dispersion analysis

	Ca, at.%	P,at.%	Ca/P	O, at.%	Ti, at.%
Before SBF	8.07	21.9	0.36	46.94	23.09
After SBF	10.26	23.53	0.43	47.19	19.02

Table 3. Chemical content of CaP coating obtained using RFMS after exposure of samples to SBF according energy dispersion analysis

	Ca, at.%	P, at.%	Ca/P	O, at.%	Ti, at.%
Before SBF	35.33	14.86	2.37	39.91	9.80
After SBF	34.69	14.47	2.39	39.87	10.97

The proposed method of forming calcium phosphate coatings allows us to form multilayer coatings on titanium implants, consisting of a combination of two methods- micro-arc oxidation and RF magnetron sputtering target. The chemical composition of the coating is the compound of calcium, phosphorus, titanium, and their oxide compounds. The composite structure of the coating allows combining the high mechanical properties of titanium with high bioactivity of calcium phosphate layer. At present, the most optimal type of coating produced by applying the combination of MAO and RFMS of target with double annealing is studied in vivo using large laboratory animals.

Acknowledgement

This work was financially supported by the Ministry of Education and Science of the Russian Federation, Federal Target Program (agreement # 14.578.21.0031, unique identifier RFMEFI57814X0031).

References

[1] ISO 23317. Implants for surgery — In vitro evaluation for apatite-forming ability of implant materials Another reference

[2] Kharapudchenko E. Hybrid calcium phosphate coatings for titanium implants/ E. Kharapudchenko, N. Bogomolova, V. Ignatov, S. Jakimov, S. Tverdokhlebov – Journal of Biotechnology and Biomaterials, 2016 – 69 p.

Surface hardening of VT-22 alloy by inductively coupled plasma nitriding and magnetron deposition of TiN films

Maxim M Kharkov, Andrey V Kaziev*, Alexander V Tumarkin, Vyacheslav E Drobinin, Tatiana V Stepanova, and Alexander A Pisarev
National Research Nuclear University MEPhI (Moscow Engineering Physics Institute), 31 Kashirskoe sh., 115409 Moscow, Russia

E-mail: kaziev@plasma.mephi.ru

Abstract. The surface of VT-22 Russian grade titanium alloy samples was modified by inductively coupled plasma (ICP) nitriding followed by magnetron deposition of TiN coatings. Different operating conditions of ICP nitriding and magnetron deposition were considered. The microhardness depth profiles were measured for samples after nitriding. The performance of TiN coatings was examined with a scratch tester.

1. Introduction

Titanium alloys are widely used in aviation industry. An attractive material for some construction applications is Russian grade titanium alloy VT-22. Certain applications require surface hardening of the alloy. The needed effect can be achieved by means of plasma nitriding or by TiN coating deposition. The techniques combining different approaches for surface modification (e. g. duplex methods) are considered to be prospective for improving the tribological performance of titanium alloys [1–5] and steels [6]. In particular, a number of studies have been focused on analyses of the duplex technology that combines plasma nitriding of titanium followed by TiN deposition [1–4]. As a rule, direct current or mid-frequency discharges are used for plasma nitriding. The coatings are mainly deposited by various physical vapour deposition (PVD) methods (e. g. arc plating [1] or electron beam evaporation [2]) as well as by plasma enhanced chemical vapour deposition (PECVD) methods [3, 4]. Papers [1–4] demonstrate that preliminary nitriding of titanium alloys aids in improvement of adhesive properties and wear resistance up to several times for TiN coatings as well as for multilayer TiN-Ti and TiN-DLC coatings.

Implementation of radiofrequency (rf) discharges for nitriding of materials including titanium and its alloys has not been studied well. It is therefore practically nothing known about performance of this technique in the combined (duplex) processing technologies. In this work, the surface hardening of VT-22 titanium alloy was carried out by nitriding in the inductively coupled plasma (ICP) followed by TiN deposition in a magnetron sputtering system. The surface hardness was studied at both steps of duplex treatment, and scratch tests of TiN coatings were performed for several regimes of nitriding and film deposition.

---

* To whom any correspondence should be addressed.

2. Experimental

The samples were cylinders of VT-22 with the diameter of 30 mm and the height of 8 mm. The elemental composition of VT-22 alloy is presented in table 1.

Table 1. Composition of VT-22 grade titanium alloy.

Fe	C	Si	Cr	Mo	V	Ti	Al	Other
0.5–1.5	< 0.1	< 0.15	0.5–1.5	4–5.5	4–5.5	79.4–86.3	4.4–5.7	< 0.7

Plasma nitriding was performed in an experimental ICP facility in mixture of Ar and N_2. The vacuum chamber was pumped down to 1×10^{-4} Pa with a turbomolecular pump backed by a dry scroll pump, and gas flow rate and composition were regulated by mass-flow controllers. The discharge was generated by a radiofrequency plasma generator RPG-250. The rf power was supplied by COMET cito 1330 series, through the automated matching network.

Prior to nitriding, the samples were sputter-cleaned in argon plasma at p_{Ar} = 1 Pa and P_{rf} = 1.5 kW. The bias voltage applied to the sample U_b = –300 V was used, and the cleaning time was 10 min.

The parameters of plasma nitriding procedures for five samples are given in table 2.

Table 2. Parameters of plasma nitriding process.

Sample no.	Pressure (Pa)	Gas mixture	Forward rf power (kW)	Bias voltage (V)	Processing time (min)	Temperature (°C)
1	0.33	N_2:Ar = 1:2.3	1.5	–300	120	< 900
2	0.33	N_2:Ar = 1:2.3	1.5	–100	30	< 700
3	0.33	N_2:Ar = 1:2.3	1.5	–100	120	< 700
4	0.33	N_2:Ar = 1:2.3	1.5	–100	240	< 700
5	0.33	N_2	1.5	–1500	120	< 700

After the experiments, the samples were cooled in argon at p_{Ar} = 130 Pa down to 200°C. Each sample was then sectioned into four parts. One part was used for measurements of depth profiles of hardness and elemental composition, as well as the cross-section image analyses. Microstructure and elemental composition were analysed by Tescan Vega 3 scanning electron microscope (SEM) equipped with energy-dispersive X-ray spectrometer (EDS) Oxford Instruments X-act. Two parts of each sample were used for coating deposition in a magnetron sputtering facility.

The vacuum chamber of the magnetron sputtering facility was evacuated to a base pressure of 1×10^{-4} Pa with a turbomolecular pump backed by a multi-stage Roots pump. Argon and nitrogen mixture was regulated with mass-flow controllers. Prior to coating deposition, the samples were sputter-cleaned in glow discharge in argon at p_{Ar} = 1.7 Pa at U_b = –1500 V for 10 min. The magnetron deposition was then performed in either of two regimes (A or B). The deposition parameters corresponding to regimes A and B are shown in table 3.

Table 3. Parameters of TiN magnetron deposition.

Regime	Pressure (Pa)	Gas mixture	Discharge power (W)	Bias voltage (V)	Deposition time (min)	Coating thickness (μm)
A	0.62	N_2:Ar = 4.50:0.30	300	–300	60	3
B	0.62	N_2:Ar = 4.64:0.16	300	–300	120	6

3. Results and discussion

After the ICP nitriding of sample no. 1 at $T \sim 900$°C, redistribution of elements in the near-surface layer was observed. The regions of higher Al (α-phase stabiliser) concentration together with lower Mo (β-phase stabiliser) concentration were found. Cross-sectional SEM image of sample no. 1 accompanied by the Mo distribution map are shown in figure 1.

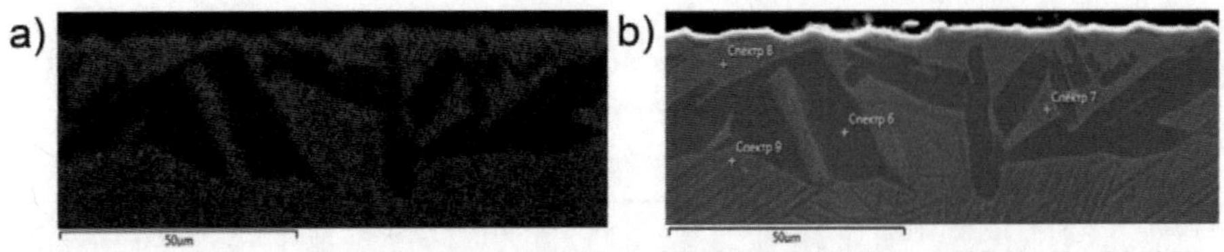

Figure 1. Cross-section of sample no. 1 after ICP nitriding (U_b = –300 V, $T \sim$ 900°C): (a) Mo distribution map; (b) secondary electron SEM image.

The observed behaviour of alloying elements is in accordance with the results of abnormal glow discharge plasma nitriding studies reported in [7]. This effect of nitriding at high temperature of 900–1000°C, as it was mentioned in [7], may lead to negative changes in mechanical properties. However, its influence on the performance of the above-deposited coatings has not been examined.

Samples nos. 2, 3, 4 were nitrided at the lower temperature ($T \sim$ 700°C) and under identical conditions except the time period (30, 120, and 240 min, respectively). Higher processing time caused improvement in surface hardness measured perpendicular to the surface (see table 4). The hardness depth profiles measured in the polished section for all samples are shown in figure 2. Increasing the processing time from 120 to 240 min did not cause substantial change in hardness profiles. However, the hardness measured on the surface noticeably increased presumably due to growth of the nitride layer. The microstructure of samples after plasma treatment (samples nos. 2, 3, 4, 5) did not exhibit noticeable changes with respect to the initial sample. Redistribution of elements was not observed in samples nitrided at $T \sim$ 700°C.

Table 4. Surface hardness of samples after plasma nitriding .

Sample no.	Hardness $HV_{0.1}$
1	1150
2	450
3	640
4	1030
5	620

Hardness of TiN coatings deposited after the nitriding procedures was measured for all samples to be nearly constant at the level of 1600–1700 $HV_{0.1}$.

The coatings were analysed by a scratch-tester Anton Paar Revetest RST with Rockwell indenter (radius 200 μm) under linearly increasing load form 0.5 to 70 N. The track length was 7 mm. Three tracks were scratched on each sample. Typical indenter track after the test is shown in figure 3.

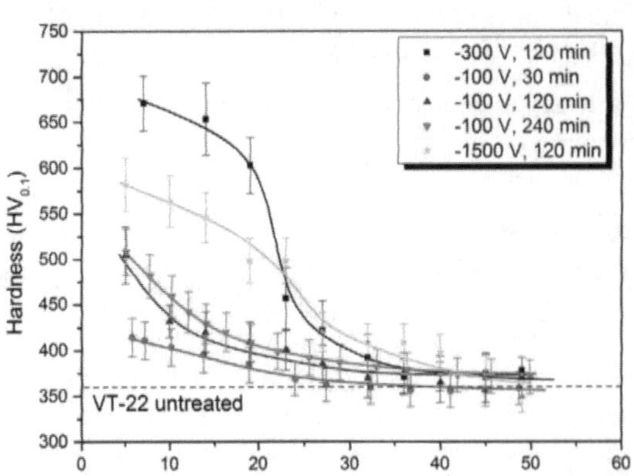

Figure 2. Microhardness depth profiles after ICP nitriding.

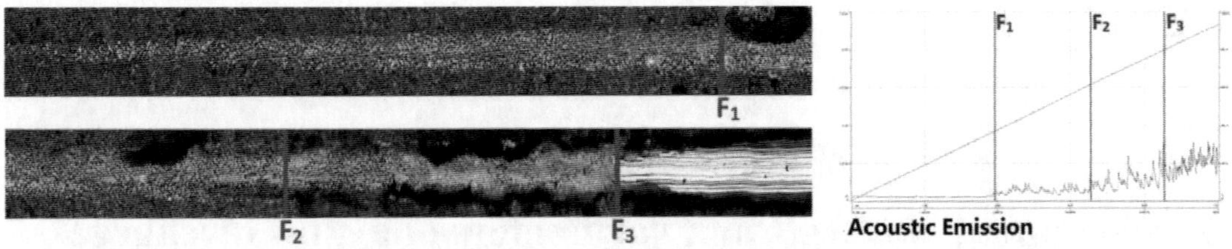

Figure 3. Typical scratch test results: F_1 — occasional cracking, F_2 — elastic deformations, F_3 — coating separation from the substrate

The indenter track demonstrated three critical load points: F_1, F_2, and F_3. Critical load F_1 is characterised by low magnitude acoustic emission that correlates with the beginning of occasional cracking of coating. The second critical load F_2 corresponds to beginning of the plastic deformations. Eventually, critical load F_3 denotes the coating separation from the substrate.

Scratch tests (table 5) demonstrate that for samples nos. 1–4, deposited TiN coatings start to crack (F_1) at 10–15 N load in regime A, and at 25–35 N load in regime B. Coating separates from the substrate (F_3) at 35–45 N in regime A, and at 60–70 N in regime B. Hence, the coatings deposited in regime B mostly demonstrate better overall adhesive performance. One must also mention that cracking and exfoliation take place easier if nitriding is performed at high temperature (sample no 1).

Table 5. Mean critical loads for TiN coatings obtained by the scratch test.

	F_1 (N)		F_2 (N)		F_3 (N)	
Sample no.	Regime A	Regime B	Regime A	Regime B	Regime A	Regime B
1	12	26	25	40	33	61
3	15	28	28	—	40	70
4	15	33	—	46	45	58
5	15	18	18	—	65	50

However, sample no. 2 nitrided in N_2 gas at high bias voltage of −1500 V demonstrates different result: regime A gives better adhesive properties than regime B.

4. Conclusion

Duplex ICP nitriding and magnetron TiN deposition technology ensures several times improvement of the surface layer hardness of VT-22 titanium alloy together with its high adhesive strength.

Acknowledgments

The work was supported by the Ministry of Education and Science of the Russian Federation (contract No. 14.Y26.31.0008) and by National Research Nuclear University MEPhI in the framework of the Russian Academic Excellence Project (contract No. 02.a03.21.0005, 27.08.2013).

References

[1] Yu T, Tian-Wen G, Jing W et al 2011 *Adv. Mater. Res.* **217–218** 1050–1055
[2] Avelar-Batista J C, Wu S, Gotman I et al 2015 *Mater. Lett.* **157** 45–49
[3] Avelar-Batista J C, Spain E, Fuentes G G et al 2006 *Surf. Coat. Technol.* **201** 4335–4340
[4] Ma S, Xu K, Jie W 2004 *Surf. Coat. Technol.* **185** 205–209
[5] Marin E, Offoiach R, Regis M et al 2016 *Mater. Design* **89** 314–322
[6] Vardanyan E L, Yagafarov I I, and Ramazanov K N 2016 *Proc. 27th Int. Symp. on Discharges and Electrical Insulation in Vacuum (Suzhou, China)* pp 722–725
[7] Borisyuk Yu V, Oreshnikova N M, and Pisarev A A 2016 *Russian J. Surf. Investig.* (in press)

Determination of the plasma parameters of a glow discharge in long tubes

A S Kiselev[1], D K Kostrin[1], A A Lisenkov[2] and E A Smirnov[1]

[1] Department of electronic instruments and devices, Saint Petersburg Electrotechnical University "LETI", 197376, Saint Petersburg, Russia
[2] Institute of Problems of Mechanical Engineering, Russian Academy of Sciences, 199178, Saint Petersburg, Russia

E-mail: alex.kiselev.epu@gmail.com

Abstract. In this work experimental current-voltage characteristics of a glow discharge occurring in long tubes for a wide range of discharge conditions (pressure, diameter of the discharge tube, nature of the gas) were obtained. On the basis of the current-voltage characteristics was calculated the longitudinal potential gradient in the positive column. With the help of the developed computer program was calculated the electron temperature for discharge conditions corresponding to the experiment. The technique is based on the use of the balance equations of ionization in gas discharges occurring in long narrow tubes, and provides the possibility for calculation of the discharge plasma parameters, both in pure gases and in multicomponent mixtures. Based on the experimental values of the longitudinal potential gradient in the positive column and the calculated values of the electron temperature was calculated the dependence of accommodation coefficient for the electrons from the discharge conditions. The compliance between the experimental and reference data was obtained.

Determination of the plasma parameters of a glow discharge is an important task in the design and analysis of the gas discharge devices [1]. Of particular interest is the determination of the parameters of the discharge occurring in long tubes, for example, in gas discharge lasers. One of the main characteristics that defines the behavior of the discharge gap in the circuit is the current-voltage characteristic (CVC). The CVC allows judging the state of the active environment, in particular, the degree of leakage of the atmospheric gases [2]. In addition, the form of the CVC has a significant influence on the dynamic properties of the discharge [3, 4].

Discharge CVC determines the behavior of the longitudinal potential gradient. The longitudinal potential gradient E_z in the positive column (PC) of a glow discharge is proportional to the square root of the accommodation coefficient χ of the electrons, which refers to the total energy of the electron W_e. The value of χ is important in the theory of a glow discharge PC: it characterizes the energy balance of the PC and connects the longitudinal potential gradient E_z with electronic temperature T_e [4]:

$$E_z = \frac{1{,}5kT_e\sqrt{\chi}}{e\lambda_e}, \quad (1)$$

where $\lambda_e = \lambda_{e0}/p$ – is the average free path length for electrons; p – gas pressure.

Function $\chi = f(T_e)$ can be found taking into account shares of energy lost by electrons in elastic and inelastic collisions:

$$\chi = \chi^* + \chi_a + \chi_w, \quad (2)$$

where $\chi^* = 2m_e/M$ – is the average fraction of the energy lost by electron in the elastic collision; χ_a – average fraction of energy loss due to the excitation; χ_w – share of energy losses at the walls.

Rigorous calculation of χ for the conditions of discharges in long narrow tubes, realized in the gas discharge lasers, is difficult because of the diversity and complexity of the processes occurring in the plasma. In the literature there are experimental data about χ relating to a number of pure gases [5, 6]. They reflect the dependence $\chi = f(T_e)$ for a narrow range of the discharge conditions, so the use of the reference values in the calculation is difficult.

To increase the accuracy of calculations of the plasma parameters of a positive column in this work was taken an attempt to refine the values of the accommodation coefficient. The idea of refining χ is based on the use of half experimental approach to the problem. Initially it was conducted a series of experiments to study the CVCs of the discharge gaps. In the experiments we used two models.

Layout 1 consisted of four discharge tubes with inner diameters of 1,1; 2; 3,4 and 5,2 mm, with movable anodes, transported by means of the magnetic actuators within the length of PC 0,1...0,25 m.

Layout 2 is a multianode tube with an inner diameter of 6,5 mm with the possible values of the active medium length 1,3; 1,12; 0,98; 0,78, 0,61; 0,47 and 0,35 m. In addition, layout 2 has an optical resonator and therefore the discharge gap can be used as the active medium of a He-Ne laser.

Layouts are connected with a vacuum pumping and filling system, which allows varying the gas content and operating pressure. In this work, were studied the characteristics of the discharge in pure helium and neon. The idea of refining χ is based on a combination of theoretical and experimental approaches to the calculation of the plasma parameters. Initially, based on experimental CVCs was obtained information about the behavior of E_z in the wide range of the discharge conditions. Since $T_e = T_e(pR)$ and $\lambda_e \sim 1/p$, from (1) it follows that $E_z/p = f(pR)$, the experimental values of the potential gradient are conveniently represented in this form (figure 1).

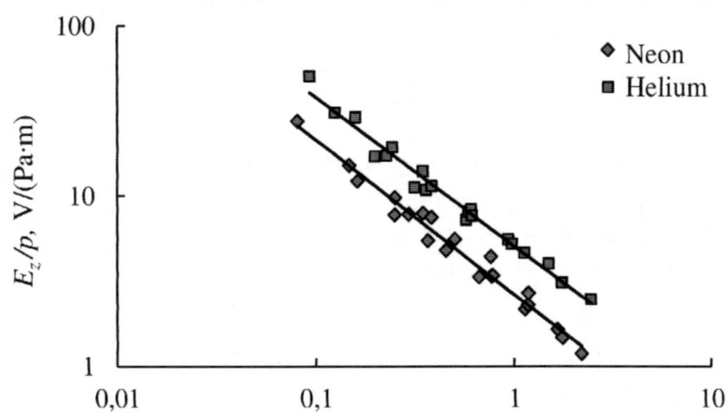

Figure 1. Dependences $E_z/p = f(pR)$ for helium and neon.

Next, using the developed computer program was calculated the function of the temperature T_e for a range of changes in the discharge conditions: pressure p; discharge channel diameter d; composition of the gas mixture corresponding to the experimental conditions. The method of calculation of T_e is as follows. Since charged particles in the plasma are connected by the Coulomb forces of interaction, their diffusion has a specific nature. Electrons with greater energy and mobility compared to the ions ahead of the last participate in the diffusion process. The spatial charge of ions left behind tends to keep the electrons and accelerates the diffusion motion of the ions. Diffusion of charged particles of both signs can occur in plasma only in conjunction and is called ambipolar diffusion.

In the experiments, the plasma is enclosed in a cylindrical tube with internal radius R. In such conditions, recombination in the volume is small, and neutralization of the charge mainly is due to the ambipolar diffusion with the subsequent recombination on the walls. The condition of the balance of charges for such a cylindrical plasma column can uniquely associate the average ionization frequency with coefficient of the ambipolar diffusion. Taking into account the expression for the ambipolar diffusion, frequency of ionization is defined as

$$\nu_i = \frac{b_i k (2,405)^2 T_e}{R^2 e}.$$

The average ionization rate can be written as

$$\nu_i = n_a \int_0^\infty q_i(U) \upsilon(U) f(U) dU, \quad (3)$$

where n_a – is the concentration of atoms; q_i – ionization cross section; $f\,(U)$ – Maxwell distribution function for electrons over energies. Approximations of the cross section of ionization for atoms and molecules were assumed linear. Thus, after integrating, was obtained an expression for the frequency of ionization:

$$\nu_i = n_\alpha \left(\frac{8kT_e}{\pi m_e}\right)^{0,5} \cdot \alpha_i \exp\left(\frac{-eU_i}{kT_e}\right)\left(U_i + \frac{2kT_e}{e}\right). \quad (4)$$

The transcendental equation (4) is applicable for the single component active media of the gas discharge lasers. For multicomponent gas mixtures the sum of ionization frequencies of all gases within the mixture should be used:

$$\sum_j \nu_i(j) = \frac{(2,405)^2 b_i k T_e}{R^2 e} = \frac{23,1 b_i k T_e}{e d^2}.$$

In these conditions, the average ion mobility of the gases can be calculated using the Blanc law.

The described above technique was used as the basis for the program to calculate T_e. The program allows the choice of the laser type or the required gas mixture, and the display mode for the calculated information, definition of the constants and ranges of changes in the diameter of the discharge channel, the total and partial pressures of the gas components and gas temperature [7].

Based on the ratio (1) between the E_z and T_e, can be obtained dependences $\chi = f(pR)$ for each gas type (figures 2, 3). For the possibility of automated calculation of χ and E_z taking into account equation (2) can be used approximation as the sum of χ^* (does not depend on discharge conditions) and a falling function in the limit tending to zero.

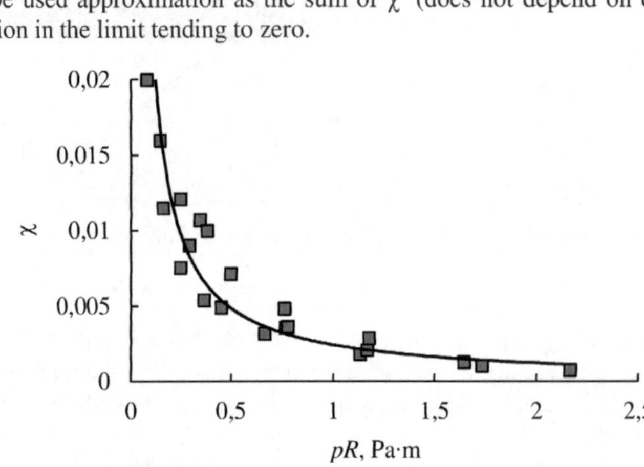

Figure 2. Dependence $\chi = f(pR)$ for neon.

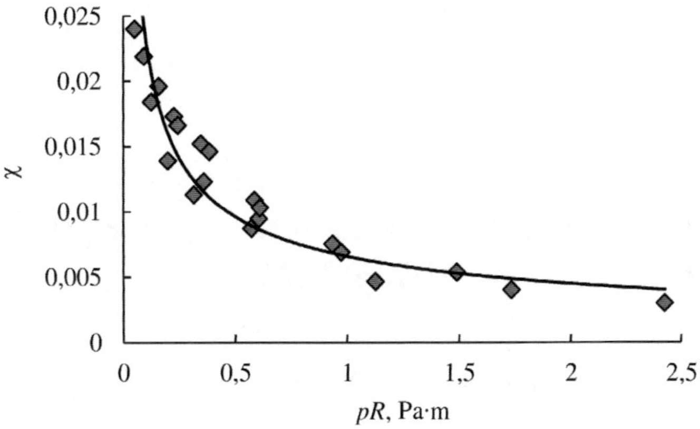

Figure 3. Dependence $\chi = f(pR)$ for helium.

Form of the dependencies $\chi = f(pR)$ is falling in nature, which is consistent with the presentation of the accommodation coefficient, that is described by expression (1). For small values of pR a significant share of the electron energy goes to heating of the walls. With the growth of pR wall loss decreases, and the main losses of energy are due to the processes in the plasma volume. The validity of a half experimental approach to the assessment of χ is confirmed by the coincidence with an accuracy of 10...15 % of these results with the reference data from the literature [5].

In further studies we intend to extend the ranges of changes of the discharge conditions and gas mixtures, including typical for glow discharge lasers, such as He-Ne and CO_2 lasers [8].

References

[1] Ramazanov A N, Kostrin D K, Goncharov V D and Lisenkov A A 2016 *Journal of Physics: Conference Series* **729** 012004
[2] Smirnov E A, Morozov S S and Timoshichev A A 2008 *Petersburg Journal of Electronics* **2–3** 39–48
[3] Smirnov E A 2005 *Proceeding of Saint Petersburg Electrotechnical University "LETI"* **2** 43–8
[4] Smirnov E A 2013 *Proceeding of Saint Petersburg Electrotechnical University "LETI"* **10** 6–10
[5] Granovsky V L 1971 *Electrical current in gas* (Moscow: Science)
[6] Levitsky S M 1964 *Collection of problems and calculations on physical electronics* (Kiev: Publishing House of Kiev University)
[7] Kiselev A S and Smirnov E A 2014 *Vacuum technique and technology* **1** 56–9
[8] Martsinukov S A, Kostrin D K, Chernigovskiy V V and Lisenkov A A 2016 *Journal of Physics: Conference Series* **729** 012023

IOP Conf. Series: Journal of Physics: Conf. Series **789** (2017) 012028 doi:10.1088/1742-6596/789/1/012028

Study on the influence of the magnetron power supply on the properties of the Silicon Nitride films.

D V Kiseleva[1], Y N Yurjev[1], Y V Petrakov[1], D V Sidelev[1], D V Korzhenko[1] and E V Erofeev[2]

[1] Department of Experimental Physics, National Research Tomsk Polytechnic University, Tomsk, 634050, Russian Federation
[2] Tomsk State University of Control Systems and Radioelectronics, Tomsk, 634034, Russian Federation

E-mail: dariyakiseleva@gmail.com

Abstract. Silicon nitride (Si_3N_4) films were deposited by magnetron sputtering of silicon target in ($Ar+N_2$) atmosphere with refractive index 1.95 - 2.05. The results of Fourier transform infrared (FTIR) spectrophotometry showed Si-N bonds in the thin films with concentration $2.41\cdot10^{23}$ – $3.48\cdot10^{23}$ cm^{-3}. Dependences of deposition rate, optical characteristics and surface morphology on rate of N_2 flow and properties of magnetron power supply.

1. Introduction

Silicon nitride (Si_3N_4) thin films are very promising due to its unique properties: high diffusion resistance to moisture and ions, high thermal stability, low level of surface state density, excellent insulating and semiconducting properties such as low level of leakage current, high breakdown voltage, satisfactory compatibility with semiconductive substrates in coefficient of thermal expansion. Due to its properties silicon nitride films are widely applied in microelectronics, e.g. as insulating layers in MIM-capacitors, masks for oxidation process and inhibitive coatings [1-6].

At present main technique to obtain silicon nitride thin films is plasma-enhanced chemical vapour deposition by decomposition of silane and ammonia (or nitrogen). But this technique has disadvantage – small fluctuation in operating parameters (pressure of gases, electrical power, gas ratio) leads to variation of another parameters. Such situation leads to difference in stoichiometric composition between Si_3N_4 and obtained thin films. Also, such thin films contain bounded hydrogen[1-6].

Magnetron sputtering is alternative technique allows obtaining stoichiometric silicon nitride thin films with good deposition rates and properties without toxic reagents (such as silane and ammonia) [1-6]. However in this case films' properties depends on parameters of power supply: pulse rate, pulse duration etc.

This article is devoted to investigation of influence of the deposition parameters and characteristics of the power supply of the magnetron sputtering systems on the properties of the silicon nitride films.

2. Experimental

2.1. Magnetron sputtering

Silicon nitride thin films coatings were deposited on ion-plasma installation (figure 1). Vacuum system (3) based on a turbomolecular pump provided a residual pressure not less than $5\cdot10^{-3}$ Pa. Pressure was measured by a wide-range vacuum gauge MicroIon Plus (1).

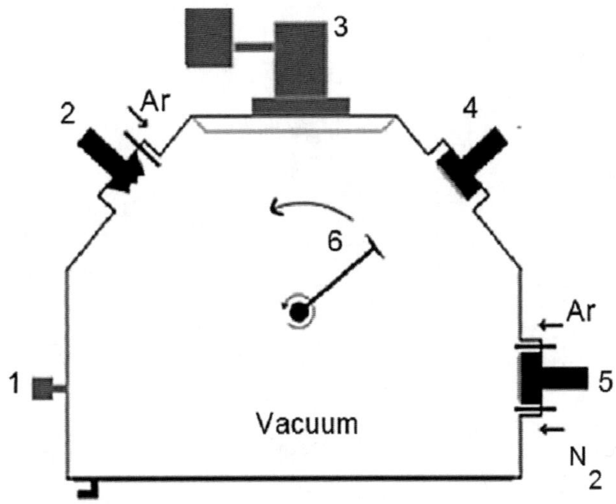

Figure 1. Scheme of ion-plasma installation: 1 - MicroIon Plus; 2 - ion source; 3 - pumping system; 4, 5 - planar magnetrons; 6 - the substrate attached to the rotary table; Ar - working gas (argon), N_2 – reactive gas (nitrogen).

Before deposition the substrates were cleaned using the ion source (2) with the parameters: voltage (U) = 2500 V, current (I) = 0.05 A. Ion cleaning time lasts 20 minutes. For film deposition magnetron was used (5) with planar silicon cathode and power supplies with different parameters:

- power source with two modes of operation: direct current (DC) and pulsed current with a frequency f = 100 kHz,
- current source with a frequency 134 kHz.

Deposition was carried out in the power discharge stabilization mode (0.5 kW) while maintaining a constant argon flow value (18 cm^3/min) and various values of the nitrogen flow (4.5 - 10.5 cm^3/min). Thin films were deposited on wafers made of monocrystalline silicon with intrinsic conduction.

2.2 Films characterization

The optical constants of films and their thicknesses were measured by ellipsometry, based on the analysis of changes in polarization state of a polarized monochromatic light reflected from the object at oblique incidence light beam (Ellips 1891 SAG device). Atoms' bonds in coatings and their concentration were investigated By FT-IR spectroscopy method based on a study of the vibrational-rotational spectra (Infralum FT-801 instrument). The surface morphology was studied by scanning electron microscopy - method of solid surface analysis of microstructure with an electron microscope, which consists of viewing the reflected "electron image" (SEM Zeiss Supra 55 with electrons' energy equal to 10 keV).

3. Results and discussion

During the reactive sputtering process, the greatest impact on the rate of growth of the thin film has a ratio between the plasma-forming (argon) gas and reactive (nitrogen) gas. Performance of the magnetron sputtering system according to the nitrogen flow rate Q (N_2) in the working chamber for different types of power supplies is shown in the figure 2.

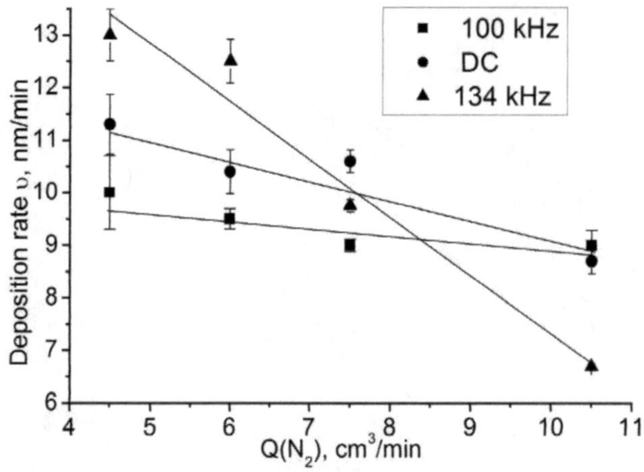

Figure 2. The dependence of the deposition rate on the nitrogen flow.

Silicon nitride deposition rate decreases with the increasing flow rate of nitrogen into the chamber. This is due to the nitriding of the target, in which a thin layer of silicon nitride is formed on its surface, whose sputtering rate is significantly lower than that of pure silicon [7]. It should be noted that the power supply with the pulse rate 134 kHz has a high deposition rate at low flow of nitrogen. This feature is associated with formation of operating current pulses, providing a high pulse power at which silicon is sputtered more efficiently. However, due to the fact that the pulses are short enough, with increasing of nitrogen flow target is nitrided more rapidly and the rate of deposition decreases more intensively [8]. Using DC power supply allows getting close to the deposition rate with respect to the pulsed mode at a frequency of 100 kHz. However, using a magnetron DC causes the frequent occurrence of microarcs on the target's surface that despite the operation arc protection system may have further effect on the quality of the coatings.
The dependences of the refractive index of thin films Si_3N_4 on nitrogen flow for different power sources are shown in figure 3.

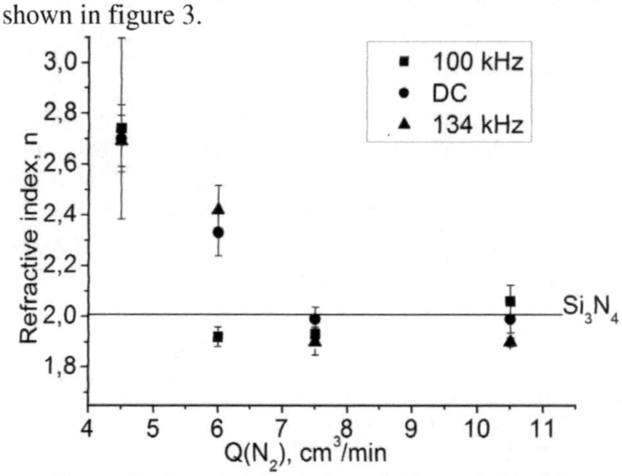

Figure 3. The refractive index of silicon nitride films

The figure 3 shows that with increasing nitrogen flow the refractive index of coating decreases, regardless the type of power source. When using a current source with a frequency of 134 kHz a more intensive decrease of refractive index is associated with the reasons described above for deposition rates.

Silicon nitride films have higher refractive index at low flow of nitrogen, which is associated with an excess of silicon. When nitrogen flow is 7-11 cm^3/min, films have refractive indices close to the stoichiometric Si_3N_4 ($n = 1.95 - 2.05$) [9].

By FTIR spectroscopy were investigated atoms' bonds presented in the coatings and their concentration. The figure 4 shows a typical spectrum for all coatings on the example of the sample obtained at a nitrogen flow of 7.5 cm^3/min and using a power supply with the magnetron pulse frequency 134 kHz. The peak with the wave number of 840 cm^{-1} indicates the presence of Si-N bond [9]. Also absence of any extraneous bonds is evident.

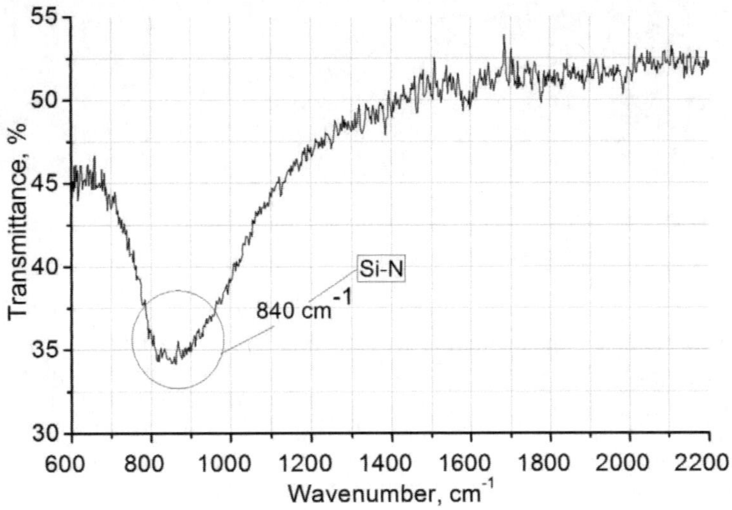

Figure 4. FT-IR spectrum of Si_3N_4 film

The concentration of the bonds (the density of the vibrating oscillators) Si-N in the films was calculated by formula:

$$C = \frac{S}{\varepsilon} \tag{1}$$

where S is the integrated intensity of the absorption band (peak area), ε is the extinction coefficient.

Table 1 shows the results of calculated concentrations of bonds depending on the reactive gas flow into the chamber to a variety of magnetron power sources.

Table 1. Concentrations of Si-N bonds.

$Q(N_2)$, sccm	Concentration of bonds, $n \cdot 10^{23}$, cm^{-3}		
	100 kHz	DC	134 kHz
4.5	2.86	3.38	2.41
6	3.07	3.24	2.58
7.5	3.52	3.11	2.76
10.5	3.28	3.13	2.61

It was found that all obtained coatings have similar values of concentrations of Si-N bonds. These values are consistent with the data for the plasma-chemical methods of Si_3N_4 films' obtaining, which is a major in microelectronics at present.

SEM images of the samples were prepared for analysis of surface morphology, some of which are shown in figure 5.

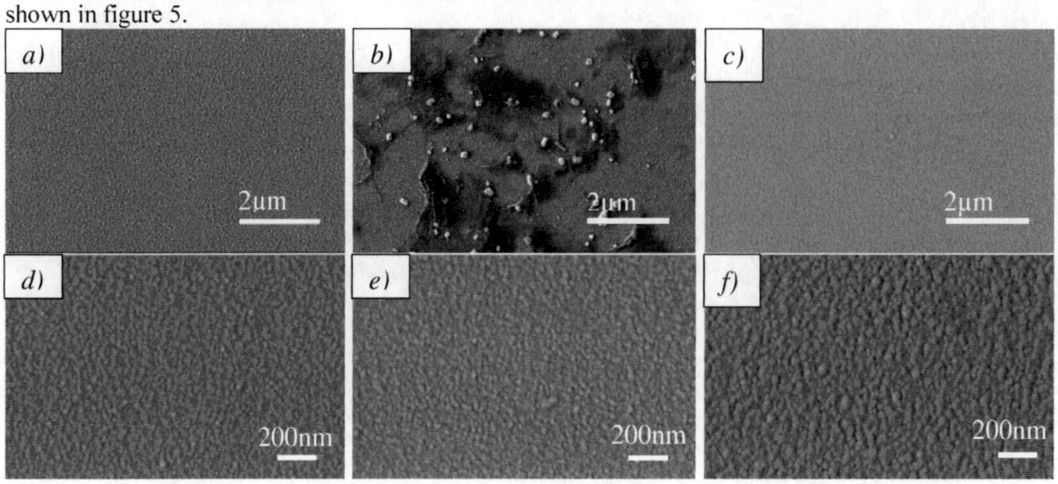

Figure 5. SEM-images of a surface test samples: (a), (d) - 100 kHz mode; (b), (e) - DC mode; (c), (f) - 134 kHz mode.

It was determined that the power source of the magnetron affects surface morphology. Thus the use of the DC power leads to the formation of the droplet fraction on the surface of the film (fig. 5b), due to the occurrence of electrical arcs on the surface of the magnetron target [10]. This is a disadvantage since it may reduce the performance of coatings. Pulse magnetron mode of operation avoids the problem of droplets (fig. 5 a, c). All obtained samples have a uniform homogeneous structure, including sections without droplets obtained at DC mode. The average grain size in the films, deposited by direct and pulsed current (100 kHz) are 20 nm, and the coatings produced by the pulsed current (134 kHz) - 50 nm.

4. Conclusions

As a result of the work, it was found that the deposition rate of Si_3N_4 films depends linearly on the flow rate of nitrogen into the chamber: increasing of N_2 flow reduces the rate. Also power supply has a negligible effect except for modes with a small flow of nitrogen.

Increasing the N_2 flow rate decreases the refractive index from 2.7 to 1.9. When the nitrogen flow rate is 7.5 cm^3/min and 10.5 cm^3/min refractive index films corresponds to the stoichiometric Si_3N_4 (n = 1,95 - 2,05).

The results of FT-IR spectroscopy showed the presence of only Si-N bonds in the films. Their concentration is practically independent on the parameters of the electric power of the magnetron sputtering system and is equal to $(2.41 - 3.48) \cdot 10^{23}$ cm^{-3}.

Scanning electron microscopy showed that all samples have a uniform homogeneous structure. The deposition of silicon nitride films in pulsed mode eliminates the problem of droplet fraction is typical for DC. The average grain size varies in the range 20-50 nm.

This work was financially supported by the Ministry of Education and Science of the Russian Federation, agreement no. 14.577.21.0204 of October 27, 2015. Unique project identifier: RFMEFI57715X0204.

References

[1] Hirohata Y, Shimamoto N, Hino T, Yamashima T and Yabe K 1994 *Thin Solid Films* **253** 425

[2] Abbas K, Ahmad R, Khan I A, Saleem S and Ikhlaq U *International Journal of Chemical, Molecular, Nuclear, Materials and Metallurgical Engineering* **10** 800
[3] Veprek S 1980 *Current ToP. Matter.Sci.* **4** 151
[4] Orfert M and Richter K 1999 *Surf. Coat. Technol.* **622** 116
[5] Nowling G R 2002 *Plasma Sources Sci. Technol.* **11** 97
[6] Zambom L S and Mansano R D 2003 *Vacuum* **71** 439
[7] Musil J. 2005 *Thin Solid Films* **475** 208
[8] Kouznetsov V A, Schneider J M and Macak K 1999 *Surface and Coatings Technol.* **122** 290
[9] Batan A, Franquet A, Vereecken J and Reniers F 2008 *The 12th European Conference on Applications of Surface and Interface Analysis* **40** 754
[10] Kelly P J, West G, Kok Y N, Bradley J W, Swindells I and Clarke G C B 2007 *J. Surface and Coatings Technol.* **202** 952

IOP Conf. Series: Journal of Physics: Conf. Series **789** (2017) 012029 doi:10.1088/1742-6596/789/1/012029

Decomposition of the hydrocarbon compounds in the vacuum arc discharge plasma

D K Kostrin[1], M I Pikus[1], S A Trifonov[1] and A A Lisenkov[2]

[1] Department of electronic instruments and devices, Saint Petersburg Electrotechnical University "LETI", 197376, Saint Petersburg, Russia
[2] Institute of Problems of Mechanical Engineering, Russian Academy of Sciences, 199178, Saint Petersburg, Russia

E-mail: dkkostrin@mail.ru, lisran@yandex.ru

Abstract. Vacuum arc discharge, maintained in the vapors of the cathode material, is used for the implementation of plasmachemical synthesis. For the formation of compounds into the plasma flux a working gas is injected. In this work the principle of decomposition of the carbon containing compounds in the low temperature vacuum arc discharge plasma is considered.

One of the commonly used methods of coating deposition is a vacuum arc discharge, existing in the vapors of a cathode material [1]. Prospect of application of the directed flows of charged particles belonging to the plasma flux is associated with the ability to control the process of their deposition during the coating formation, and also high density of the energy released by the substrate, and possibility to regulate its temperature. Despite all these advantages of the technological process, it should be noted their complex relationship shown in figure 1 [2, 3].

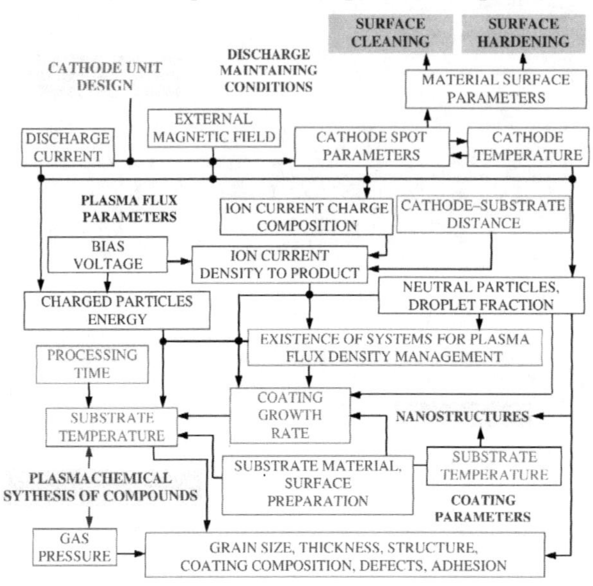

Figure 1. Relationship and influence of the parameters of the plasma flux on the formed coating.

Technology using vacuum arc plasma sources allows both to create clean coatings and to carry out the plasmachemical synthesis of the compounds [4]. To obtain carbide compounds, possessing high melting temperature and wear resistance, into the flux of metal plasma the hydrocarbon compounds are introduced.

Introduction of a carbon containing compound into the vacuum chamber from the side of the sputtered cathode (figure 2) ensures mixing of the plasma flux and gas atoms, and also the active dissociation of the gas molecules into the excited molecules, molecular ion formations, individual atoms and particles of different valences (radicals).

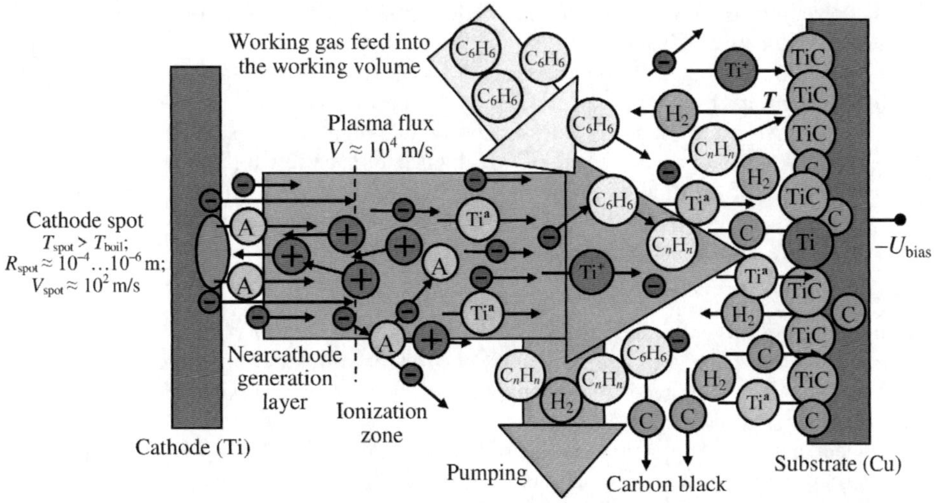

Figure 2. Plasmachemical process of the carbon based compounds formation.

The ground state of a molecule is a condition with the most stable electronic configuration. In this case, the molecule is in an energy minimum and has fairly high activation barriers for the changes in its structure. In the periodic table for carbon (C – $1s^2\ 2s^2\ 2p^2$) and hydrogen (H – $1s^1$) the number of the valence electrons corresponds to the number of the valence orbitals, which explains the high stability of the chemical bonds C–C (351.8 kJ/mol) and H–H (432.1 kJ/mol). Compounds of carbon with hydrogen differ in the nature of bonds, number of atoms in the molecules and their structure. Hydrocarbons constitute a homologous series, in which two consecutive hydrocarbons differ in exactly one same group of atoms. With increasing number of carbon atoms in homological series of hydrocarbons or with an increase in molecular weight generally grows their melting and boiling point and also their state changes.

As the working carbon containing compound commonly used is benzene (C_6H_6) – aromatic hydrocarbon (C_nH_{2n-6}, $n \geq 6$) with the first ionization potential 9.24 eV. Benzene contains the stable cyclic group of atoms (benzene nucleus) with a closed system of conjugated bonds. The bond length between the carbon atoms in benzene is 0.139 nm, which is an intermediate value between the length of a single bond in the alkanes (C_nH_{2n+2}, 0.154 nm) and a length of double bonds in alkenes (C_nH_{2n}, 0.133 nm). Circular conjugation allows gaining energy 150 kJ/mol, what is the magnitude of the conjugation energy – the amount of energy required for the destruction of the aromatic system C_6H_6.

Noticeable decomposition of the aromatic hydrocarbons in the gas phase with long bonds into molecules of shorter length is observed already at temperatures around 800 K, under these conditions the molecule becomes thermodynamically unstable.

The process of decomposition of the hydrocarbons is a complex process, consisting of the consecutive and parallel chemical reactions. Forming intermediate particles are characterized by altered electronic structure of an atom or molecule, and because of the continuous participation in the occurring chemical reactions have a short life time. Molecules and intermediate particles decay into

IOP Conf. Series: Journal of Physics: Conf. Series **789** (2017) 012029 doi:10.1088/1742-6596/789/1/012029

the fragments with sequential transformation of one free radical into the other: $\dot{R}_1 \rightarrow \dot{R}_2 \rightarrow \dot{R}_3 \rightarrow \ldots$, and also into separate atoms of hydrogen and carbon, as well as into excited states of molecules and molecular ion formations [5].

In the gas discharge plasma to the chemical processes of pyrolysis is added the decomposition reaction of hydrocarbons due to the interaction with the plasma flux components: the formation of CH_x radicals is in the process of direct dissociation by an electron impact. The structure of the formed radicals depends on the rate of separation of the hydrogen atoms, while the less hydrogen atoms are contained in the molecule, the easier is the decomposition of the hydrocarbon. The speed of course of the chemical reactions is proportional to the multiplication of the current concentrations of the reactants (law of Guldberg and Vahe).

Thermal transformations of benzene occur in a wide range of temperatures. At low temperatures (800...1100 K), is observed a breakage of C–H bond, accompanied by the formation of biphenyl ($C_{12}H_{10}$ – twin phenyl radical) and to a lesser extent of polyphenyls. Radical C_6H_7 is formed by attaching the atomic hydrogen to benzene. At higher temperatures the flow of complex and nonselective processes of splitting of the phenyl nucleus are observed. The decay rate of C_6H_5 and C_4H_3 is so great that the kinetics of the decay of C_6H_6 in these terms and conditions and in the assumption of the direct reactions of dissolution can be written as:

$$C_6H_6 \rightarrow C_2H_2 + C_4H_2 + 2H.$$

In the plasmachemical synthesis the sputtered flux of metal ions of the cathode material dn_i/dt reaching the substrate is calculated using the current of the vacuum arc discharge I_{dis} selectable from the conditions for stable burning of the discharge and the operating temperature of the cathode. A balanced flow of the gas molecules dn_{C6H6}/dt is determined based on the ratio of a chemical reaction between the metal and reactive gas and the size of the receiving surface [6].

The formation of a coating on a solid surface follows simultaneously with the process of carbon black formation occurring in the working volume. These reactions proceed in parallel.

Chemical conversion processes of the different carbon compounds into the solid particulate carbon are accompanied by the physical processes of the emergence of new dispersed phase. Carbon black is formed during nonequilibrium spontaneous condensation of carbon due to the inelastic collisions of particles. At the initial stage occurs decomposition of the carbon substances and formation of the clusters in a form of a radical mesh with torn bonds, after which on the condensation nuclei there is an intensive buildup of the carbon particles.

The formation of the compound flowing on the surface of condensation may occur according to the following scheme of splitting of hydrocarbons with long bonds into the molecules of a shorter length:

$$C_6H_6^{\uparrow} \xrightarrow{W} {}'C_nH_{Xn}^{\uparrow} + C_xH_{Yn}^{\uparrow} + H_2^{\uparrow} \xrightarrow{W} C + C_xH_Y^{\uparrow} + H_2^{\uparrow};$$

$$Me_{\text{substrate}} + (Me_{\text{plasma}} + C_6H_6)_V \xrightarrow{T} (M_{\text{substrate}} + Me_{\text{substrate}}C + Me_{\text{plasma}}C + C) + C_{\text{carbon black}} \downarrow + C_nH_n^{\uparrow}.$$

Using as the processed substrate the transition metal with unpaired electrons [7], allows to additionally apply it as a catalyst, which increases the rate of decomposition of the hydrocarbon and influences the structure of the formed carbonaceous material. The catalytic reaction proceeds on the surface and is caused by the activation of molecules of the reagents in the interaction with the surface. The evolved carbon is adsorbed on the surface and enters into a chemical reaction with it.

The advantage of using vacuum arc plasma sources is the formation of the coating by deposition of the ions with high kinetic energy. The speed of the plasma flux is mainly formed in the cathode region [8], and in the area of transportation it does not change and is about 10^4 m/s. The plasma flux is moving in a diverging magnetic field. It has significant spatial heterogeneity due to the presence of actuators with different geometry, and motion of charged particles is performed at a substantial distance from the axis of the system.

Analysis of the plasma flux composition was carried out using emission spectral analyzer ISM3600 [9]. Figure 3 shows emission spectra of a vacuum arc discharge in the visible range, obtained in a substrate region at a current of 80 A during sputtering of a titanium cathode in vacuum at a pressure of $7.8 \cdot 10^{-3}$ Pa and in benzene vapors at a pressure of $8.7 \cdot 10^{-2}$ Pa. On spectra lines corresponding to excited neutral titanium atoms (Ti I), singly (Ti II) and doubly (Ti III) charged titanium ions and singly charged carbon ions (C II) are marked.

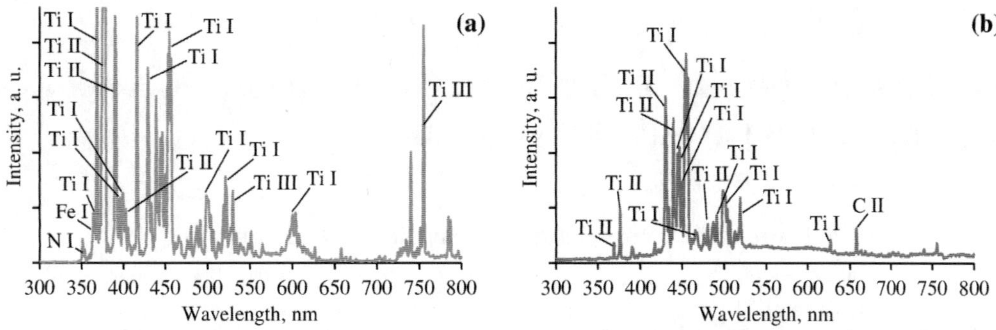

Figure 3. Radiation spectra in the process of titanium (**a**) and titanium carbide (**b**) deposition.

Introduction of a benzene vapor into the plasma flux significantly modifies the emission spectrum, affects the charge composition of the titanium plasma and the concentration of titanium in the implementation of the plasmachemical synthesis. The resulting titanium carbide has a well formed crystalline structure with all the lines of maximum intensity recorded on X-ray diffraction patterns.

Changing the value of the accelerating potential *U*, defined on the workpiece, makes possible to regulate the energy of the deposited ions and hence to control the flow of the process of the coating formation. Preliminary condensation of the sprayed cathode material (Mo, W) to a face of a similar metal contributes to obtaining of a structural based coating, which in a subsequent diffusion reaction reduces the interfacial tensions related to the difference of the crystal lattices, and improves the strength of the adhesive bond. Plasmachemical synthesis of carbide compounds in the flux of a metal plasma of a vacuum arc discharge was carried out in the vapors of benzene (C_6H_6).

Because of the peculiarities of the electronic structure and complex covalent-ionic-metallic nature of the interatomic interactions the phase diagram of the carbide systems for Mo and W have several structural modifications, stable in different temperature and concentration intervals [10]. Under the chosen conditions of the coatings formation have been produced layers composed, respectively, of the phases of molybdenum carbide – MoC or tungsten carbide, formed from the inner layer of W_2C and polyamide monocarbide surface layer WC with a hexagonal structure.

It is thus shown that the composition of the original carbon containing gas determines the yield of carbon and the nature of course of the chemical reaction of formation of the carbide compounds.

References

[1] Kostrin D K and Lisenkov A A 2016 *Materials Science Forum* **843** 278–83
[2] Barchenko V T, Vetrov N Z and Lisenkov A A 2013 *Technological vacuum arc plasma sources* (Saint Petersburg: SPbETU)
[3] Bystrov Yu A, Vetrov N Z, Lisenkov A A and Kostrin D K 2014 *Vakuum in Forschung und Praxis* **5** 19–23
[4] Lisenkov A A and Valuev V P 2011 *Vakuum in Forschung und Praxis* **6** 32–6
[5] Bystrov Yu A, Vetrov N Z and Lisenkov A A 2008 *Technical Physics Letters* **9** 734–6
[6] Bystrov Yu A, Vetrov N Z and Lisenkov A A 2013 *Technical Physics Letters* **10** 914–6
[7] Kostrin D K and Lisenkov A A 2016 *Materials Science Forum* **870** 371–6
[8] Bystrov Yu A, Kostrin D K, Vetrov N Z and Lisenkov A A 2015 *Vakuum in Forschung und Praxis* **2** 22–5
[9] Kostrin D K, Uhov A A and Lisenkov A A 2016 *Vakuum in Forschung und Praxis* **3** 34–7
[10] Bystrov Yu A, Vetrov N Z and Lisenkov A A 2010 *Technical Physics Letters* **6** 570–3

IOP Conf. Series: Journal of Physics: Conf. Series **789** (2017) 012030 doi:10.1088/1742-6596/789/1/012030

Peculiarities of the structure formation of nanoscale coatings from the vacuum arc discharge plasma

D K Kostrin[1], M I Pikus[1], E A Smirnov[1] and A A Lisenkov[2]

[1] Department of electronic instruments and devices, Saint Petersburg Electrotechnical University "LETI", 197376, Saint Petersburg, Russia
[2] Institute of Problems of Mechanical Engineering, Russian Academy of Sciences, 199178, Saint Petersburg, Russia

E-mail: dkkostrin@mail.ru, lisran@yandex.ru

Abstract. In this paper features of the structure formation of nanoscale coatings (TiN)–(AlN)–(Ti–Al–N) in the vacuum arc discharge plasma are considered. The composition and structure of the formed nanoscale coatings are studied. The main factors influencing the quality of the formed coatings are shown.

Nanoscale and nanocomposite coatings represent a new type of materials obtained by ion-plasma technological devices. Nanocomposite materials, due to the small dimensions of the grains, exhibit very different performance properties compared to the conventional materials.

As the first nanostructured multilayer coating can be considered a combination of layers Ti:TiN, because titanium nitride is the most used reinforcing coating. For multilayer coatings was obtained a hardness level of 3700 kg/mm^2, which is higher than the hardness of homogeneous films.

The wear resistance of the workpiece increases due to the layer-by-layer deposition of nitrides of various metals [1, 2] generated using the plasma of a vacuum arc discharge while sputtering the cathode material by a cathode spot [3, 4]. Coatings consisting of multiple layers have a high hardness while maintaining the viscosity as a result of brittle fracture energy dissipation on the intergranular and interlayer interfaces formed by the layers.

Currently, one of the possible types of such coatings is alternating layers of nanocrystalline nitrides of titanium (TiN) and aluminum (AlN), with an intermediate layer of complex composition (Ti–Al–N). Nitride compounds are metal-like formations which have high hardness, melt without decomposition at a high temperature and are able to possess superconductivity.

For the plasmachemical synthesis as the reaction gas is used a chemically reactive molecular nitrogen N_2 ($2s^2\ 2p^3$), which has a strong triple bond with $d_{N\text{–}N}$ = 1.095 Å [5]. The destruction of the triple N–N bond prevents unusually strong first of the breaking bonds in the N_2 molecule (~ 1250 cal/mole).

Due to the small values of the threshold energies of excitation for vibrational and rotational levels of molecules particles are already excited at small values of the average energy, and by the electron collisions the processes of vibrational relaxation and dissociation of molecules not only through the main, but also through the electron excited states are accelerated.

Therefore, in the discharge gap charged particles not only cause the dissociation of N_2 molecules, but also excite higher energy vibrational levels of N atoms contributing to the occurrence of the reverse reaction N + N → N_2 with the allocation of the energy spent on decomposition. Chemical transformations of the neutral particles affect the ionization balance in the plasma flux and the distribution function of the electron energy and as a result all the plasma characteristics.

Increasing the strength of the films is possible due to the structural changes associated with the variation of the deposition conditions and also can be obtained by reduction of their thickness. Films with nanometer thicknesses have hardness several times higher than the micrometer films. Their durability is influenced by the structure of the free surface and the interface film–substrate, formed in the process of condensation of atoms.

The increase of the negative bias potential that is set on the substrate and consecutive increase in the power of the sprayed particles, and also the variation of the product temperature lead to a change in the phase structure of the formed coating and the percentage content of the elements in the coating due to the sputtering of the light elements with increasing energy of ions bombarding the substrate. For deposition of the coating can be used technological devices, which enable simultaneous spraying of different materials (figure 1), and also the spraying of the alloy cathode consisting of various materials in different proportions.

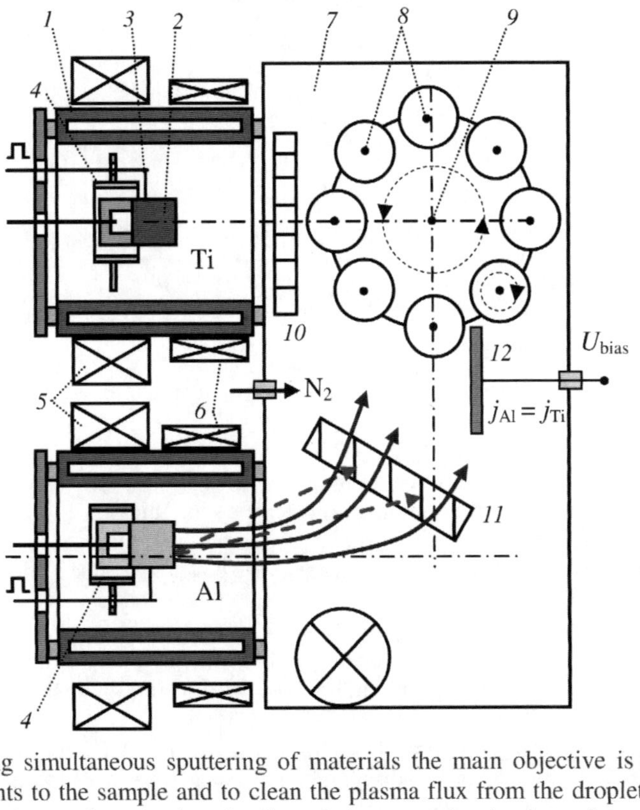

Figure 1. Scheme of the technological apparatus for producing complex composite coatings: *1* – anode; *2* – cathode; *3* – igniting electrode; *4* – screen; *5* – stabilizing solenoid; *6* – focusing solenoid; *7* – working volume; *8* – workpiece; *9* – planetary rotation mechanism; *10* – plasma attenuator; *11* – plasma separator; *12* – substrate.

During simultaneous sputtering of materials the main objective is to harmonize the density of ion currents to the sample and to clean the plasma flux from the droplet formation of the aluminum. For the purpose of weakening the flux of, for example, titanium plasma was used an attenuator *10*, and for cleaning – the separator *11*, creating a solid, impenetrable barrier to droplet formation, having the straight path of movement. In this case, their deposition is carried out on the separator surface facing the working surface of the cathode.

The change in the set bias voltage leads to a change in the content of nitrogen in the composition of the nitride coatings MeN_{1-x} in the range $x = 0.5…0.9$ and the change in the lattice of the nitride phase, which is associated with an increased likelihood of the nitrides formation at growing energy of nitrogen and titanium ions [6].

When applying combined coatings it was taken into account that Ti and Al have different melting and boiling points, sputtering yields and atomic masses. The last condition leads to the fact that the plasma flux contains particles with different kinetic energy. Therefore, the main task was to harmonize ionic currents of titanium and aluminum coming to the sample.

Depending on the conditions of deposition in the coating was observed the presence of both phases of titanium and aluminum, and intermetallic compound (Ti–Al). It was noted that with growth of aluminum ion current reaching the substrate, the percentage of the intermetallic compound in the coating is increased, and the decrease of the current I_{iAl} leads to the predominance in the coating of the titanium nitride phase. Figure 2 shows X-ray diffraction pattern for Ti–Al–N system obtained using bias voltage –130 V.

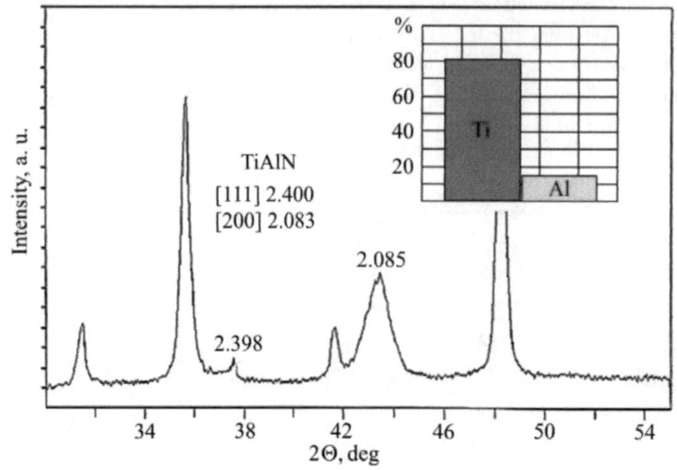

Figure 2. X-ray diffraction pattern for Ti–Al–N system.

For the Ti–Al–N system were obtained the following ratios for content of titanium and aluminum in the coating depending on the set bias voltage: 80 % Ti and 13 % Al for U_{bias} = –130 V; 63 % Ti and 31 % Al for U_{bias} = –100 V; 55 % Ti and 45 % Al for U_{bias} = –20 V.

The use of the multilayer systems in the coating formation requires to investigate and simulate not only the structure of each of the layers individually, but also to determine the thickness, type of the material and number of layers in the period, create artificial structures with unique functional properties, as the use of multilayered structures leads to the increase of the coatings hardness.

However, for multilayer nanoscale films based on Ti and Al was found that with the increase in the percentage of more solid component (Ti), the hardness of the coating decreases and, conversely, with the increase of the percentage content of the more plastic component (Al), the hardness of the coating increases. This fact shows that materials in the coating and thickness of the films must be picked up in a special way. Main parameters of the deposition process are shown in figure 3.

Conducted research of the technological peculiarities of the formation of TiN, AlN and Ti–Al–N films from a flux of metal plasma of the vacuum arc discharge showed that the main parameters influencing the structure and properties of coatings are:

- gas pressure, that is responsible for elemental and phase composition of the coating;
- discharge current, that determines the composition of the plasma flux and the growth rate of the coating;
- distance from the cathode to the workpiece, that sets a relation between the discharge current and gas pressure, determining the phase relations metal–nitrogen (Me:N) in the formed coating;
- bias potential, that provides the energy of deposited particles, and thereby the coating growth rate, and its microstructure and hardness;

- substrate temperature, that is responsible for the adhesion, microstructure and residual stress in the obtained coating.

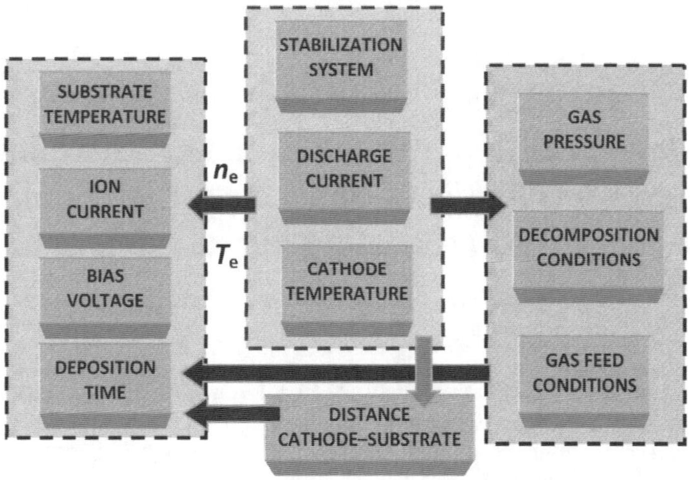

Figure 3. Factors that determine the quality of the applied coating.

It was received, that providing the cleaning of the plasma flux from the droplet formation, by varying the process parameters it is possible to effectively manage the composition and structure of the formed multilayered coating.

References

[1] Barchenko V T, Vetrov N Z and Lisenkov A A 2013 *Technological vacuum arc plasma sources* (Saint Petersburg: SPbETU)
[2] Vinogradov M L, Kostrin D K, Smelova V V, Trifonov S A and Lisenkov A A 2016 *Proceedings of the 2016 IEEE North West Russia Section Young Researchers in Electrical and Electronic Engineering Conference (Saint Petersburg)* pp 729–30
[3] Lisenkov A A and Valuev V P 2011 *Vakuum in Forschung und Praxis* **6** 32–6
[4] Bystrov Yu A, Kostrin D K, Vetrov N Z and Lisenkov A A 2015 *Vakuum in Forschung und Praxis* **2** 22–5
[5] Bystrov Yu A, Vetrov N Z and Lisenkov A A 2012 *Technical Physics Letters* **10** 938–40
[6] Bystrov Yu A, Vetrov N Z, Lisenkov A A and Kostrin D K 2014 *Vakuum in Forschung und Praxis* **5** 19–23

Effects of low pressure radio frequency discharge on the physical and mechanical characteristics and chemical composition of diffusion coating on a surface of complex configuration details

V I Ladianov[1], F Z Gilmutdinov[1], R M Nikonova[1], N F Kashapov[2], M F Shaekhov[3], V I Khristoliubova[3]
[1] Physics Technical Institute, Ural Branch of the Russian Academy of Sciences, 426000, Russian Federation, Izhevsk, 132 Kirova st.
[2]Kazan Federal University, 420008, Russian Federation, Kazan, 18 Kremlyovskaya street.
[3] Kazan National Research Technological University, 420015, Russian Federation, Kazan, 68 Karl Marks st.

E-mail: valllerrriya@mail.ru

Abstract. The work deals with the influence of low-pressure radio frequency (RF) discharge on the surface properties of metals and their alloys. As objects of research to study the interaction of the jet low pressure RF discharge into the surface of the material the following materials were chosen: tungsten cobalt alloy, high speed steel, structural steel. In the presence of the materials energy parameters of low pressure RF discharge flows in the discharge chamber and the electrode gap were studied. A quantitative assessment of the gas composition inside the chamber to determine the characteristics of the plasma flow, making the major contribution to the modification of the surface was carried out. The influence of the input parameters of the plasma unit on the discharge characteristics was held. Identification of the main processes responsible for the modification of the surface of metals and alloys with the metal sample in the plasma jet and the effect of samples of products complex configuration on its properties is determined.
The results of studies of physical and mechanical characteristics and chemical composition of the surface layers of high-speed steels, alloys and steel before and after treatment by low pressure radio frequency discharges with the instrumental indentation methods and X-ray photo-electron spectroscopy. With the help of the quality control system of the inner surfaces tubular products were studied.

1. Introduction

For the set technological and operational properties of mechanical engineering products numerous technologies are widely used, allowing to modify the surface properties of materials.

Depending on the physical and chemical processes occurring during the modification of materials, hardening methods can be divided as follows: the methods of forming a coating or film on the surface of the product; methods that convert the chemical composition of the surface; methods of modifying microgeometry surface layer (surface roughness); methods that alter the structure and energy supply of the surface; methods that alter the structure of the material throughout.

There are several methods of modifying the surface by which the individual nanolayers properties can be changed, while other surface properties remain unchanged. One of such methods is a magnetic pulse treatment, which can be used to harden the layers of 3-10 nm thickness, however, the surface roughness of this layer does not change, fissured and embossed layers remain.

For surface cleaning and intensifying the process of chemical-thermal treatment various methods of electro physical effects are used: nitriding in an ultrasonic field at high pressures in a fluidized bed, ion nitriding in a glow discharge plasma, heating high-frequency currents, finishing treatment in a glow discharge ion bombardment, plasma-chemical treatment in active gas environment. However, the processes are very long, it takes time for at least one hour. At glow discharge plasma surface of the material is bombarded by charged particles. Plasma in an oxygen environment is effective for the cleaning of organic contaminants, since carbon is oxidized and removed in the form of CO and CO2 [1, 2].

Plasma phase condensation method in the conditions of ion bombardment is the most widely used in the domestic industry. As a result of erosion of material in the cathode spots, the burning at the cathode, which material corresponds to the composition of the coating material the plasma stream is generated. The anode is the body of the vacuum chamber. The pressure in the chamber while maintaining the discharge is 10-1 - 10-5 Pa [3]. Deposition of titanium nitride occurs by the evaporation of titanium cathode in nitrogen at pressures of 0.1 - 0.4 Pa [4]. Addition of methane or acetylene into the vacuum space or chamber leads to a condensation of carbide coating. To achieve high density of ion flux values directed to the surface plasma-specialized devices to control both the rate of plasma flow and its physical characteristics are used [5].

The ion-plasma methods are carried out by condensation of the plasma state of matter on the surface of a product [6,7]. These techniques are now widely used to produce heat-resistant and wear-resistant coatings of compounds such as carbides and nitrides of metals of IVa, Va and VIa groups of the Mendeleev periodic system, oxides of aluminum, silicon, beryllium. These methods permit to obtain coatings of not more than 0.1 mm thickness. Product temperature during the deposition process is 500 - 800 K, deposition rate - not more than 0.03 m / s. The hardness of the coating is more than 14 GPa, the roughness parameter degrades no more than 1 micron.

Plasma promotes chemical reactions between the atoms of the reactive gas and metal, formation of the gas ions, which affects the flow of the substrate, and also stimulates the neutral particles that are deposited on the substrate. Exposure to high-energy ion flux to the substrate influences the coating adhesion strength to a substrate, the coating structure, etc. By ion bombardment of the substrate the surface layer defects are formed, spraying, change of morphology and surface composition, crystallographic structure, the absorption of gas, heating the substrate are possible.

Based on the analysis of sources [8-11], there are many scientific papers among the PVD methods devoted to the studies of nanostructured materials with unique physical and mechanical properties, obtained by magnetron sputtering arc evaporation. Nanostructured carbide, nitride and boride coating (TiC, TiN, TiB2, Ti (C, N) etc.) are used in industry in many countries as a wear-resistant coatings on metal-working tool.

Electric spark alloying (ESA) based on the pre-emptive destruction of the anode material with the spark discharge in the gaseous environment and the subsequent products transfer to the cathode erosion [12]. On the surface of the cathode the layer with anti-corrosion properties, high values of hardness and wear resistance, change of thermodynamic state and phase composition is formed. Erosion of the alloying of the electrode depends on the processing conditions, material structure and other factors.

Efficiency of electric spark alloying process, the morphology and thickness of the coating depends on the phase composition of products of erosion, which in turn depends on the technology processing conditions: duration of doping, discharge energy, form of tool electrode movement. Phase and structural transformations are associated with the distribution of chemical elements in the surface layer, dependent on the duration of the treatment, discharge parameters, the properties of the electrode material and other factors.

The most important feature of the electric spark alloying process - the possibility of forming a coating on the working surfaces of parts with a wide range of physical, mechanical and chemical properties by varying the electrode material, electrical characteristics, transconductance environment. The formation of the hardened layer by doping parts improves their durability, which depends on the structure of the coating, the magnitude of residual stress and other factors. The electric spark alloying process disadvantages include limit thickness (50 micrometers), high roughness of the coating, the possibility of applying only to the conductive coatings.

The coatings obtained by different methods of deposition and sputtering, have a significant drawback - it is a problem of an adhesive strength, in order to achieve high levels of quality that requires pre-finishing and surface preparation. The main drawback of the method - the presence in the coating of microdroplet phase.

Using the method of ion-plasma processing of structural materials inner surface of the articles can be processed. A method of vacuum processing of the inner tube surface includes a coaxial arrangement of the discharge electrodes. As one of the electrodes of the workpiece is used, the initiation of the vacuum electrical discharge between the electrodes is created by the potential impact of bit-discharge plasma onto the inner surface of the product. The potential difference is created between the cathode -product and the anode made of non-magnetic material that is not shorter than the length of the workpiece. Exposure to plasma is carried out with applied to the discharge zone crossed electric magnetic fields, and the discharge current density is maintained at 15-20 mA / cm2, nitrogen pressure 25 Pa, and the magnetic flux density equal to 10-20 mT [13]. With this method the corrosion resistance of the inner surface of the tubular article is increased. The disadvantage of the method is the processing time. Vacuum ion-plasma method does not allow to speed up the process by increasing the density of the ion flow, since there is overheating of parts as a result surface hardness is reduced [14].

The above drawbacks are not observed in the modified nanodiffuzive nanolayers surface including coatings, which are obtained by ion implantation by exposure to RF discharges [15-18]. The method consists of gas saturation of the surface, gas atoms and molecules are introduced into the surface of the metal layer, reacting with the atoms of the material to form compounds of nitrides or carbides of the corresponding metal. Direction of nanodiffuzive layers creation on the surface of metals and alloys is poorly studied. These shortcomings have led to the development of long-term process, which combines the operations of the materials evaporation, directional particle flow formation and introduction of them to the treated surface, the ionization and excitation of atoms, cleaning and polishing of surfaces using radio frequency plasma torch operating at low pressures [19,20]. The advantages of this method for processing materials in order to form the directional properties of the surface are: short duration of treatment process, coating uniformity, the possibility of modifying the configuration of complex products and different materials, forming of a high-density flow of ions, the use of different types of electrodes, including electrolytic [21-27].

2. Experiment

For the processing of metals and their alloys by the flow of low pressure radio frequency (RF) plasma radio frequency capacitor installation with symmetric planar electrodes developed at the Department of "Plasma chemical and nanotechnologies high molecular materials" is used (Figure 1) [28,29]. RF voltage is applied from the radio frequency generator AECesar 1330. For the automatically matching of the generator with a load the matching device AEVarioMatch 5000 is used.

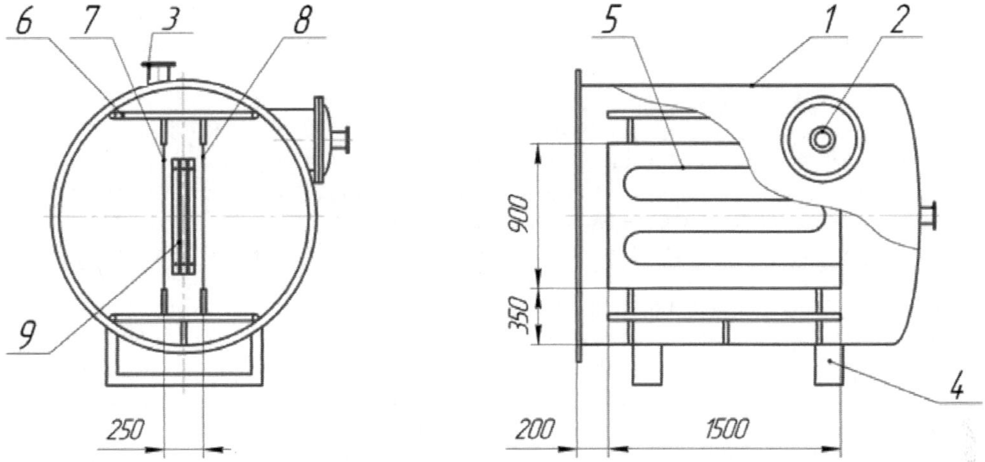

Figure 1 – RF plasma unit: 1 - vacuum chamber, 2 - flange vacuum system 3 - RF input, 4 - bearing, 5 - circuit cooling, 6 - dielectric rack 7 - RF electrode, 8 - ground electrode, 9 - workpiece.

The camera body is welded, made of stainless steel. The chamber walls are equipped with ribs. On the top radio frequency inputs, the unit of gas flow controllers, RF generator to the load matching system are placed. Hydraulic collectors of cooling systems are situated on the left side of the camera. There are nozzles for connecting the pumping system on the back wall of the chamber. is A collector system of process gases inlet is inside the discharge chamber in the door flange plane between the RF electrodes.

Vacuum pumping system is comprised of:

- Vacuum spool assembly;
- Vacuum double-rotor pumps;
- Vacuum pneumatic valve;
- Vacuum pneumatic valve;
- Gas flow regulator;
- Locking the gauge;
- Thermal sensor;

Spool vacuum pump AVZ-180 is designed to produce a pressure of 5000-6000 Pa depending on plasma gas flow rate or a mixture thereof (0.004 - 0.12 g / s) in a vacuum chamber and maintaining the pressure at the outlet the vacuum double-rotor pump DVN-500 during the operation of the first pump is contemplated. Pneumatic shutter is designed to prevent the ingress of atmospheric air into the vacuum chamber during emergency situations. Valves KVP-100 and KVP-63 are used for switching and bypass pumping foreline. Valve DPC-63 serves for air inlet after the manufacturing cycle.

Plasma torch with a flat electrodes for plasma discharge represents two water-cooling copper plates located at a distance of 40 cm. The electrodes are placed in a vacuum chamber with the special means for securing samples located between the electrodes.

Closed-circuit cooling system is designed to cool the unit AVZ-180 and RF electrode. System is based on the installation of refrigeration TDC-20.

Process parameters of RF capacitive discharge in a vacuum chamber with planar symmetrical electrodes are the following: gas pressure 20 - 25 Pa, the gas flow rate 0.01 - 0.1 g / s, the power delivered to the electrodes 1,5 - 1,8 kW. Products treated in pure argon plasma for 20 minutes are followed by 20 minutes in a mixture of argon and methane. Oscillator frequency - 13.56 MHz.

Tungsten cobalt alloys, high-speed steel as the most common materials in the tool industry, intended for the manufacture of cutting tools and machining, and structural steel designed for the manufacture of hydraulic actuators used in mechanical engineering were chosen as objects of study for the researches.

Process parameters of the discharge of the tubular article inside are the following: gas pressure 20 - 25 Pa, gas flow 0.001 - 0.01 g / s, the power delivered to the electrodes 1,5 - 1,8 kW. Argon was used as the plasma gas in the processing of tubular products with the aim of pre-polishing and cleaning the surface from dirt, a mixture of argon and nitrogen in a ratio of 80:20 was used for carrying out the processes of hardening of the surface layers. The duration of treatment in the argon plasma was 20 minutes, then followed by 20 minutes in mixture of argon and nitrogen. Oscillator frequency - 13.56 MHz.

The workpiece is set and sealed between the upper and lower vacuum sealing flanges. The tubular product is part of the oscillating circuit and is analogous to inductance. RF power (13.56 MHz) of the radio frequency generator is supplied to the oscillation circuit. In the vicinity of the electrode vacuum tuning capacitor forming an oscillating circuit with electrode and allowing the best possible transmission of RF energy to produce accurate contour adjustment to the working frequency of the generator are arranged. Matching device is used for the automatically matching the generator with a load.

To equalize the temperature along the length of the product an additional supply of DC voltage is provided. Pumping equipment - vacuum unit ATS-150, consisting of a backing pump AVZ-20D and booster DVN-150. Vacuum gauges - membrane capacitive sensor MKS 627B. Working gas supply is held via the mass flow controller MKS 1179A.

The tubular product is isolated from the main body of the vacuum plant with quartz inserts. To the lower and upper parts of products electrical contacts are provided through which the product are fed with the RF voltage from the LC - circuit with a frequency of 13.56 MHz. Thus, the generated radio frequency electric fields are closed to the grounded to the portions tooling followed after insulators.

There is a potential difference between the electrodes that creates an electric pulling field within a tube channel that allows penetration of the plasma over the entire length of the workpiece for uniform processing.

Researches of the chemical composition of the surface layers of metals and alloys in plasma were carried out for the estimation of the impact effect of low pressure jet RF discharge on the details placed to the discharge chamber for the modifying processes. X-ray photo-electron spectroscopy (XPS) on SPECS electron spectrometer using A1Kα radiation (1486.6 eV) combined stratified surface etching of argon ions with an energy of 4 keV and a current density of 30 mA / cm2 (~ etching rate of 1 nm / min) was used.

Analyzes of the changes in the chemical bonds of the surface layers and the analysis of the depth of penetration of atoms of alloying elements into the metal was carried out by X-ray photoelectron spectroscopy. Experimental data are processed using the software package CasaXPS. The relative error in determining the concentration of elements was ± 0.5% of the measured value. The concentrations are calculated by the integral intensities of the respective spectra taking into account the element of sensitivity coefficients that reflect the photoionization cross section of the main electronic levels. Decomposition of multicomponent spectra of C1s and O1s were carried out using the software package CasaXPS. Some peaks are approximated by a mixed function of the Lorentz-Gauss. Their parameters are taken from reference spectra measured under identical conditions.

Mechanical properties were determined on the samples before and after treatment of the surface by low pressure RF plasma in methane by the indentation in an integrated measuring system NanoTest 600, which is designed to study the physical and mechanical properties of materials on the microscale measurements. The researches were performed according to the method of Oliver - Pharr [Oliver W., Pharr G. An Improved Technique for Detemining Hardness and Elastic Modulus Using Load and Displacement Sensing Indentation Experiments // J. Mater. Res. 1992. № 7 (6). P.1564-1583], using a Berkovich indenter (triangular diamond pyramid with an apex angle of 65.30 and a radius of curvature of about 200 nm). This technique consists of the selection of the parameters of the strength function which describes the experimental dependence of the immersion depth of the indenter and the contact area of the applied force, and calculating the hardness and elastic modulus from the data. In accordance with the procedure Oliver - Pharr plastic indentation depth is determined by the expression [30]:

$$h_c = h_{max} - m^*(C^*F_{max}) \ ,$$

where C – contact compliance, defined by the slope of the unloading curve at the point of application of maximum force. The value depends on the geometry of the indenter, which for a Berkovich indenter $m = 0.75$.

Herdness H is defined by the Vikkers method according to the maximum load F_{max} applied to the indenter contact area A with the sample: $H = F_{max} / A$.

To determine the reduced Young's modulus it is necessary to determine the angle of unloading curve inclination in accordance with the ratio, which depends on the contact area

$$C = \sqrt{p} / \left(2E_r\sqrt{A}\right) .$$

Reduced Young's modulus is determined by the expression:

$$1/E_r = \left(1 - n_S^2\right)/E_S + \left(1 - n_I^2\right)/E_I \ ,$$

where n_S – Poisson's ratio of the sample material, n_I – indenter Poisson coefficient (0.07), E_S – Young's modulus of the sample, E_I – Young's modulus of the indenter (in the experiment E_I = 1141 GPa).

Measurement of the mechanical characteristics of the surface layers of the specimens were limited with depth and load forces, they were 150 and 1.53 mN. Lloading time and unloading point of the indentation was 20 seconds, the holding time at maximum stress – 10 seconds.

The distance between the points of the indentation corresponds to 40 microns, which was set automatically by software integrated measuring system NanoTest 600. After each introduction of the indenter in to the sample at the transition to the next point of the indenter measuring retracted from the surface to a distance of 20 mm to avoid contact with the surface.

The degree of roughness plays an important role in the measurements of physical and mechanical properties of the surface layers. In this regard, in order to increase reliability of the measurement results of the procedure was performed at least 30 times. After the measuring cycle, defined automatically, manually were deleted those data points that were in doubt.

The quality control system is intended for inner surfaces of the tubes for measuring surface roughness and hardness by non-destructive measurement of the inner surface of the 10 mm in diameter and 1,000 mm in length tubular product. Climatic performance is fullfilled according to GOST 15150-69.

Evaluation of surface changes, roughness, structure, research and measurement of geometrical parameters were carried out by means of confocal laser scanning microscopy OLYMPUS LEXT 4000 and OLYMPUS LEXT 4100. Relative measurement samples roughness error does not exceed 5%.

General increases range is in the frame from 50 to 17280 fold and depends on the lens that are used. The horizontal resolution is up to 120 nm, vertical up to 10 nm. Accuracy of measurement values are 0,2 + L / 100 mm, wherein L - value measured in microns. Microscope Olympus OLS LEXT 4000 has the functions of height measurement, area and volume, roughness parameters, automatic processing of the received images.

Analysis of samples crystal structure is performed by the X-ray diffraction at 0-9 diffractometer D8 Advance (BrukerAXS) using SuKα radiation. In order to identify the differences in the structural

IOP Conf. Series: Journal of Physics: Conf. Series **789** (2017) 012031 doi:10.1088/1742-6596/789/1/012031

condition of the surface layers and a matrix (in volume) analysis was carried out in a moving beam geometry (fixed angle a = 0.5 °) and parallel beam geometries. A parabolic Gobel mirror (60 mm), horizontal Soller slit (0.12 °) and a semiconductor Si (Li) Sol-XE detector were used. The diameter of the goniometer is 500 mm. Shooting in the range of angles 2θ of 35^0 to 90^0 degrees, 0.02 degrees increments. Time at each point - 5 seconds.

The crystal structure was investigated by parallel beams circuit mode grazing incidence (angle of incidence = 3^0) at 2θ angles ranging from 30^0 to 55^0 0.02^0 increments and record time at each point is equal to 3 seconds with the help of multifunctional X-ray diffractometer Rigaku SmartLab. The slots of the incident beam: $5,0^0$ (Soller slit), 0.7 mm (a gap divergence). Diffracted beam slits: $5,0^0$ (Soller slit), 0.7 mm, anti-scatter slit 0.7 mm. Software processing of diffraction patterns was performed taking into account of the overlap $K\alpha 1$ and $K\alpha 2$ lines and without taking into account the instrumental function. Wave length $CuK\alpha 1 \approx 1.5406$ Å.

Optimization of the process by scheduling a three-factor experiment with the following independent variables - the processing time, power level and pressure of the plasma gas – argon was carried out. The relative change in hardness as one of the defining physical and mechanical characteristics of the material is selected as a response function. Processing of the data was performed using STATISTICA 6.0 software package. Optimization results are shown in Fig. 2.

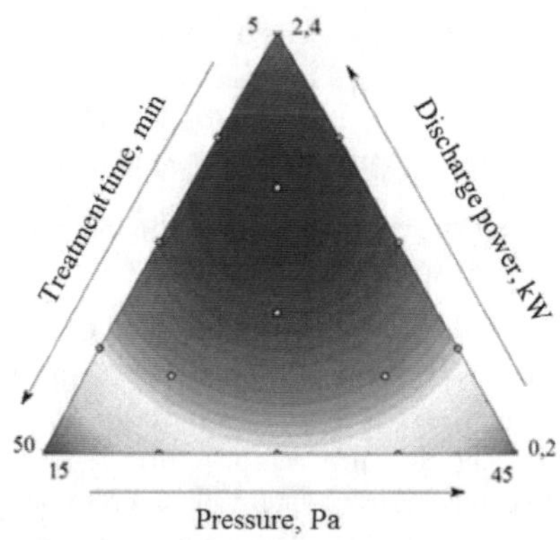

Figure 2 – Optimization of the process parameters of the hardening of metals and alloys in the low pressure RF discharge.

As seen it is seen in Fig. 2, the optimal conditions of the process of metals and their alloys hardening by a low pressure RF discharge are: processing time 20 minutes, the discharge power - 1.5-1.8 kW, argon gas pressure in the chamber - 25-28 Pa.

3. Results and discussion

The dynamic quadrupole mass spectrometer monitoring the gas composition was done in a vacuum chamber. Mass spectra of the plasma in a mixture of an inert (argon) and plasma chemical (methane, nitrogen) gases are given in Figures 3a and 3b.

The method is based on determining the mass to charge ratio of ions formed in the ionization of the gas supply during the generation of high-frequency jet low pressure discharge, it allows to make a quantitative assessment of the composition of the gas inside the chamber to determine the characteristics of the plasma flow, making the major contribution to the modification of the surface. In all cases in the chamber after evacuation, water is presented in an amount of 5-7% and a large amount of hydrogen - 23% in a mass spectrum by methane treatment. Ratio of argon and methane, argon and nitrogen from Figures 3a and 3b respectively corresponds to the flows of feed gases (G_{Ar} = 0,03g / c, G_{CH4} = 0,006g / c; G_{Ar} = 0,06g / c, G_{N2} = 0,009g / c) . Maximum decomposition of methane, which can be estimated from the amount of hydrogen in the chamber corresponds to the ratio of argon gas mixture with methane 80:20.

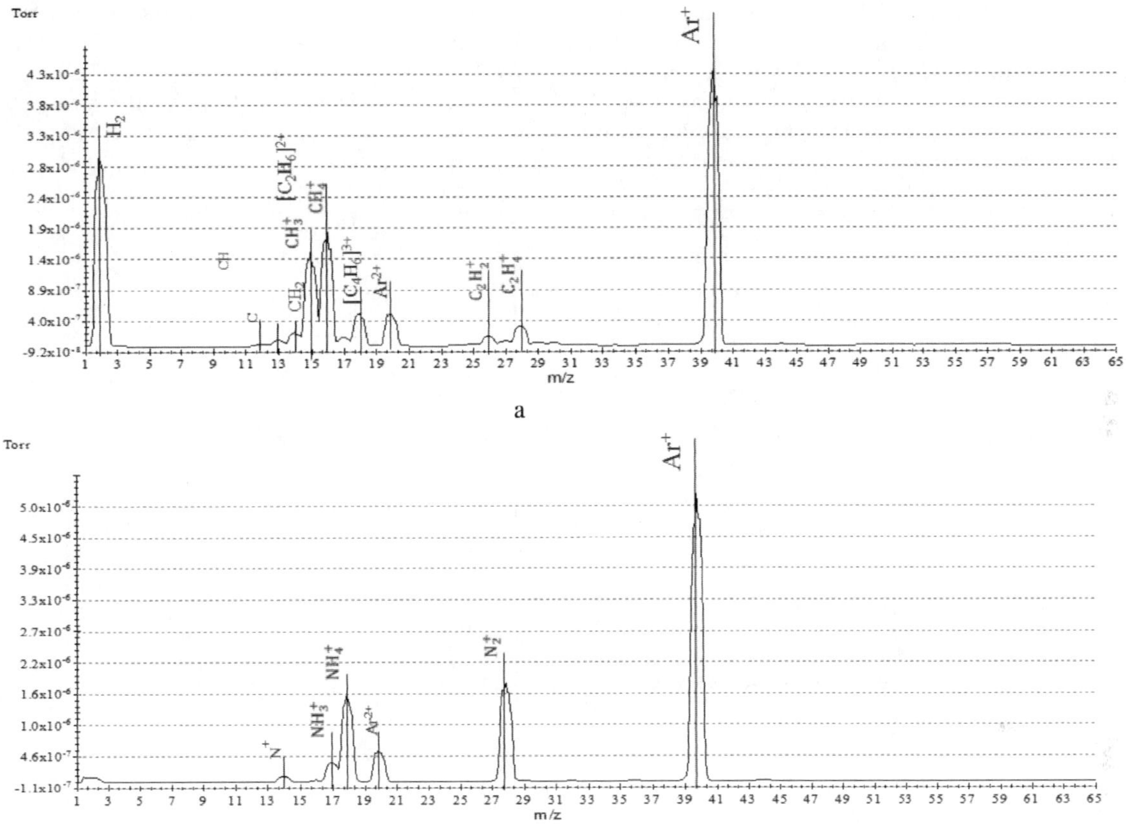

Figure 3 - Low pressure RF discharge mass spectra in the electrode gap in the vacuum chamber in a medium: a) Mixtures of reactive and inert gases (argon and methane in a ratio 80:20) b) Mixtures of inert and reactive gases (argon and nitrogen in a ratio of 80:20).

Figure 4 shows the X-ray photoelectron spectra of the untreated and treated high speed steels samples in methane. At the ion etching it is established that according the the depth of the sample the component 707 eV appears and grows corresponding unoxidized (metal) iron, but the iron oxide component (710 - 711 eV) is dominant. At the same time it is seen that in the treated sample unoxidized peak intensity of the iron is above compared with the untreated sample at the same depths. Apparently, there is a partial reduction of oxides in the processing in methane. Fe2p oxide component spectrum in its parameters corresponds Fe3O4 or complex oxide on its base with minor impurities of other alloying elements of the alloy. In the structure of the oxide present as two or three linkage ferric iron. In particular, the "shoulder" at 716 eV in the spectra Fe2p3/2 (sattelite) is associated with the presence of two linkage ferrous iron. The peak position of unoxidized iron is shifted by 0.3 eV towards higher binding energies relative to the position of pure (metal) iron, which may be due to the formation of carbide bonds Fe-C.

Carbon spectra are shown in Figure 5. One difference of C1s spectra of the treated and untreated samples is shoulder relative intensity spectrum in 288-290 eV field. In the sample without processing the spectra of the shoulder it is more pronounced because of the higher concentration of oxygen-containing compounds with carbon participation. After treatment with the methane low pressure RF discharge shoulder is substantially smaller or absent. In depth from the surface the iron unoxidized peak intensity increases in intensity in the spectrum of C1s 283.5 eV to appearance and growth synchronously, which corresponds to the C-Fe bonds. On the whole the integral spectrum of C1s peak is shifted from the position characteristic of the C-H (on the surface) in the direction of lower binding energies (C-C). The value of the binding energy for the reference graphite is 284,3-284,6 eV, the width of the standard graphite peak is ~ 1.4 eV. In the case of samples of high speed steel there are several non-equivalent chemical states of carbon in the sample layer that form a diffuse broad peak and are identified by the decomposition of the spectra into components.

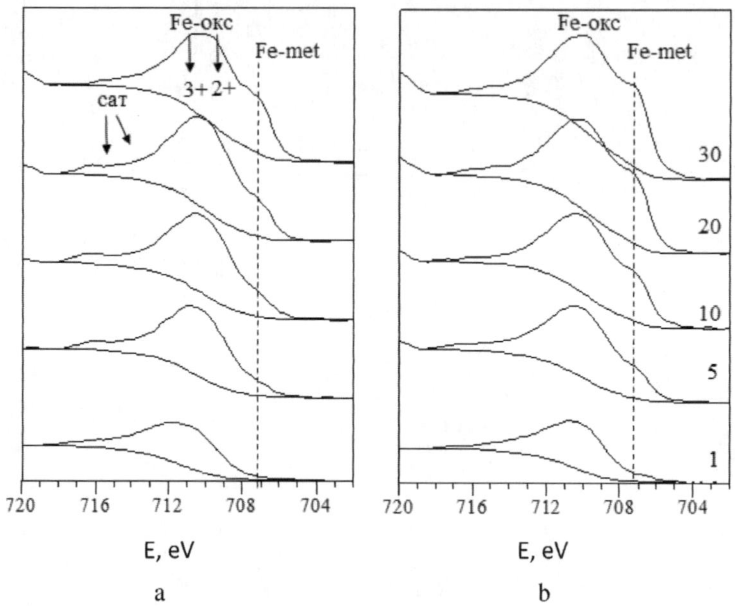

Figure 4 - Fe2p3/2 spectra of the treated (a) and untreated (b) high speed steel samples by the loew pressure radio frequency methane plasma. The numbers on the right - the depth in nm.

The spectra C1s component with Eb = 284.6-284,8 eV corresponds to the C-C bond. This peak is shifted slightly in the depth in the direction of the high-energy bonds. It is seen that in the treated sample relative intensity peak is significantly higher compared with the untreated sample, which may be due to the presence of a carbon or graphite structure. Accordingly, in the spectra of the treated sample peak intensity of hydrocarbon (C-H 285.6 eV) decreases. Processing in low pressure RF discharge leads to a decrease in the relative intensity of peaks in the field of energies Eb 287 eV, typical for carbon bonds with oxygen, which is consistent with a decrease in the oxygen concentration in the surface layers of the treated sample. At a depth of 30 nm in the spectra of the two samples a peak with Eb = 283.6 eV was revealed corresponding to carbon in an iron chemical. Thus, it can be argued that the treatment in methane leads to an increase in the proportion of carbon in the C-C bond (sp2-hybridized carbon).

Thus, according to the data obtained by XPS surface layers of high speed steel test samples contain high concentration of carbon. Carbon is chemically nonequivalent in several states, including chemical bonds C-C and C-Fe. The plasma treatment improves the carbon in methane concentration in the C-C bond, lowers the amount of C-H. According to the spectra characteristic losses, the surface layer structure, which includes carbon, forming C-C bond is characterized by a reference structure of graphite. The surface layers of both samples contain a significant amount of oxygen. Iron is oxidized throughout the analysis depth (not less than 30 nm). Processing in methane leads to partial reduction of iron oxides and oxygen concentrations decrease.

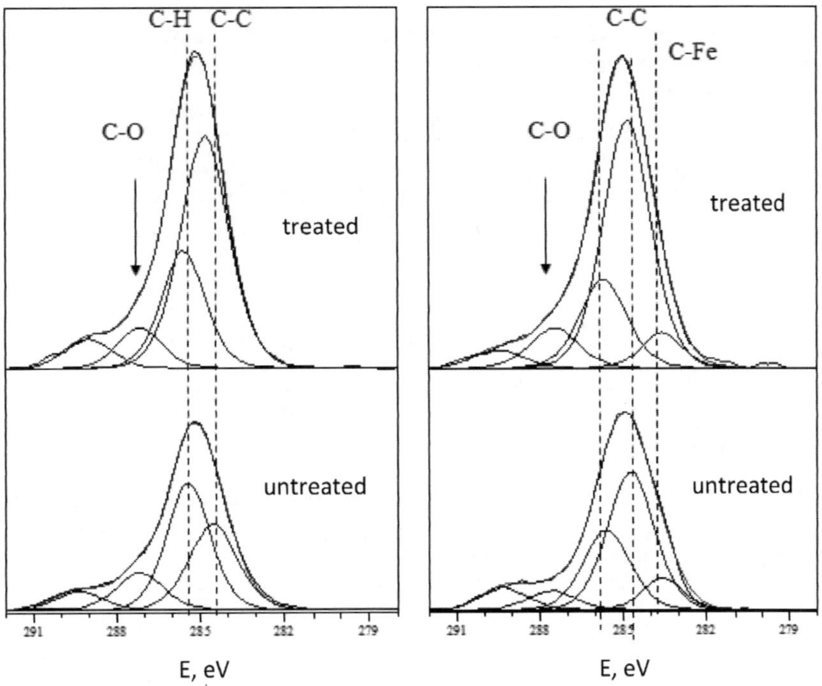

Figure 5 - Cls spectra of the treated and untreated samples at the 5 nm (a) and 30 nm (b) depth.

The concentrations of the elements in the surface layers of the tungsten cobalt alloy of untreated and treated samples are given in Tables 1 and 2 respectively. The absolute concentration of carbon in the treated sample is higher than in the sample without treatment. This - the same result as in the case of samples of high speed steel.

Structure analysis was carried out using the multi X-ray diffractometer Rigaku SmartLab. As a result of diffraction investigations revealed that the WC phase is present in both samples. Peak positions in the diffraction patterns are different - in untreated sample the interplanar distance is approximately 0,001-0,002 angstroms greater than in the treated sample, it can be assumed that changes were formed due to the introduction into the metal lattice structure of carbon ion because of the plasma chemical effect on the surface of the material.

Table 1 - The concentrations of elements in the surface layers of the tungsten and cobalt alloy sample without processing by low pressure radio frequency discharge.

Depth, nm	C	O	W	Co	Cu	Zn
1,0	61,5	23,8	0,8	1,9	5,7	6,3
3,0	43,3	31,2	1,4	3,4	9,8	10,8
5,0	40,3	30,0	1,7	4,3	10,9	12,7
10,0	37,2	29,2	2,1	4,8	13,0	13,7
20,0	30,5	26,6	3,0	4,9	20,3	14,6
30,0	25,5	22,6	3,9	4,4	29,2	14,4

Low pressure RF discharges processing in methane plasma chemical gas leads to the saturation of the surface layers by carbon and, simultaneously, to a partial recovery of cobalt and tungsten oxides and reduce the oxygen concentration.

Table 2 - The concentrations of elements in the surface layers of the tungsten and cobalt alloy sample after the low pressure radio frequency discharge processing.

Depth, nm	C	O	W	Co	Cu	Zn
0,5	74,2	19,0	0,5	1,0	1,5	3,7
1,0	73,0	15,4	0,5	1,0	5,6	4,5
3,0	69,7	15,1	0,8	1,5	7,2	5,7
5,0	61,5	21,4	0,9	2,1	7,6	6,6
10,0	58,0	18,2	1,8	3,5	10,4	8,1
15,0	54,3	19,2	2,4	3,8	12,2	8,1
20,0	50,8	17,9	2,8	4,0	16,3	8,3
30,0	44,0	18,0	3,5	4,1	20,7	9,7
40,0	36,1	15,4	4,4	3,9	29,9	10,4
50,0	32,1	12,4	5,2	4,2	34,8	11,2

Fig. 6 shows the spectra N1s nitrogen after background subtraction of inelastically scattered secondary electrons. Maximum nitrogen has a binding energy spectra in the 397-398 eV, which corresponds to its chemical interaction with metals.

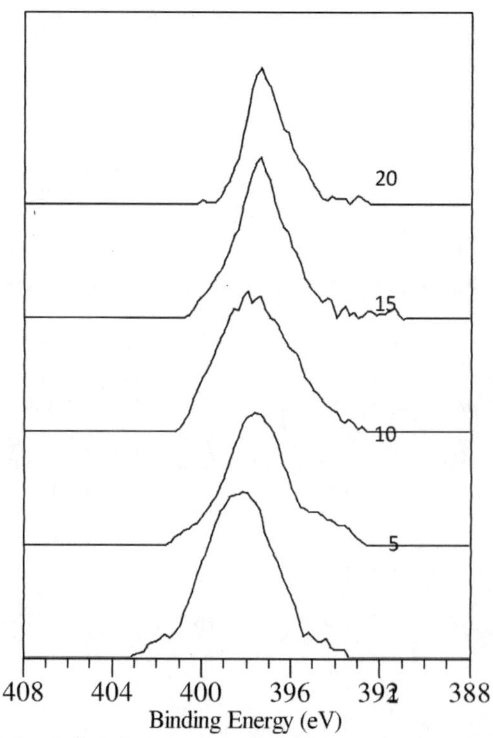

Figure 6 - N1s spectrum of the surface layers of steel 30HN2MFA after plasma treatment in nitrogen.

Characteristic diagrams "loading - unloading," which were included in the estimates for the high speed steel samples to determine the physical and mechanical characteristics of the surface layer before and after treatment by low pressure radio frequency discharges are shown in Fig. 7. Dependence of the applied force F, mN the penetration depth of the indenter h, nm in samples of high speed steel are shown in Fig. 8 (chart loading - unloading, for the forces of 1,53 nH).

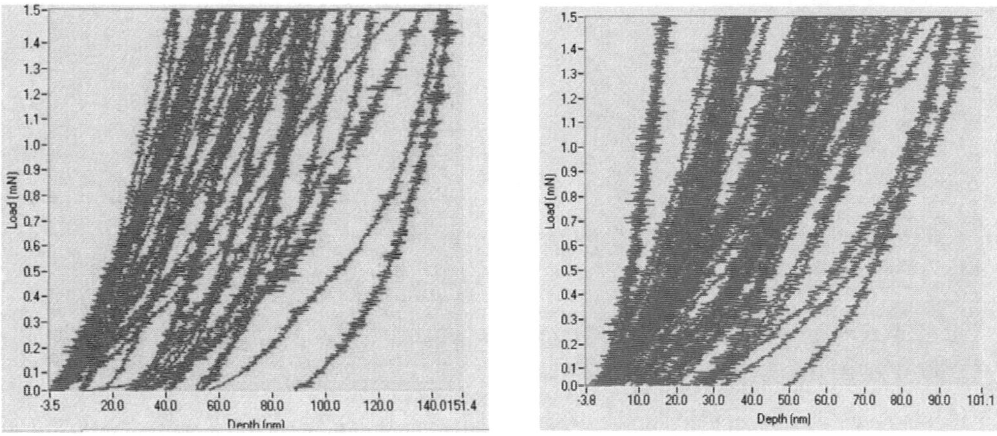

a b

Figure 7 - Typical diagrams "load - unload" - according to the applied force F, mN penetration depth of the indenter h, nm. High speed steel samples before (a) and after (b) treatment of the surface by low pressure radio frequency plasma. The strength Fmax 1,53 nH.

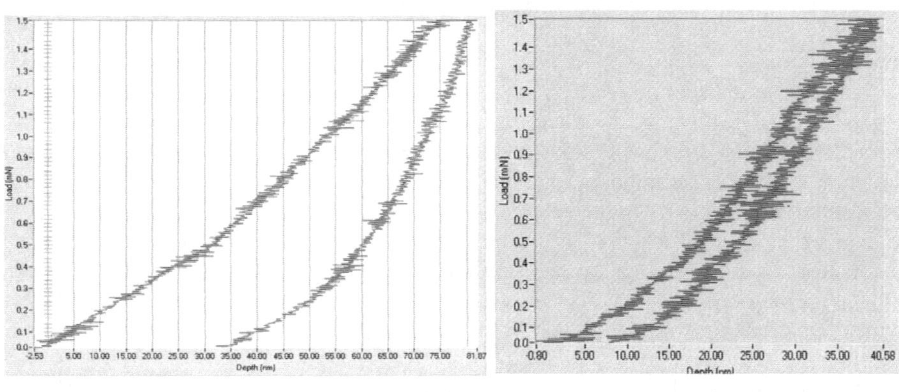

a b

Figure 8 - The dependence of the applied force F, mN on the penetration depth of the indenter h, nm in high speed steel samples (chart loading - unloading) for strength Fmax 1,53 nH. Sample pre-treatment (a, 16 sample points indentation) and after treatment after treatment of the surface by low pressure radio frequency plasma (b, 20 sample points indentation).

The results of the studies are summarized in Table 3. Analysis of the table shows that after treatment of the high speed steel samples surface by low pressure RF plasma there is an increase in microhardness of 30-40%, thereby the modulus of elasticity also increases, as a result reduced plastic depth hc and maximum indentation depth hmax decrease. The differences are fixed by limiting the maximum load Fmax 1.53 mN at a depth of 60-80 nm.

Table 3 – Physical and mechanical properties of high speed steel surface layers before and after low pressure radio frequency plasma treatment.

Physical and mechanical characteristics	High speed steel	
	Before treatment	After treatment
Maximum depth h_{max}, nm	90,06	60,75
Maximum load F_{max}, mN	1,53	
Plastic identation depth h_c, nm	67,04	40,18
Microhardness H , GPa	2,30	3,66
Reduced Young's modulus of elasticity E_r , GPa	55,20	84,91
Young's modulus of elasticity E_S	0,42	0,61

4. Conclusions

Complex studies of influence of effective parameters of the low pressure radio frequency discharges processing on the chemical composition and physical and mechanical properties of the surface layers of high speed steels, tungsten cobalt alloys and steels made by X-ray photoelectron microscopy and mechanical tests using instrumental indentation and scanning probe microscopy have allowed the following conclusions:

- It is found that the RF treatment in the jet discharge at low pressure causes a change in the chemical composition of the surface layers of metals and their alloys. Surface modification of metal is possible in the RF discharge generated between the symmetrical parallel plate electrodes. The effect of low pressure RF discharge on the properties of metals and their alloys depends on the plasma parameters, and the structure of the material.
- It is experimentally found that alteration of the chemical properties of high speed steel and alloys following after the exposure of tungsten cobalt alloys and steels by low pressure RF jet plasma occurs with a change of the crystal lattice parameters and increase of the physical and mechanical properties of the material. Diffusion Nano layers formation on the metal surface increases the wear resistance of the final product, increase operating characteristics of equipment.
- Studies have shown that the use of an ionized gas stream as a processing tool (plasma) allows the modification of the surface of cutting tools, parts of complex configuration. Gas saturation occurs with surface layers of metals and alloys at a depth of 500 nm for the processing time to 40 minutes, resulting in an increase in strength properties, durability, and product service life.

Ladianov Vladimir Ivanovich - doctor of sciences, head of the department of structural and phase transformations, Physics Technical Institute, Ural Branch of the Russian Academy of Sciences;

Gilmutdinov Faat Zalalutdinovich – Ph.D., head of the department of physics and chemistry of the surface, Physics Technical Institute, Ural Branch of the Russian Academy of Sciences;

Nikonova Rose Muzafarovna - Ph.D., senior researcher, Laboratory of phase transitions, Physics Technical Institute, Ural Branch of the Russian Academy of Sciences, +7 (3412) 216955, e-mail: RozaMuz@ja.ru;

Kashapov Nail Faikovich - doctor of sciences, provost of Kazan Federal University;

Shaekhov Mars Farifovich - doctor of sciences, professor of department Plasma Technology and Nanotechnology of High Molecular Weight Materials, Kazan National Research Technological University, shaekhov@kstu.ru;

Khristoliubova Valeriia Igorevna – assistant teacher, department of “Garment and footwear desigh”, Kazan National Research Technological University, +7(906)113-56-36, valllerrriya@mail.ru

References

[1] Saifutdinov A.I. Dynamics contraction of DC glow discharge in argon /A.I. Saifutdinov, A.A. Saifutdinova, N.F. Kashapov, SA. Fadeev //(2016) Journal of Physics: Conference Series, 669 (1). – art. no. 012045.

[2] Lopatin I.V. Independent glow discharge of low pressure hollow cathode at a current of tens of amperes /I.V. Lopatin P.M. Schanin, Y.K. Akhmadeev, S.S. Kowalski, N.N. Koval // Plasma physics.-2012.-№7.-V.38.-P. 639-643.

[3] Zhang, C.S. Experimental and theoretical study on interaction between lanthanum and nitrogen during plasma rare earth nitriding/C.S. Zhang, M.F. YanZ. Sun//Applied Surface Science1.–2013. –v.287.–№15.–p.381–388.

[4] Semenov, A.P. Wear-resistant coatings applied by vacuum ion-plasma methods / A.P. Semenov, A.I. Grigoriev // Technology of mechanical engineering. -1973.-№7.-P. 15-20.

[5] Patent. 110088 RU. A device for vacuum coating / I.I. Vasiliev; the applicant and the patentee "Kazan National Research Technological University" .- Application №2011113754; application date 08.04.2011;publishing date 10.11.2011.

[6] Brzhozovskii, B. Composite ion-plasma coatings with nanodisperse reinforced phase: scientific and practical aspects of synthesis / B Brzhozovskii, V Martynov, E Zinina and M Brovkova // IOP Conference Series: Materials Science and Engineering. – 2016. – V. 116. – 01022007.

[7] Smallman, R.E. Oxidation, Corrosion and Surface Engineering/R.E. Smallman//Modern Physical Metallurgy.–2014.–v.8.–p.617–657.

[8] Marin, E. Diffusive thermal treatments combined with PVD coatings for tribological protection of titanium alloys / E. Marin, R. Offoiach, M. Regis, S. Fusi, A. Lanzutti, L. Fedrizzi//Materials & Design.–2016.–v.89.–p.314–322.

[9] Catalogue of Vacuum technology NPF "Alan Practitioner» PVD vacuum coating application methods. Website URL: http://www.elanpraktik.ru/technology/pdf/pvd_%20metody.pdf (reference date 03/10/2014)

[10] German S Fox-Rabinovicha Hierarchical adaptive nanostructured PVD coatings for extreme tribological applications: the quest for nonequilibrium states and emergent behavior /German S Fox-Rabinovicha, Kenji Yamamotob,Ben D Beakec,Iosif S Gershmand,Anatoly I Kovaleve,Stephen C Veldhuisa,Myriam H. Aguirref, Goulnara Dosbaevaa, Jose L Endrinog//Science and technology of advanced materials.–2016.–V.13.–№4.

[11] Pelletier, J. Plasma-based ion implantation and deposition: A review of physics, technology and applications /J. Pelletier, A. Anders//IEEE Transactions on Plasma Science.–2005.–v.33.–№ 6.–p.1944–1959.

[12] Belyakov A.V. The technology of electric spark for the rotor blades of steam turbines and equipment for its realization resource improvement / A.V. Belyakov, V.V. Sarantsev, A.N. Gorbachev, F.I. Panteleenko, E.L. Azarenka, B.F. Reutov // Electrical staitions.-2016.- №1.

[13] Patent. The method of the vacuum tube inner surface processing, RU 2039845.

[14] Calanda, N.A. Magnetic and magnetoresistive properties of nanoheterostructures Al2O3-Sr2FeMoO6-δ-Al2O3 / N.A. Calanda, G.G. Peas, M.V. Yarmolich, A.A. Lozovenko, E.J. Buzzards // Physics of solid substance.-2016.-№2.-V.58.-P. 341-348

[15] Frederico Augusto Pires Fernandes Microstructure of nitrided and nitrocarburized layers produced on a superaustenitic stainless steel/Frederico Augusto Pires Fernandes, Luiz Carlos Casteletti, Juno Gallego//Journal of Materials Research and Technology.–2013.–v.2.–p.158–164.

[16] Abdullin, I.Sh. Surface Layers Modification of Tungsten-Cobalt Alloy by Low Pressure RF Plasmas/I.Sh. Abdullin, V.S. Zheltukhin, I.R. Sagbiev, R.F. Sharafeev//Proc. of 61st annual gaseous electronics conference, Oct. 13-17, 2008.–Dallas, Texas, USA./Abstract: FTP1.00021.

[17] Rosiana Boniatti The influence of surface microstructure and chemical composition on corrosion behaviour in fuel-grade bio-ethanol of low-alloy steel modified by plasma nitro-carburizing and post-oxidizing/Rosiana Boniatti, Aline L. Bandeira, Ângela E. Crespi, Cesar Aguzzoli, Israel J.R. Baumvol, Carlos A. Figueroa//Applied Surface Science.–2013.–v. 280.–p.156–163.

[18] Zheltukhin, V.S. Low Pressure Radio-Frequency Plasmas in the Nanolayers Formation on the Surface of Construction Materials/V.S. Zheltukhin, I.Sh. Abdullin, I.R. Sagbiev, R.F. Sharafeev//Proceedings of International Conference «Micro- and Nanoelectronics-2007».–Moscow-Zvenigorod, Russia.–2007.–P. 32–39.

[19] Kashapov N.F. Hardening of the surface plasma jet high-frequency induction discharge of low pressure /N.F. Kashapov and S.N. Sharifullin //IOP Conference Series: Materials Science and Engineering. – 2015. – V. 86. doi:10.1088/1757-899X/86/1/012021.

[20] Sharifullin S.N. Plasma method of increasing the roughness class and hardening surfaces of products /S.N. Sharifullin, N.R. Adigamov, S.V. Slavnina // Bulletin of Samara Scientific Center of the Russian Academy of Sciences. Special. Issue "Actual problems of mechanical engineering", 2009. - P. 15 - 18.

[21] Makhotkina , L.Yu. Design of special purpose products made of nanomodified leather / L.Yu. Makhotkina, V.I. Khristoliubova, L.R. Khannanova-Fakhrutdinova // IEJME – Mathematics education. – 2016. – V. 11, № 6. – P. 1495-1503.

[22] Sagbiev, I.R. Low pressure radio frequency Jet discharge in the process of surface modification of structural materials nanolayers: Dis. Doctor. Sciences /I.R.. Sagbiev.-Kazan, 2009.-332p.

[23] Sharafeev, R.F. The flow of low pressure radio frequency plasma in the processes of interaction with the surface materials: Abstract. Ph.D. Dis. / R.F. Sharafiev.- Kazan, 2010.- 20 p.

[24] Shaehov, M.F. Physics of low pressure radio frequency discharge in the processing of capillary-porous and fibrous materials: Abstract. Doctor. Science Dis. / M.F. Shaehov.-Moscow, 2006.-32 p.

[25] Gaysin, Al.F. Features of the transition of low-current radio frequency capacitive discharge with an electrolytic electrode to the high-current discharge / Al.F. Gaysin // High temperature. -2015.-№1.-V.53.-P. 18-22.

[26] Markov, A.V. Heating of a gas and a polymeric material in low-temperature plasma of a high-frequency discharge//A.V. Markov, Y.P. Yulenets//High Temperature.–2015.–№2.–T.53.–C.167–170.

[27] Makhotkina, L.Yu. Design of special purpose products made of nanomodified collagen-containing materials with radio-frequency discharge / L.Yu. Makhotkina, S.N. Sharifullin // IOP Conference Series: Materials Science and Engineering. – 2016. – V. 134. – 012020.

[28] Abdullin, I. Formation of diffusive nanostructured layers on a surface of metals and their alloys/I. Abdullin, A. Khubatkhuzin, V. Khristoliubova// State-of-the-art Trends of Scientific Research of Artificial and Natural Nanoobjects: abstracts for 4th International Scientific conference, 22-25 April 2014. –Saint-Petersburg. –P. 161–163.

[29] Khristoliubova, V.I. Gas and plasma dynamics of RF discharge jet of low pressure in a vacuum chamber with flat electrodes and inside tube, influence of RF discharge on the steel surface parameters / V.I. Khristoliubova, N.F. Kashapov, M.F. Shaekhov // IOP Conference Series: Materials Science and Engineering. – 2016. – V. 134. – 012017.

[30] Oliver, W.C. An improved technique for determining hardness and elastic modulus using load and displacement sensing indentation experiments / W.C. Oliver, G.M. Pharr // J. Mater Res.-1992, V. 7.- no. 6.- pp. 1564–1580.

Studies of the influence of nonequilibrium plasma thermal exposure on the characteristics of the capillary-porous polymer material

L Yu Makhotkina[1], **V I Khristoliubova**[1]
[1]Kazan National Research Technological University, 420015, Russian Federation, Kazan, 68 Karl Marks st.

E-mail: valllerrriya@mail.ru

Abstract. Capillary-porous materials, which include natural macromolecular tanning material, are exposed to a number of factors during the treatment by a nonequilibrium plasma. Plasma particles exchange the charge and energy with the atoms of the material during the interaction of the plasma with the surface. The results of treatment are desorption of atoms and molecules from the body surface, sputtering and evaporation of material's particles, changes of the structure and phase state. In real terms during the modification of solids by nonequilibrium low-temperature plasma thermal effect influences the process. The energy supplied from the discharge during the process with low pressure, which is converted into heat, is significantly less than during the atmospheric pressure, but the thermal stability of high-molecular compounds used in the manufacture of materials and products of the tanning industry, is very limited and depends on the duration of the effect of temperature . Even short heating of hydrophilic polymers (proteins) (100-180 ^{0}C) causes a change in their properties. It decreases the collagen ability to absorb water vapor, to swell in water, acids, alkalis, and thus decreases their durability. Prolonged heating leads to a deterioration of the physical and mechanical properties. Higher heating temperatures it leads to the polymer degradation. The natural leather temperature during plasma exposure does not rise to a temperature of collagen degradation and does not result in changes of physical phase of the dermis. However, the thermal plasma exposure must be considered, since the high temperatures influence on physical and mechanical properties.

1. Introduction

Plasma consists of active and inactive particles with high kinetic and potential energy, that determine the physical and chemical nature of their interaction with the surface. Furthermore, physical and chemical transformations on the treated surface depend not only on the characteristics of the plasma stream but also on the charging state of surface [1-3].

Capillary-porous materials, which include natural macromolecular tanning material, are exposed to a number of factors during the treatment by a nonequilibrium plasma. Plasma particles exchange the charge and energy with the atoms of the material during the interaction of the plasma with the surface.

The results of treatment are desorption of atoms and molecules from the body surface, sputtering and evaporation of material's particles, changes of the structure and phase state [4,5].

In real terms during the modification of solids by nonequilibrium low-temperature plasma thermal effect influences the process. The phenomena of physical and chemical nature can take place at the same time. Physical processes include transfer of thermal energy between the plasma gas and the processed natural capillary-porous material, resulting in a substantial increase in temperature and, as a result, physical transformations. The chemical process is accompanied by the occurrence of one or more chemical reactions in the gas phase and on the polymer surface.

Due to the structure of the polymers and the macromolecules flexibility capillary-porous materials do not require a big amount of energy to carry out structural changes of macromolecules, their geometric shape changing, changing of the polymer physical state.

The energy supplied from the discharge during the process with low pressure, which is converted into heat, is significantly less than during the atmospheric pressure, but the thermal stability of high-molecular compounds used in the manufacture of materials and products of the tanning industry, is very limited and depends on the duration of the effect of temperature. Even short heating of hydrophilic polymers (proteins) (100-180 ^{0}C) causes a change in their properties. It decreases the collagen ability to absorb water vapor, to swell in water, acids, alkalis, and thus decreases their durability. [6] Prolonged heating leads to a deterioration of the physical and mechanical properties. Higher heating temperatures lead to the polymer degradation.

Analysis of elementary processes of low pressure RF plasma discharges interaction with a capillary-porous materials shows that the largest contribution to the modification of the surface layer in the non-equilibrium low-temperature plasma processing is made, first of all, by ion bombardment, the kinetic energy of which ranges from 10 to 100 eV, the second, is the recombination of ions (argon - 15.76 eV, oxygen - 15.6 eV, the nitrogen - 14.8 eV), the third, the deactivation of excited atoms at the surface, the energy is transferred to 3-5 eV, the fourth, thermal effect.

Changing of the parameters of the plasma torch (discharge power, pressure in the discharge chamber, the gas flow rate) allows to obtain the concentration of particles with a certain energy. The nature of the particles is determined by the effects of interaction with the polymer molecules of atoms. Time of macromolecule excitation depends on the molecular weight, structure of polymer, the packing density of the atoms. Due to the special properties of polymers related to their conformational mobility as a result of interaction with ions, there is a high probability of realization of the process of structural transformations in the polymer without changing the chemical composition [7].

2. Results and discussion

An important mechanism of the the physical exposure of plasma is ion sputtering. Despite the fact that the energy of the sputtering threshold of 10-40 eV, and the energy of the atoms on the surface of various materials is in the range of 2-8 eV, an unambiguous answer to the question whether there is sputtering in the interaction of the low-temperature RF plasma with high-molecular materials in the investigated range of characteristics of the plasma torch is impossible, as polymer systems have strong forces within - and intermolecular interaction energy and cohesion in polymers are often much more than energy of chemical bonds of individual atoms.

Thermal resistance of capillary-porous hydrophilic polymers depends on their moisture content. Momentary heating of collagen at temperatures near the welding point or prolonged heating at temperatures well below the sealing temperature causes a change in a number of properties. When heated in the dehydrated state and under certain conditions in the presence of water collagen can sequentially move to the highly elastic, and then in a plastic state.

The electron temperature in the jet was evaluated by measuring the value of the current density in a nonequilibrium plasma $j_\varphi = (0,4 - 1,4)\cdot 10^6$ A/m^2. Since the electron mobility is several orders higher ion mobility conduction current in the plasma is determined by the movement of electrons by the formula:

$$j_e = n_e \overline{V}_e e \;;$$

where $\overline{V}_e$ - the average velocity of ordered motion of electrons. The kinetic energy of the ordered motion of electrons:

$$K_e = \frac{m_e \overline{V}_e^{\,2}}{2} = \frac{m_e j_e^{\,2}}{2 n_e e^2} \;;$$

At the $n_e = 10^{15} - 10^{16}$ m^{-3}, the kinetic energy of the electrons K_e = 63 - 0.63 eV.

For the jet the thickness of the double layer is $n_e = 10^{16}$ m^{-3}, T_e = 10000 K, r_{de} = 7,64 mkm, at $n_e = 10^{15}$ m^{-3}, T_e = 10000 K, r_{de} = 8,23 mkm.

Thermal processes do not make a significant contribution to energy processes of modification.

Theoretical calculations confirmed by experimental data.

To determine the influence of thermal plasma on investigated high-molecular compounds temperature measurement of the sample was performed during plasma processing. The objects of study were natural tanning materials at various stages of their production.

Processing was carried out in air, argon. The discharge current density (j_φ) and plasma exposure time (τ) were ranged. The samples were dried before treatment to constant weight in an oven at 107 ^{0}C (method1), by evacuation (method 2), by extracting the water by the alcohol (method 3). The temperature change of leather material are shown in Fig. 1-2.

The temperature of leather samples increases during the plasma treatment, but does not rise more than 90 ^{0}C, thus it remains below the welding temperature.

Skin temperature depends on the degree of processing and the amount of water contained in it: so pelts is heated up to 70 ^{0}C (welding temperature90 ^{0}C); colored semi-finished up to 95 ^{0}C (welding temperature 120 ^{0}C). Increasing of the current density in the discharge above $1{,}6{\cdot}10^5$ A/m2 in the plasma flow results in degradation of the samples.

The higher the moisture content of the sample, the lower the temperature rises, since the water filling the capillary and pore volume reduces the effectiveness of the plasma exposure. A sample with an initial moisture of 60-80% is heated up to 50-60 ^{0}C, the skin, dehydrated beforehand, is heated up to 70-100 ^{0}C [3]. This is happened due to the fact that for the ionization of water molecules the energy 12.60 eV is necessary, and while for air - 13.62 eV. Consequently, part of the energy that was input to the discharge is spent on the ionization of water molecules and the heat capacity (enthalpy) of water is much greater than air or argon. Consequently, for the heating of one water molecule to the same temperature as air or argon molecules is necessary to spend more energy.

The temperature depends on the pre-dewatering methods of samples (Tg <Tc <Tks). The difference between TV and TCS is 8 to 13 ^{0}C and depends on the degree of processing of the dermis.

The presence of dyes in samples promotes greater heating of tanning samples if thermal conductivity and heat capacity of the samples is higher than the gas phase and substances in the pores. Leather-painted semi-finished product is heated more than unpainted 8-10 % and much greater than raw pelts and 20-40 % [8]. The presence of metal ions in the samples and other chemicals (dyes) results to the temperature increase.

Analysis of the results leads to the conclusion that the temperature of the wet sample is increased less than dehydrated because water, filling the pore volume reduces the effectiveness of the nonequilibrium low temperature plasma impact by reducing the recombination energy of ion flux to the inner surface of the material. Increasing of the samples temperature due to the higher thermal conductivity of water than air is insignificant, as in the range of investigated plasma parameters deposited energy to the discharge used to increase the energy of the ions, and the proportion of the heat flow is negligible. Desorption of water in the plasma jet is not more than 1.5 – 3 % of the initial weight of the dehydrated sample.

Table 1 Maximum temperature of materials heating
(air, G = 0,04 g/s, τ = 300 s, , j_{φ}= $1,2\cdot10^5$ A/m^2, P = 53.3 Pa)

Dehydration process	Temperature, ^{0}C			
	Leather materials			
	raw	pelts	tanned unpainted semi-finished	tanned painted semi-finished
1	56	70	83	93
2	48	61	73	80
3	53	67	79	88
Without dehydration	22	52	62	71

The change in dermis samples welding temperature and its processed products after plasma exposure in the test range of the technological parameters of the plasma torch is investigated. Raw materials, pelts, tanned unpainted and painted semi-finished leather products in watered and in the dehydrated state are taken as samples. Samples were dried by vacuum, hot air and alcohol.

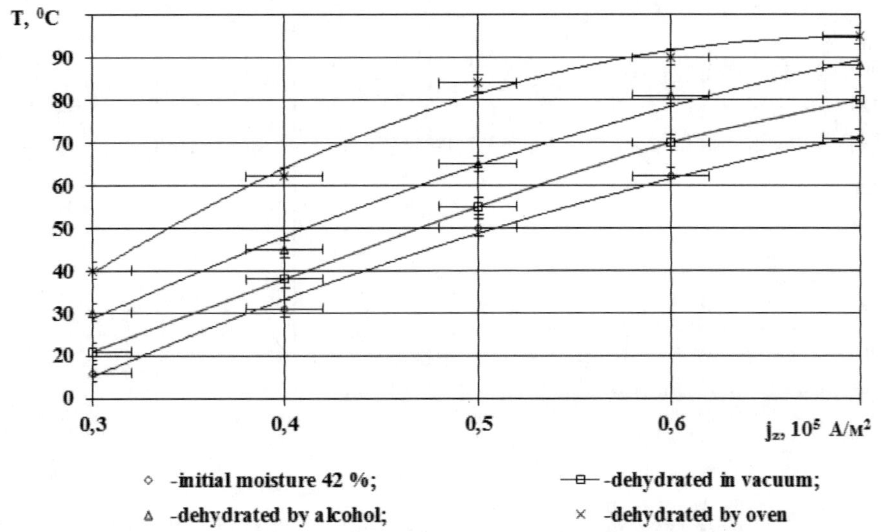

Fig. 1. Changing of the raw material temperature on the current density in the discharge

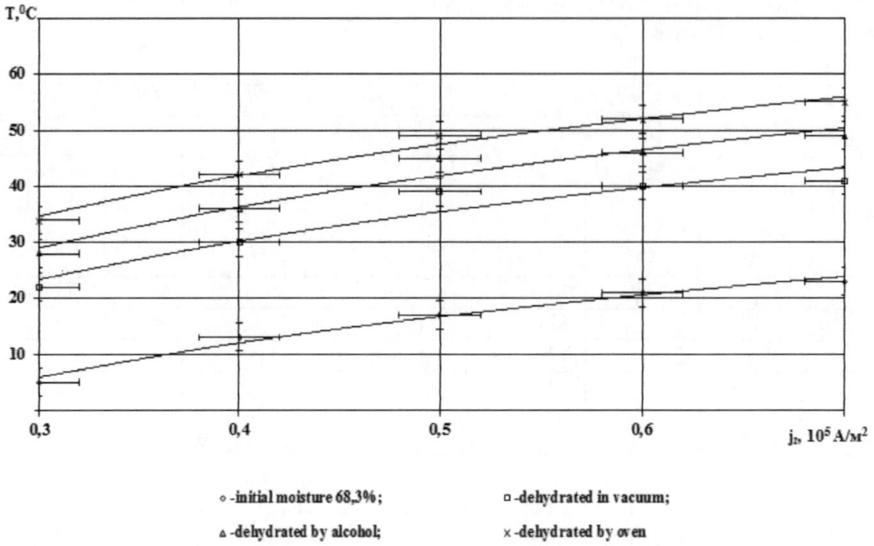

Fig. 2. Changing of the painted semi-finished temperature on the current density in the discharge

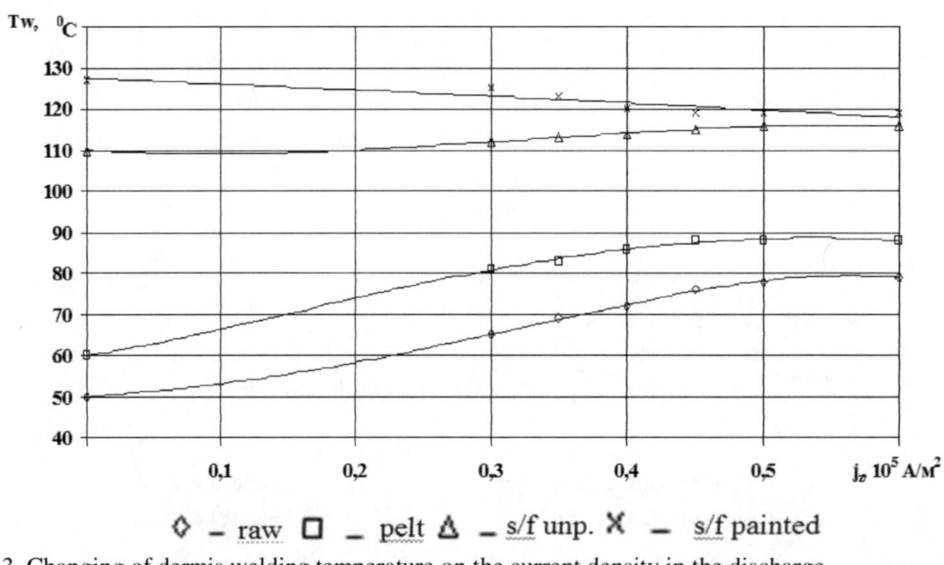

Fig. 3. Changing of dermis welding temperature on the current density in the discharge

Figure 3 shows the curve s of the welding temperature of the dermis and processed products on the current density in the plasma. According to information received, the nature of the gas has little effect on T_w. T_w in the argon plasma is only 2-3 ^{0}C higher than in the air. A marked increase in the welding temperature is observed for samples of raw materials and pelt with increasing of current density in the discharge of about 30 ^{0}C. Welding temperature of tanned unpainted semi-finished product is increased by 5-10 ^{0}C and painted is decreased by 3-5 ^{0}C. Welding temperature change of all samples is registered in the range of discharge current density from 0 to $1{,}0 \cdot 10^5$ A/m^2. With further j_φ increase the temperature does not vary.

3. Conclusions

On the basis of the experiments it can be concluded that the nonequilibrium plasma flow has no thermal effect on the material being processed in these modes of the installation (P = 53.3 Pa, τ = 300 s, G = 0,04 g/s, the gas - air) and at the same time it increases the welding temperature - the temperature of leather phase changes.

Thus, the natural leather temperature during plasma exposure does not rise to a temperature of collagen degradation and does not result in changes of physical phase of the dermis. However, the thermal plasma exposure must be considered, since the high temperatures influence on physical and mechanical properties. Exposure to low pressure RF plasma in studied range does not lead to polymer etching. Therefore, the non-equilibrium low-pressure plasma can be used in modifying of the collagen materials properties.

References

[1] Osin Yu.N. SEM and X-ray analysis of surface microstructure of a natural leather processed in a low tempraeture plasma / Yu.N. Osin, L.Yu Makhotkina, L.N. Abutalipova, I.Sh. Abdullin // Vacuum: volume 51, №2. - P.221-225.

[2] Makhotkina, L.Yu. Design of special purpose products made of nanomodified collagen-containing materials with radio-frequency discharge / L.Yu. Makhotkina, S.N. Sharifullin // IOP Conference Series: Materials Science and Engineering. – 2016. – V. 134. – 012020

[3] Khristoliubova, V.I. Gas and plasma dynamics of RF discharge jet of low pressure in a vacuum chamber with flat electrodes and inside tube, influence of RF discharge on the steel surface parameters / V.I. Khristoliubova, N.F. Kashapov, M.F. Shaekhov // IOP Conference Series: Materials Science and Engineering. – 2016. – V. 134. – 012017.

[4] The impact of environmental factors on materials of light industry. Monograph Zhiharev A.P., Fukin O.V., Abdullin I. Sh, Makhotkina L.Yu., Kazan, 2011

[5] Makhotkina L.Y., Khristoliubova V.I. VARIATION OF FOOTWEAR WEAR RESISTANCE DEPENDING ON METHODS OF TREATMENT. International Journal Of Applied And Fundamental Research. – 2015. – № 1 – URL: www.science-sd.com/460-24749 (11.03.2015).

[6] Dispersing additives for waterborne paint materials: review of the literature (123100, Moscow, 2nd Zvenigorod, 12, JOURNAL-LKM @ MTU-NET RU) Sirazieva L.F., Stepin S.N., Makhotkina L.Yu. Coating materials and their application, 2004, № 10, P. 25.

[7] Processing of integrated footwear materials by RF plasma of reduced pressure Makhotkina L.Yu. Leather and footwear industry. 2006. № 5. P. 36-37

[8] Makhotkina , L.Yu. Design of special purpose products made of nanomodified leather / L.Yu. Makhotkina, V.I. Khristoliubova, L.R. Khannanova-Fakhrutdinova // IEJME – Mathematics education. – 2016. – V. 11, № 6. – P. 1495-1503.

About a possibility of increasing the adhesion strength between mineral glass and polymeric binder under radio-frequency induction plasma treatment

I S Miftakhov, A V Trofimov, A I Nagmutdinova, E F Voznesensky, F S Sharifullin, I V Krasina and G R Rakhmatullina

PNTVM Department, Kazan National Research Technological University, Kazan, 420015, Russia

fortmayn@mail.ru

Abstract. The paper investigated influences of radio-frequency induction plasma treatment on the surface of sheet mineral glasses for household purpose. Discussion for casting the most suitable treatment modes and theirs substantiation is shown. During the investigation the most productive plasma treatment modes for applied binders have been found. It is shown that the durability of adhesive joints between mineral glass and polymeric binder under low-temperature plasma treatment increase to 65%.

1. The Introduction

One of the most important properties of glass materials is a possibility to change smoothly their features, depending on driven additives, in case of making, and a possibility of non-stoichiometric structure formation, what allow obtaining materials with properties different from properties of the crystalline material [1].

It may be noted that glasses have such properties as chemical and biological inertness, low adhesive durability of the surface, which can produce negative effect on the technical and operational properties of finished stock. For glasses are used as constituent element in composite, the activity of surface, adhesive properties, wettability will show the main role, as the processes which are proceeded on limit of the section of diverse materials are responsible for quality of the final product in the course of production [2].

It has been concluded that theory of durability of adhesive joints between mineral glass and polymeric binder lies in the diffusion of adhesive into substratum and bonding a strong connection. Theory specifications should be in the mechanical interaction: coupling of adhesive and substratum of having microcracks, roughnesses, hollows on the glass surface. Though, chemical theory of adhesion has been concluded that active formations on surface can form bonds with adhesive, while forming the reaction products.

The most common possible methods for increasing the adhesive durability include: chemical modifications of the glass surface, for example, by using salt of metakrilovy acid, hydrochloric acid [3], and hromoksilkhloridy, or another one chemical compound, which could interact in a complex with both: polymer and glass molecules, for example — four-chloride silicon or non-hardening epoxy [4]. Also, adhesive durability could increase by using the reagents which generate chemically bonded layer with glass surface, on which one further interaction with adhesive is followed [5].

However, chemical treatment can cause damage to the technical properties of glass and glass materials and produce negative effect on ecology.

Alternative method is high-frequency plasma treatment. The advantages of the method, such as speed, one-staging, nondestructive on the internal structure influence, ease of a variation of parameters, preservation of optical and strength properties of glasses, may to be noted.

2. Materials, methods and equipment

The objects of the investigation are samples of sheet silicate glasses. An influence of the radio-frequency induction plasma discharge on the adhesion value of the researched objects to road enamel KO-525 was considered. Glass processing was carried out in the experimental radio-frequency plasma equipment [6].

3. Results

Technical parameters of the radio-frequency plasma treatment are: ion energy Ei =15–30 eV; ionic current density Ji = 5–23 A/m^2 (current intensity on the anode of a generating lamp Ia=1,0-2,5A); plasma-forming gas – Argon; gas rate 0,06 g/s; pressure in the working chamber P =40–90 Pa; processing time T=300 s. Samples were treated at various height over a plasmatron cut: 60, 50, 40, 30 mm. A series of experiments for different height and ion energy was conducted.

The adhesion strength was estimated by ISO 4624 by using Neurtek KN-10 equipment; the results of the measurements are shown in the table 1 and in the figure 1.

Table 1. Average values of adhesion strength, kg/m2

Ion energy, eV (Ia)	Samples height over a plasmatron cut, mm				
	60	50	40	30	The control sample
15 (1,0 A)	140	473,9	599,2	941,6	577,8
20 (1,5 A)	310	371,3	834,6	663,4	
25 (2,0 A)	337,1	684,8	941,6	735,1	
30 (2,5 A)	627	642	863,5	613,1	

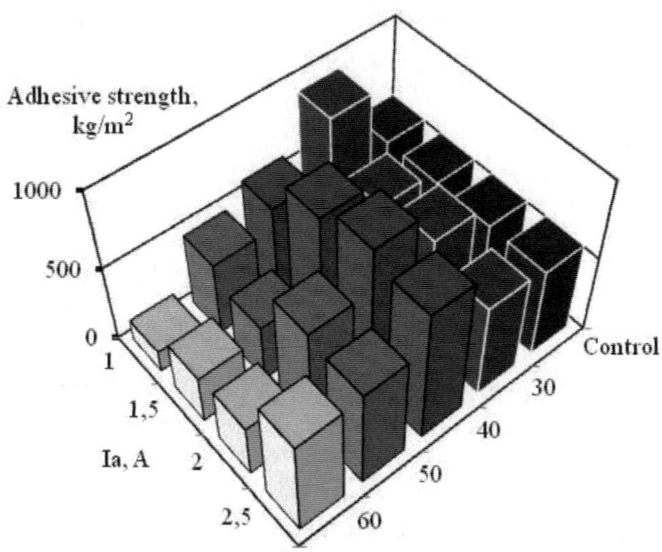

Figure 1. Average values of adhesion strength, kg/m^2

4. Conclusions

Thus, found results provide evidence for increasing the adhesion strength of glass to polymer binders under effect of the radio-frequency induction argon plasma discharge in cases with different height samples, lying over a plasmatron cut and different ion energy. At the same time, the most effective technical parameters are height over a plasmatron cut h=40 mm and h=30 mm.

Presumably, the surface of sheet silicate glasses is exposed to microetching and regrouping of silicate network, can be ascribed to deep adhesive penetration into glass surface and its chemical bond with silicate groups.

References

[1] Pavlushkin N M 1983 *Himicheskaya tehnologiya stecla e sitalov* [Chemical technology of Glasses and Glass-ceramics]. Moscow, (Izdatel'stvo StroyIsdat) 247 p (In Russian)

[2] Kuznecov V P, Baumgarten M I, Nevzorov B P, Fadeev U A 2014 *Adkheziya v kompozicionyh materialah: terminy e fizhicheskaya sush'nost'* [Adhesion in the composite materials: terms and physical essence]. Kemerovo (*Vestnik Kemerovskogo universiteta* vol 1) pp. 173-177 (In Russian).

[3] Kardashow D A 1976 *Sinteticheskiye klei* [Sintetic adhesives]. Moscow (*Izdatel'stvoHimiya)* 504 p. (In Russian).

[4] Sergeev L V, Baygojin A, Sergeev L V, Fattahov S G 1962 *Adgeziya organicheskih polimerov k silikatnomu steklu. II. Obrazovanie molekulyarnyih organosiloksanovyih plenok i ih vzaimodeystvie s poverhnostyu opticheskogo stekla* [Adhesion of organic polymers to silicat glass. II. Molecular organic siloxane films formation and its iteraction with opticl glass surface]. Moscow, (*Vyisokomolekulyarnyie soedineniya* vol 4) pp. 977-981. (In Russian).

[5] Korol A N 1985 *Nepodvizhnyie fazyi v gazozhidkostnoy hromatografii: spravochnik* [Non-subvisible phases in gas-liquid chromatography: directory]. Moscow (*Izdatel'stvoHimiya)* 240 p. (In Russian).

[6] Abdullin I S, Zheltukhin V S, Kashapov F N 2000 *Vyisokochastotnaya plazmenno-struynaya obrabotka materialov pri ponizhennyih davleniyah. Teoriya i praktika primeneniya* [High-frequency plasma-jat of low pressure materials tretament. Theory and using practice]. Kazan (Izdatel'stvo Kazanskogo tekhnologicheskogo universiteta) 348 p. (In Russian).

IOP Conf. Series: Journal of Physics: Conf. Series **789** (2017) 012034 doi:10.1088/1742-6596/789/1/012034

Electrical discharges with liquid electrodes in optical materials processing

D N Mirkhanov[1] **Al F Gaisin**[1]

[1]Faculty of Physics and Mathematics, Kazan National Research Technical University named after A.N.Tupolev - KAI, Kazan, 420111, Russia

E-mail: mirhanov@list.ru

Abstract. In this work the experimental research of high frequency capacitive electric discharge in the processing of semiconductor material surface is presented.

1. Introduction

Currently the electrical discharges with metal and liquid electrodes at constant current are studied well enough. Such discharges are widely used to produce low-temperature plasma and have proven themselves in the plasma technique and technology. They are used in the technology of cleaning, sanding, coating of metals and have a wide practical application in industry. In recent years, they show special interest in high-frequency capacitive electric discharges (HFCED), as the processes occurring in these discharges are more complex and are consequently poorly studied. The significant complexity of the HFCED plasma means that a fundamental understanding of their physics is still not available. Studies of the HFCED with liquid electrodes will lead to new scientific results and open new technological opportunities. The high-frequency plasma of the HFCED with liquid electrodes can be used to modify the surface of metals, dielectrics, semiconductors, as well as materials of different nature. One of the advanced ways to use the HFCED discharge is a processing of optical semiconductor lenses surface, allowing to achieve the required parameters of surface roughness, for further application of various optical coatings.

2. Experiment

The experimental facility is designed to study the HFCED discharge with different types of electrodes [1-8]: solid metal, flowing and stream-droplet electrolytic. The facility is designed for experimental studies of the HFCED discharge, its occurrence, peculiarities of combustion, mechanics and form. Also on the experimental facility a technology of modification of optical semiconductor lenses surface was developed that allows to improve the quality of application of various hardening and antireflection optical coatings. The distinctive feature of the facility is that by means of simple modifications it can be used for research the HFCED discharge with various electrodes and characteristics in the wide range of pressures. The structural scheme of the experimental facility is shown in Fig.1a-b. As it was noted earlier, the facility can be used for research of the HFCED discharge with various forms and types of electrodes, for this purpose it is necessary to replace the system of the electrolyte supply to the electrode of the necessary form. The facility allows to study electrical discharges in the range of discharge voltage $U = 0 \div 5$ kV, current $I = 0{,}01 \div 5$ A, the length of the electrolyte jet $l_c = 0{,}5–40$ mm, diameter of the jet $d_c = 2–8$ mm and a speed of the electrolyte movement $v \approx 0.4$–1 m/s For the generation of the discharge as a power supply system a high frequency generator 8-60/13 is used. In the experiment, the sample of the semiconductor optical lens from optical germanium GMO-1 (Fig.2), which is widely used in infrared optics, was taken. This lens is treated, after which a wear-resistant diamond-like antireflection coating is applied.

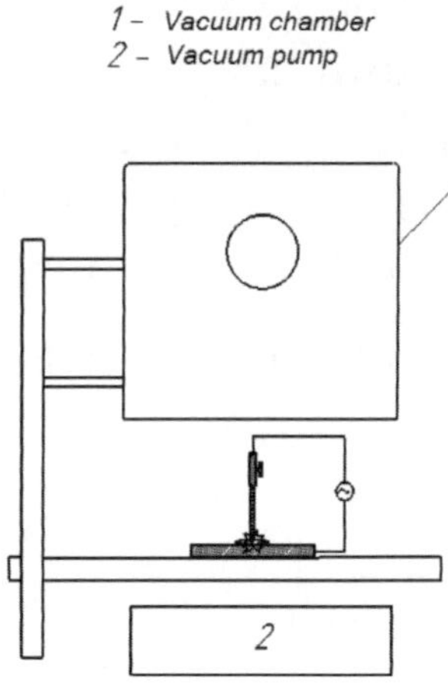

Figure 1a. Structural scheme of the experimental facility with vacuum system

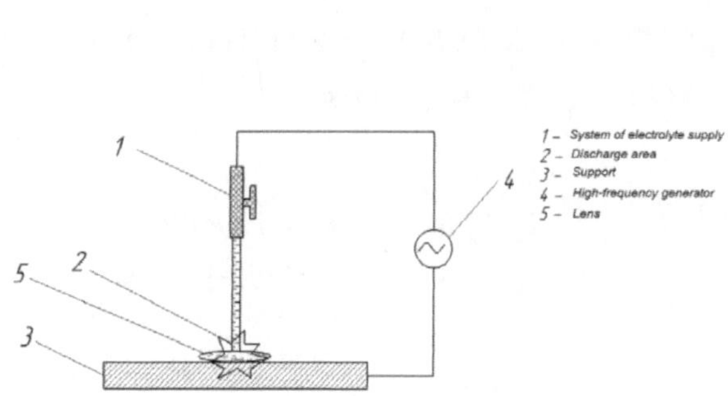

Figure 1b. Structural scheme of the experimental facility inside vacuum chamber

However, due to the low quality of microrelief preparation the coating peels off quickly. In the experiment, the sample of the semiconductor optical lens from optical germanium GMO-1 (Fig.2), which is widely used in infrared optics, was taken. This lens is treated, after which a wear-resistant diamond-like antireflection coating is applied. However, due to the low quality of microrelief preparation the coating peels off quickly.

Figure 2. Optical germanium GMO-1

During the experiment, for the sample the optimal mode of operation of the experimental facility was identified, as an electrolyte a 3% iodized NaCl in the service water was chosen. Experimentally for the sample from germanium the following optimal characteristics of the experimental facility was chosen: the processing time t = 10 s, the jet length l = 4 cm, the generator voltage U = 2 kV, anode current generator Ia = 0.4 A, the pressure P ≈ 100000 PA. The processing is shown in Fig. 3.

Figure 3a. Processing of optical germanium lens GMO-1 between stream electrolytic and optical liens on the metal electrodes.

Figure 3b. Extension of electric discharge on the surface of optical germanium lens GMO-1

Fig. 3 shows that at the initial moment of time t = 0.1-3 s the discharge combusts locally in the form of spot (Fig. 3.a-b), at the moment of time t = 3-10 s, the discharge in the form of a ball is distributed over the entire surface of the lens, whereby the processing occurs throughout the area. Fig. 4 shows the surface after processing in the selected mode.

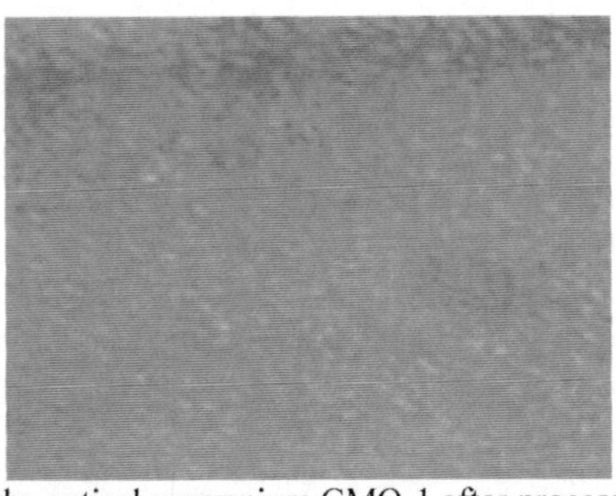

Figure 4 The surface of the optical germanium GMO-1 after processing in the selected mode

After the treatment a significant improvement of surface roughness was observed, which will significantly improve the adhesion of further applied optical coating.

3. Results

The mechanism of the surface modification was as follows: the electrolyte is applieв on the surface of optical lenses, the high frequency electric field is superimposed, as a result of the secondary emission the formed electrons interact with the surface microrelief, improving its roughness. the steam-gas envelope is formed on the treated surface of the lens where the microdischarges appear, with increasing of the voltage the discharge can go into an arc.

References

[1] Reiser Yu.P., Physics of Gas Discharge (Edition 3, updated and revised). Dolgoprudny, 2009. - 734 p.
[2] Reiser Yu.P., Schneider M.N., Yatsenko N.A. High-Frequency Capacitive Discharge: Physics.
[3] Experimental Technique. Appendixes. M.: Nauka, 1995. – 320 p.
[4] Fortov V.E., Son E.E., Gaysin F.M., Bromberg L., Son K.E., O John Khe, I Khe Iong. Plasma Technology (in Korean) Moscow Institute of Physics and Technology, KOFST, 2006. -135 p.
[5] Gaisin A.F. 2013 High Temperature **6** 863 – 866
[6] Akishev Yu.S., Grushin M.E., Karalnik V.B., Monich A.E., Pnkin M.V., Trushkin N.I., Kholodenko V.P., Chugunov V.A., Zhirkova N.A., Irkhina I.A., Kobzev E.N. Plasma Physics. 2006. V. 32. No. 12. P. 1142.
[7] Gaisin A.F., Sarimov L.R. 2011 Plasma Physics Reports **6** 535 – 540
[8] Dautov G.Y., Dautov I.G, Fayrushin I.I, Kashapov N.F. 2013 Journal of Physics: Conference Series **1** 012014

Structure and properties of nitrides on the surface of collagen

M M Mironov and M M Grebenshchikova

Department of Plasma Technology and Nanotechnology of High Molecular Weight Materials, Kazan National Research Technological University, K.Marx str., 68, Kazan, 420012, Russia

Abstract. A study of the structure and composition of titanium and hafnium nitride coatings, deposited by condensation from plasma phase, on the leather by methods of scanning electron microscopy and x-ray fluorescence analysis. Nanostructured nitride coatings don't degrade the properties of the leather and also inhibits the growth of pathogenic microflora and slow the migration of chromium ions.

Nanostructured, superhard nitride coatings are condensed from the plasma phase by magnetron sputtering and cathodic arc deposition. Some condensates, for example, hafnium inhibit the growth of the pathogenic microflora, and may be used in medicine [1].

Of particular interest are hafnium nitride coatings on a natural collagenic substrate. They could be used to produce orthopedic products.

Collagen is the main structural protein. It is polymer with NH_2 and COOH active groups. Collagen forms the fiber by aggregation of microfibrils. Porous nanostructure of the dermis is the foundation of natural leather materials which tanning with aldehyde, titanium, zirconium, and chrome compounds.

Research material is the chrome leather for artificial limbs and parts of musical instruments according to national standards. The leather contains to 16 % chemically combined and adsorbed moisture. Reduction of moisture content to 10% or less and heating to 100 °C or more leads to changes in structure of collagen. And the leather loses their shape and all properties permanently. The leather is used in the production of artificial limbs, orthopedic insoles, stump sockets and others. These products must meet the requirements of national standards. Leather material is in the contact with the human skin directly. Therefore it must meet the medical, technical and sanitary-chemical requirements, including durability, elasticity, hypoallergenic, no toxicity. The chrome leather is not fully responsible this requirements, because it contains chromium. Because of secretion of the sweat glands, chrome migrates to human skin and causes allergic reactions. Collagen is ideal place for the development of pathogenic microflora such as bacterial and microfungal infections. Antiseptics don't solve problems of toxicity and allergic reactions.

The new nanostructured coatings are known to inhibit the growth of pathogenic microflora. They are based on hafnium nitrides and condensed from the plasma phase in nitrogen atmosphere.

Cathodic arc deposition is widely used in all over the world to synthesize extremely hard film to protect the surface of cutting tools and extend their life significantly. A wide variety of coatings can be created by this technology including TiN, HfN, aluminium, chromium, etc.

Properties of collagen materials are fundamentally different from the properties of metals. Primarily, they have a low thermostability (up to 100 °C), high moisture content in the structure and dielectric properties. That limits the use of vacuum technology for modification of the leather and allows the use of only the elements of this technology. Before condensation the leather is kept at a pressure of 0.01 Pa. It's necessary to prevent loss of vacuum. Then the synthesis and condensation are carried out at higher pressure of nitrogen. Arc evaporator was turned on periodically to create a low temperature condensing mode. The coating was formed on the obverse and underside of the leather by rotation in front of evaporator.

Defective samples were determined by organoleptic method. The consumer characteristics, surface topography, composition of the condensates at the macro and micro level were measured on the rest of samples. The end result of working off of the technology was to evaluate the properties in accordance with national standard requirements.

The surface topography of nitride coating on the leather with maximal condensation time is shown in figure 1. It is fundamentally different from the relief of coatings on metals. Relief of coating on the leather resembles a mountain landscape in microsizes with elements having a nominal diameter of 1-3 μm. Projections of balls with a diameter of 0.3-3 μm are dropping phase. They are formed by arc evaporation cathode metal. Large drops with diameter of 2-4 μm spontaneously fall from the surface to the bottom of the vacuum chamber. But small ones (200-300 nm in diameter) stick to the surface. Then they are overgrown nitride phase.

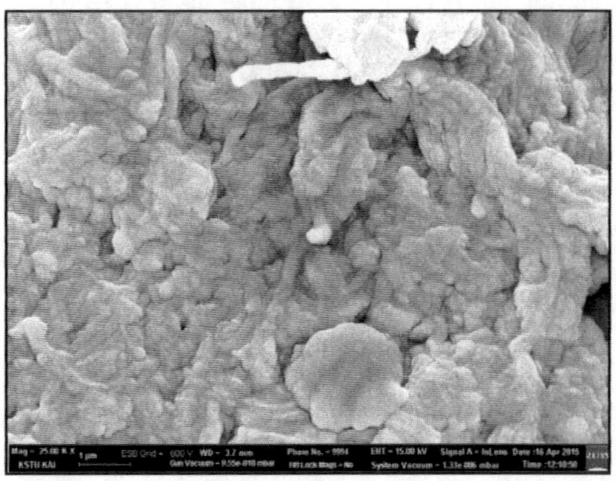

Figure 1. SEM image of the surface topography of nitride coating on the leather.

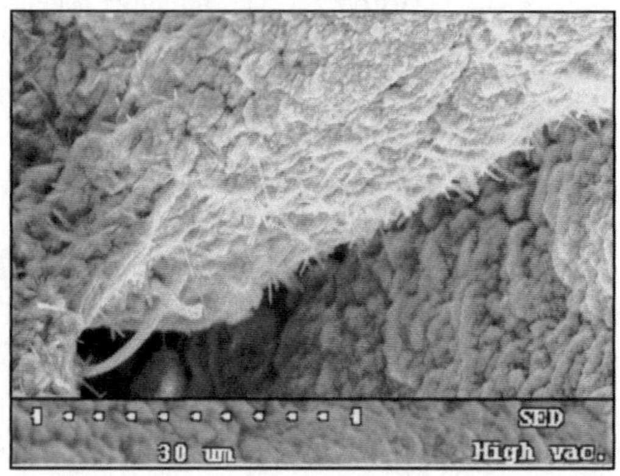

Figure 2. SEM image of nanowhiskers.

The reason of this hillocky relief of coating was revealed by changing the condensation regime. Figure 2 shows the nanowhiskers with length of 5-6 μm and diameter of 40-80 nm. When there is no shaking of rotation, they grow with abnormal speed. And then they fall randomly to the surface and adhere to it due to the Van der Waals forces. Further, they are also overgrown nitride phase and form the characteristic relief.

Crystals grow fragmentary and form scales (dendrites) under condition of limited time of condensation. They are not connected with each other. They are fixed only at the elastic collagen substrate and can move relative to each other (figure 3). Such coatings don't crack and don't hinder the permeability and elastic properties of the leather.

The scales are ordered and range in size from 0.1 to 1.0 μm. They cause diffraction of the incident light. It gives the surface a golden color. Also the iridescent color pattern can appear. It has the characteristic metallic luster. And it appears when sizes of the scales are changed monotonically.

The scaly structure of the coating on the leather was visualized by scanning electron microscopy at workstation Auriga CrossBeam using secondary electrons (figure 4) with the energy dispersive spectrometer INCA. It allowed to more accurately examining the structure and composition of the individual fragments.

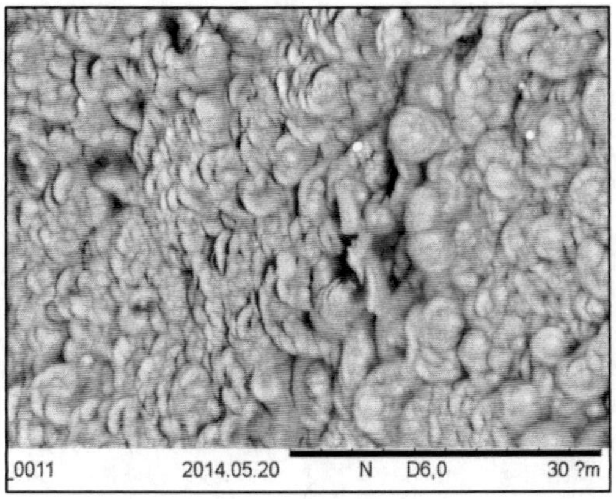

Figure 3. SEM image of the scaly structure of nitride coating on the leather.

Figure 4. SEM image of the scaly structure of nitride coating on the leather.

The distribution of elements on the electron energy spectrum is shown in figure 5. Composition of outer layer of scaly coating at depths of several atomic layers showed the presence of hafnium oxide and titanium in a ratio of 1:4.

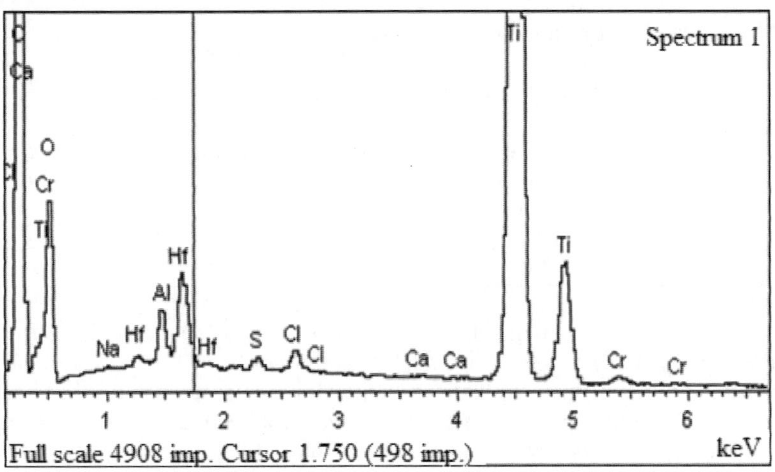

Figure 5. The distribution of elements on the electron energy spectrum.

Probably the atmospheric oxygen and water from collagen are oxidized the outer layers from nitrides to oxides. This is due to the large difference in the value of the standard enthalpy of formation of 88 and 265 kcal/mol, respectively.

Hafnium-titanium coatings on the polycorundum substrate were studied by X-ray fluorescence analysis on such spectroscopes as Bruker «Tornado» and «Picofox». Studies on depths up to 60 μm showed the ratio of hafnium and titanium in equal parts with ± 20% variation.

The columnar crystals are viewed in scales structure. It is typical for nitride condensates. The crystals have a length equal to the thickness of the coating and a diameter of 20-50 nm. Growth of crystals in the columnar direction is not interrupted by a transition of metal vapor-plasma phase to another. However, a thin interface between the phases TiN and HfN is viewed in the electronic image (figure 6).

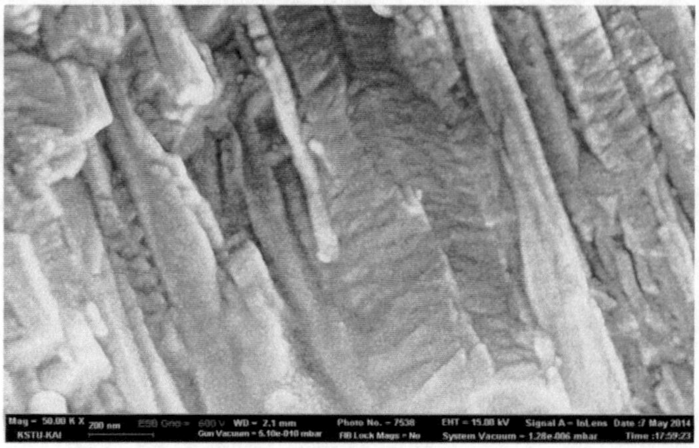

Figure 6. SEM image of the structure of scales.

It is found that the coating of hafnium nitride doesn't degrade the properties of the leather. It gives to the front surface of biocompatibility with human tissues and inhibits the growth of pathogenic microflora and microfungi [2]. Main characteristics of the leather with coatings are given in table 1. They satisfy the requirements of national standards [3].

Table 1. Properties of the leather with coatings.

	Welding temperature (°C)	Tensile strength (MPa)	Relative extension (%)	Moisture content (%)	Biological activity
HfN	92.5	12.3	28.0	10.6	Bioactive, non-toxic
TiN	92.0	12.0	42.5	8.8	Bioinert

Characteristic relief of nitride coating with thickness of 1-3 μm is formed on the leather during the growth of whiskers with a length of 5-6 μm a diameter of 40-80 nm.

The composition of the nanostructured titanium-hafnium nitride coating on depths up to 60 μm has the ratio of hafnium and titanium in equal parts with ± 20% variation. Unlike the composition of outer layer of scaly coating at depths of several atomic layers showed the presence of hafnium oxide and titanium in a ratio of 1:4.

The scaly structure of covering with thickness in 200 nm and the presence in the composition of Ti and Hf allow saving gas permeability, elasticity of the leather, and also inhibits the growth of pathogenic microflora and slow the migration of chromium ions 35 times.

The use of leather with titanium-hafnium nitride coatings is planned in the production of leather-metal prostheses of lower extremities.

References

[1] Patent RU 2 554 773 (2015).
[2] Veinov V P, Bayazitova L T, Haldeeva E V, Mironov M M and Grebenschokova M M 2016 *Sientific session 2016* (Kazan: KNRTU) p 489
[3] Mironov M M, Khramov E N and Grebenschikova M M 2016 *Bulletin of the Technological University* vol **12** pp 108-110

Calculation of the film flowing over horizontal tube surface

I V Morenko

Institute of Mechanics and Engineering, Kazan Science Center, Russian Academy of Sciences, Kazan, Russian Federation

E-mail: morenko@imm.knc.ru

Abstract. Hydrodynamic processes taking place during the motion of liquid droplets and the formation of a liquid film flowing over surface of a horizontal tube are considered. Mathematical modelling of motion of two-phase medium is based on the Lagrangian approach. The results of numerical calculations of the liquid film thickness over surface of the tube are presented.

1. Introduction

The flow of liquid films over three-dimensional surfaces is presented in nature and some technological processes. Research of the deposition process of liquid functional coatings is relevant in medicine, engineering, energy, chemical industry. The deposition of metal, dielectric and semiconductor coatings is used in the manufacture of integrated circuits. Ceramic coatings protect the details and mechanisms of corrosion, oxidation and wear, is increased resource cutting instruments. The use of modern computer technology and mathematical modelling allows you to select technological equipment and its modes of operation rationally and scientifically.

The regime of the film flow over surface depends on the Reynolds number, the physical properties of the fluid, surface roughness and other factors. At low Reynolds numbers the film flows over the smooth surface in the laminar regime, the film surface is smooth, no waves [1]. When the Reynolds number increases the gravitational waves are formed on the film surface. The ripple, short capillary waves are observed on the surface of the film in other cases. At high velocity of liquid film the flow becomes turbulent. Furthermore, thin films are often is torn, the dry regions are formed [2]. The droplet flow observed in the case of low liquid flow rates. The transverse instability of the contact line leads to separation of wet and dry regions of the flow [3].

In this paper the problems of directed motion of liquid droplets from the spray device, the subsequent formation of a thin liquid film and falling film over horizontal surface of the tube are considered. D=0.1 m is diameter tube. The part of the tube length L=0.1 m is considered. The axis Ox_3 of the Cartesian coordinate system $x_1x_2x_3$ is parallel to the axis of the tube (figure 1). The spray device is located on top at a distance 0.1 m from the tube.

It should be noted that the liquid atomization process on the cylindrical surface of the body includes a large number of interconnected physical processes: the formation of droplets, the movement of droplets in a gaseous environment, interaction droplets with the wall or the liquid film, the flow of thin liquid film, spraying, condensation, evaporation of the liquid and others.

2. Governing equations

Taking into account all factors fully at mathematical modeling is very difficult, therefore, makes a number of assumptions. The droplets have a spherical shape, the same diameter d_p, the drop rotation, the split, the merge, the evaporation, the condensation is not considered. The investigated processes are isothermal. It is believed that the gas velocity is zero.

General equations are written for the problems of calculating the trajectories of the droplets and the gravitational falling thin liquid film.

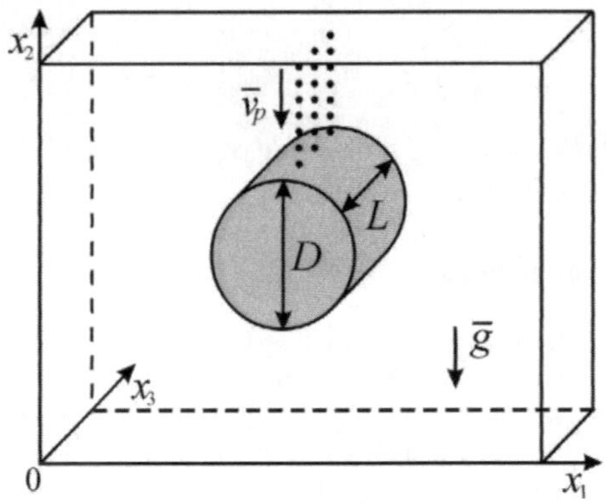

Figure 1. Sketch of computational domain.

The trajectory of the droplets is described by equations

$$\frac{dx_i}{dt} = v_{pi}, \tag{1}$$

where t is time; v_{pi} is the the components of the velocity vector $\overrightarrow{v_p}$ of the drop. The motion equation of the drops is represented as

$$m_p \frac{d\overrightarrow{v_p}}{dt} = \overrightarrow{F_g} + \overrightarrow{F_D}, \tag{2}$$

where $m_p = (1/6)\rho_p \pi d_p^3$ is the mass, ρ_p is the drop density.
In equation (2) takes into account gravity and buoyancy

$$\overrightarrow{F_g} = \frac{\pi}{6} d_p^3 (\rho_p - \rho)\vec{g},$$

and the drag force acting on the droplet from the gas medium

$$\overrightarrow{F_D} = \frac{3}{4}\frac{\rho m_p}{\rho_p d_p} C_D \left(\vec{v} - \overrightarrow{v_p}\right)\left|\vec{v} - \overrightarrow{v_p}\right|,$$

where ρ, $\vec{v}$ is the density, the air velocity, $\vec{g}$ is the acceleration of gravity. The drag coefficient of drop C_D depends on the Reynolds number of the drop $\mathrm{Re}_p = \frac{\rho d_p \left|\overrightarrow{v_p} - \vec{v}\right|}{\mu}$:

$$C_D = \begin{cases} \dfrac{24}{\mathrm{Re}_p}, & \mathrm{Re}_p \le 1; \\ \dfrac{24}{\mathrm{Re}_p}\left(1+0.15\mathrm{Re}_p^{0.687}\right), & 1 < \mathrm{Re}_p < 1000; \\ 0.44, & \mathrm{Re}_p \ge 1000. \end{cases}$$

The energy loss associated with overcoming the drag force at the laminar motion of a spherical body. In the case the turbulent flow past of the sphere the inertia forces are significant.

Numerical simulation of the flow of the liquid film is performed under the following assumptions. The laminar film flow is assumed. The velocity profile of the film in the normal direction is parabolic. The tangent component of the film velocity is significantly more than the normal component velocity. The normal component of the film velocity set to zero. The pressure is constant in the wall-normal direction.

Let δ is the film thickness, $\vec{u} = \frac{1}{\delta}\int_0^{\delta} \vec{u^*}\, d\delta$ is the mean tangential velocity of the film. The movement of the liquid film is described the continuity and momentum equations [4]

$$\frac{\partial \rho\delta}{\partial t} + \nabla \cdot \left(\rho\delta\vec{u}\right) = S_{imp}, \tag{3}$$

$$\frac{\partial \rho\delta\vec{u}}{\partial t} + \nabla \cdot \left(\rho\delta\vec{u}\vec{u}\right) = -\delta\nabla p + \vec{S}_{\rho\delta u}, \tag{4}$$

where $S_{imp} = \sum_j m^*_{p,j}$ is the mass source per unit wall area due to impingement, $m^*_{p,j}$ is the mass flux of j-th impinging droplet.

In equation (4) the pressure $p = p_{imp} + p_\sigma + p_\delta + p_a$ is comprised the impingement pressure of droplet $p_{imp} = \sum_j m^*_{p,j}\left(\vec{v}_{p,j} \cdot \vec{n}\right)$, capillary effects $p_\sigma = -\sigma\nabla_s^2\delta$, where $\sigma = 0.072$ N/m is the surface tension, the hydrostatic pressure $p_\delta = -\rho\left(\vec{n} \cdot \vec{g}\right)\delta$, the local gas medium pressure p_a.

The term $\vec{S}_{\rho\delta u} = \vec{S}_{\rho\delta u,imp} + \vec{\tau}_a + \vec{\tau}_w + \rho\vec{g}_t\delta + \vec{\tau}_\theta$ is comprised the momentum source per unit wall area due to impingement droplet $\vec{S}_{\rho\delta u,imp} = \sum_j m^*_{p,j}\vec{v}_{p,t,j}$, the viscous shear stresses on the film surface $\vec{\tau}_a$, the friction force $\vec{\tau}_w = -\mu 3\vec{u}/\delta$, the gravity force $\rho\vec{g}_t\delta$, $\vec{g}_t$ is the gravity components tangential to the wall, the stress related to contact-angle force $\vec{\tau}_\theta$.

The computation domain is subdivided using tools blockmesh into a nite number of control volume cells over which the governing equations are integrated. The numerical solutions of problem (1)–(4) are implemented in OpenFOAM licensed GNU GPL. Impose restrictions on Courant number $C_O \le 0.3$.

3. Calculation results

In the numerical experiments the following input parameters are given: the density of gas $\rho = 1.2$ kg/m³, the dynamic viscosity coefficient of gas $\mu = 1.79$ kg/(m·c), the density of liquid $\rho_f = 10^3$ kg/m³, the dynamic viscosity coefficient of liquid $\mu_f = 10^{-3}$ kg/(m·c), droplet diameter $d_p = 10^{-3} \div 5\cdot 10^{-3}$ m, the drop velocity vector at the outlet from the spray device is directed downwards $v_{p1}^0 = 0$ m/c, $v_{p2}^0 = -0.5$ m/c, $v_{p3}^0 = 0$ m/c. The liquid spraying begins at time $t = 0$ c and ends at time $t = 0.3$ c. The mass flow rate is 0.023 kg/c.

The hydrodynamic regime of film motion is determined by the Reynolds number $\mathrm{Re} = \dfrac{4\Gamma}{\mu_f}$, Γ is the flow rate. The Reynolds number of the film $\mathrm{Re} = 460$, which corresponds to the laminar flow regime with a wavy air-fluid interface according to the given classification [1].

The calculations show that on the surface of the thin liquid film the waves observed when the film flowing under the gravity force over surface of the horizontal tube. The wave motion is due to the perturbations. These perturbations occur at the interaction of the falling liquid droplets with the thin film surface and the effect of surface tension forces. The results of numerical calculations of the film thickness are presented on figure 2.

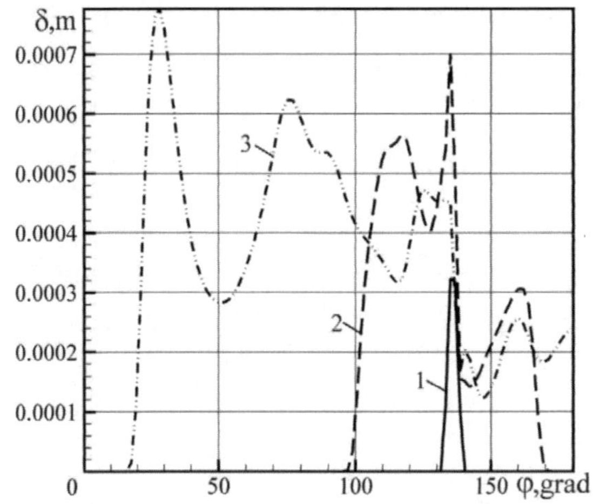

Figure 2. The thickness of the liquid film over tube surface at different times: 1 – $t = 0.13$ c, 2 – $t = 0.18$ c, 3 – $t = 0.26$ c.

It is shown that the film thickness δ less than $8 \cdot 10^{-4}$ m. At time $t = 0.13$ c the maximum value of the film thickness is achieved in the area of the tube surface $\theta = 130 \div 140°$. The angle θ measured from the top of the tube. At times $t = 0.18$ c, 0.26 c the film thickness δ is not constant on the tube wall (curves 2, 3). The values of film velocity less than the velocity of drops at the outlet of the spray device $|v_p^0|$.

4. Conclusions

In conclusion, we note that for the given parameters of the spray device the thin liquid film is not completely covered the surface of horizontal tube. On the moving front of the liquid film the transverse instability of the contact line leads to the formation the rivulets.

References

[1] Kasatkin A G 2004 *Basic processes and apparatuses of chemical technology* (Moskow: Alliance) 753
[2] Maltsev L I and Zavarzin D S 2012 *Thermophysics and Aeromechanics* **19**(6) pp 711–718
[3] Alekseenko S V, Bobylev AV, Guzanov V V, Markovich D M and Kharlamov S M 2010 *Thermophysics and Aeromechanics* **17**(3) pp 371–384
[4] Meredith K V Heather A, Vries J and Xin Y 2011 *Computational Methods in Multiphase Flow* **70** pp 239–250

The interaction of acoustic waves with a three-layer barrier at different angles of incidence

A A Nikiforov[1]

[1]Institute of Mechanics and Engineering, Kazan Science Center, Russian Academy of Sciences, 2/31 Lobachevsky str., Kazan 420111, Russia

E-mail: anikiforov1@yandex.ru

Abstract. In this paper influence of the bubble layer on the transmission of the acoustic waves through the three-layer barrier placed in the water at various angles of incidence is investigated. Barrier consists of a layer of gel with polydispersed air bubbles limited polycarbonate layers. It is shown that the presence of bubbles in the layer substantially lowers the transmission of the acoustic waves through the multi-layer barrier for frequencies close to the resonant frequency of the bubbles. For these frequencies the values of the reflection and transmission coefficients are almost independent of the angle of incidence and determined by the parameters of the bubble gel.

Introduction

The acoustic multilayer material properties which can effectively control transmission acoustic waves appears a significant subject to be investigated, given its wide-ranging possible applications. Important parameter of any acoustic waves is speed of propagation of them in the medium. At the study of acoustics of liquids it is necessary to take into account the bubbles contained in her. Bubbles increases compressibility of liquid, considerably reducing speed and increasing attenuation of acoustic waves. The fundamentals of mechanics and thermal physics of bubble liquids and various basic aspects of acoustics of multiphase flows are considered in [1]. Systematic exposition of theory of distribution of acoustic waves in the layered media is given in [4]. The theoretical and experimental results of interaction of acoustic waves with the layer of bubbly liquid are presented in [3-6]. In this paper influence of the bubble layer on the transmission of the acoustic waves through the three-layer barrier placed in the water at various angles of incidence θ is investigated. Barrier consists of a layer of gel with polydisperse air bubbles limited polycarbonate layers. A bubble layer thickness $d_3 = 1$ mm, polycarbonate layers thickness $d_2 = d_4 = 1.6$ mm. The function of the distribution of bubbles over sizes was determined as

$$N(a) = \frac{n_t}{\sqrt{2\pi\varepsilon}} exp\left\{ -\frac{[ln(a/a_0)]^2}{2\varepsilon^2} \right\}, \quad a_{min} \le a \le a_{max}. \tag{1}$$

Where $a_0 = 0.08$ mm, $\varepsilon = 0.05$, $n_t = 50$ is the distribution parameters [4]. Bubbles distribution range $a_{min} = 0.04$ mm, $a_{max} = 0.16$ mm. The dependence of the complex wave number K_* on the vibration frequency ω according to [6] has the form

$$\left(\frac{K_*}{\omega}\right)^2 = \frac{1}{C_f^2} + \frac{3\alpha_{g0}\alpha_{l0}\rho_{l0}\langle Q\rangle}{3\gamma_g p_0 - \langle QS\rangle}, \tag{2}$$

$$C_f = C_l/\alpha_{l0},\quad Q = 1 - 3(\gamma_g - 1)(y\cdot coth(y) - 1)/y^2,\quad y = \sqrt{i\omega a^2/\kappa},$$

$$S = i\omega a^2 h_a \rho_{l0}/(1 + h_a t_a),\quad h_a = -i\omega + 4\nu_l/a^2,\quad t_a = a/(C_l(\alpha_{g0})^\beta),\quad \beta = 1/6,$$

$$\langle Q\rangle = \frac{1}{\rho_{g0}}\int_{a_{min}}^{a_{max}} N(a)g_0(a)Q\,da,\quad \langle QS\rangle = \frac{1}{\rho_{g0}}\int_{a_{min}}^{a_{max}} N(a)g_0(a)QS\,da,$$

$$\rho_g = \int_{a_{min}}^{a_{max}} N(a)g_0(a)\,da,\quad g_0(a) = \frac{4}{3}\pi a^3\rho_{g0},$$

where ρ°, ρ are the density and the average density, and p is the pressure, γ_g is the adiabatic exponent of gas in bubbles, C_l is the sound speed in liquid, ν_l is the kinematic viscosity of liquid, $N(a)$ is function of bubble size distribution, a is the bubble radius, κ_2 is the thermal diffusivity of gas and α is the volume fraction. The subscripts l and g refer to the parameters of liquid and gas phases, respectively, and the subscript 0 denotes the unperturbed state.

According to [2], the result of transmission of a plane monochromatic wave $\sim exp[i(Kx - \omega t)]$ through a multilayer sample is a plane wave $\sim Wexp[i(Kx - \omega t)]$, where W is the wave-transmission coefficient, which is determined through the layer impedances Z_i and the input layer-boundary impedances Z_i^{in}:

$$W = \frac{2Z_1}{Z_1 + Z_2}\frac{Z_2^{in} + Z_2}{Z_2^{in} + Z_3}exp(iK_2 d_2\cos\theta_2)\frac{Z_3^{in} + Z_3}{Z_3^{in} + Z_4}exp(iK_3 d_3\cos\theta_3)\frac{Z_4^{in} + Z_4}{Z_4^{in} + Z_5}exp(iK_4 d_4\cos\theta_4), \tag{3}$$

$$Z_j = \rho_j\frac{\omega}{K_j\cos\theta_j},\quad Z_j^{in} = \rho_j\frac{Z_{j-1}^{in} - iZ_j tg(K_j d_j\cos\theta_j)}{Z_j - iZ_{j-1}^{in}tg(K_j d_j\cos\theta_j)}Z_j,\quad Z_1^{in} = Z_1.$$

Here, θ_j the angle between the direction of wave propagation in a layer with the normal to the boundary layer, d_j and K_j are, respectively, the thickness and wavenumber of the *j*-th layer (for the bubble layer, $K_3 = K_*$). The water parameters marked by the subscripts 1 and 5, the layers of polycarbonate marked by the subscripts 2 and 4, the bubbly layer marked by the subscript 3. The reflection coefficient is determined by

$$R = \frac{Z_4^{in} - Z_5}{Z_4^{in} + Z_5}. \tag{4}$$

Results

The calculations are performed with the following values: $T_0 = 293$K, $p_0 = 0.1$MPa. For polycarbonate sound velocity $C = 2280$m/s, density $\rho^\circ = 1190$kg/m^3; for water $C = 1480$m/s, $\rho^\circ = 1000$kg/m^3; for gel $C = 1450$m/s, $\rho^\circ = 1000$ kg/m^3 the viscosity of the gel $\nu = 2\cdot 10^{-5}$ m^2/s.

The results of calculations dependencies absolute values of the reflection $|R|$ and transmission $|W|$ coefficients from the dimensionless frequency f/f_o ($f=\omega/2\pi$, f_o is the Minnaert resonance of the bubble) are shown on figures 1 and 2. The lines *I* and *II* is the calculation for the three-layer barrier comprising a layer of bubbly gel, the lines *III* and *IV* is the calculation without bubbles in gel ($\alpha_g=0$). Accordingly, the lines *I* and *III* are calculated with $\theta=0$, *II* and *IV* – with $\theta=\pi/6$.

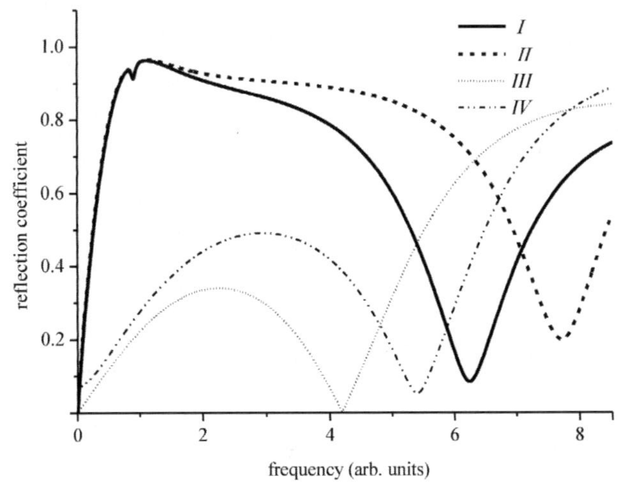

Figure 1. Frequency dependences of the reflection coefficient absolute value

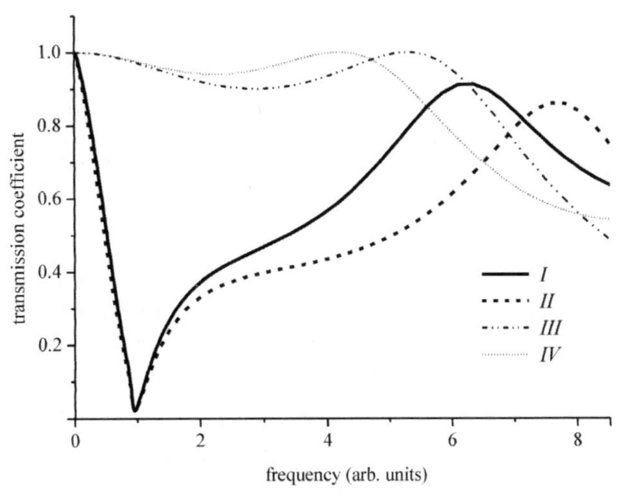

Figure 2. Frequency dependences of the transmission coefficient absolute value

For the lines *I* and *II* in figures 1 and 2, when bubbles are present in the gel, significant difference from the lines *III* and *IV* is achieved in the frequency range acoustic opacity stripe ($f/f_0 < 6$), where the values of the phase velocity and attenuation coefficient taking high values [2, 5]. In this area, the influence of the angle of incidence on the nature of the curves and is not essential. In area $f/f_0 = 1$ the influence of the angle of incidence on the reflection and transmission coefficients is not essential. At $f/f_0 = 1$ influence is already noticeable. Figure 3 displays the transmission coefficient as a function of the angles of incidence for the frequency values $f/f_0 = 2.5$ (line I), $f/f_0 = 10$ (line II) and $f/f_0 = 17.5$ (line III).

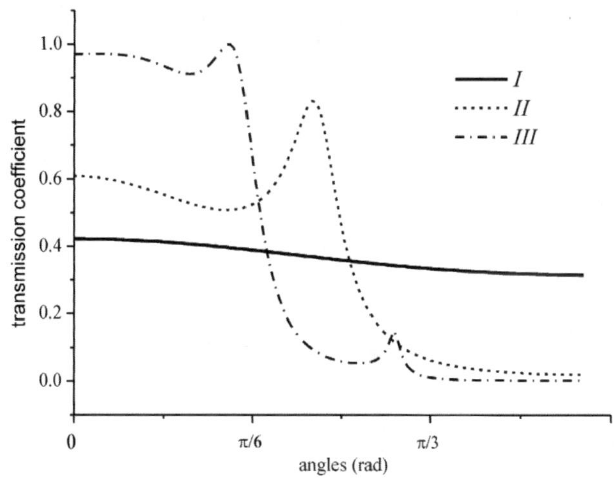

Figure 3. The angular dependence of the transmission coefficient

Thus shown, for a three-layer barrier with a bubbly layer at different angles of incidence of the acoustic wave transmission coefficient in area Minnaert resonance of the bubble remains practically unchanged. When changing the angle of incidence may be possible to control transmission of acoustic waves in said frequency range.

References

[1] Nigmatulin R I 1991 *Dynamics of Multiphase Media* (New York: Hemisphere)
[2] Brekhovskikh L M 1980 *Waves in layered media* (New York: Academic Press)
[3] Shagapov V Sh, Sarapulova V V 2015 *Acoustical Physics* **61** 37
[4] Leroy V, Strybulevych A, Page J H and Scanlon M G 2008 *J. Acoust. Soc. Am.* **123** 1931
[5] Nigmatulin R I, Gubaidullin D A, Nikiforov A A 2014 *Doklady Physics* **59** 286
[6] Gubaidullin D A, Fedorov Yu V 2016 *Doklady Physics* **61** 499

Heating of heavy oil reservoirs using high-temperature heat agent recirculation

V B Osnos[1], V V Kuneevsky[1], V M Larionov[1], E R Saifullin[1], A V Gainetdinov[1], Yu V Vankov[2], I V Larionova[3]

[1]Kazan Federal University, 18 Kremlyovskaya str., Kazan 420008, Russian Federation
[2]Kazan State Power Engineering University, 51 Krasnoselskaya str., Kazan 420066, Russian Federation
[3]Kazan National Research Technical University named after A.N. Tupolev, 10, K.Marx str., Kazan 420111, Russian Federation

E-mail: mr.emilsr@gmail.com

Abstract. The prospects of using high-temperature heat for heating oil reservoirs are estimated. Considered one of the possible schemes of heat agent recirculation flow implementation. The conditions of the developed scheme practical using are analyzed.

1. Introduction

The demand for oil over the last decade, the world has increased significantly. However, light oil reserves diminish. Therefore, improvement of methods for heavy oil field development is an urgent task.

For the production of heavy oil methods of heat stimulation are promising. There are deposits of such occurrence and properties of the oil, in which the effects of thermal methods may be the only ones that allow development.

Most thermal stimulation methods based on the using of heat agents for heat transferring to the oil reservoir. The best heat agent among the technically possible - water and steam due to their high enthalpy. In general, the heat content of steam is higher than the water, but with pressure increasing benefits of steam are reduced, if they are assessed only in terms heat entered into the formation. This also indicates that the greatest efficiency is achieved when steam is injecting in shallow wells that require low pressures [1].

With steam injection occurs temperature front lagging behind the displacement front. However, due to the vaporization heat of steam condensation warmed reservoir area increases by 3-5 times (depending on the dryness of the injected steam, and pressure) as compared with injection of hot water [2]. This is one of the advantages of steam using as a heat agent. However, the use of steam with high pressure due to the lag of the temperature front from displacement front by 8-10 times increases the chance of a breakthrough in the production wells or aquifers. [3] With increasing temperature of the injected heat agent the efficiency of heating the reservoir increases, however, the use of superheated steam at a pressure of 2-3 MPa and a temperature of 400-500 °C for bitumen deposits does not provide sufficient warm-up, as at a pressure of 2 MPa and 400 °C, the enthalpy is 2834 kJ/kg, and at a pressure of 1.5 MPa and a temperature of 200 °C – 2792 kJ/kg [2], i.e. the values are almost equal. However, heating to a temperature of 400 °C and maintaining a pressure of 2 MPa requires equipment that is 2 times more expensive, and the cost of energy 1.5 times larger than a steam at pressure of 1.5 MPa and a temperature of 200 °C. There is a high risk of breaking the producing formation bottom, oil spills and steam on the surface of the ground or in the upper aquifers if shallow bedding of productive layers at injection pressure of more than 2 MPa [4].

Of particular interest is the use of high-temperature heat agent (HTHA) with temperatures ranging from 175 to 450 °C for heating the heavy oil reservoirs. HTHA advantage over other heat agents is the ability to use at high temperatures with noncritical pressures. To use the HTHA in practice necessary equipment providing HTHA supply with required parameters in the oil reservoir and realizing the coolant recirculation.

The aim of work is development of the implementation scheme of oil reservoir heating using HTHA.

2. Description of the scheme

Implementation scheme of oil reservoir heating using HTHA shown in Figure 1.

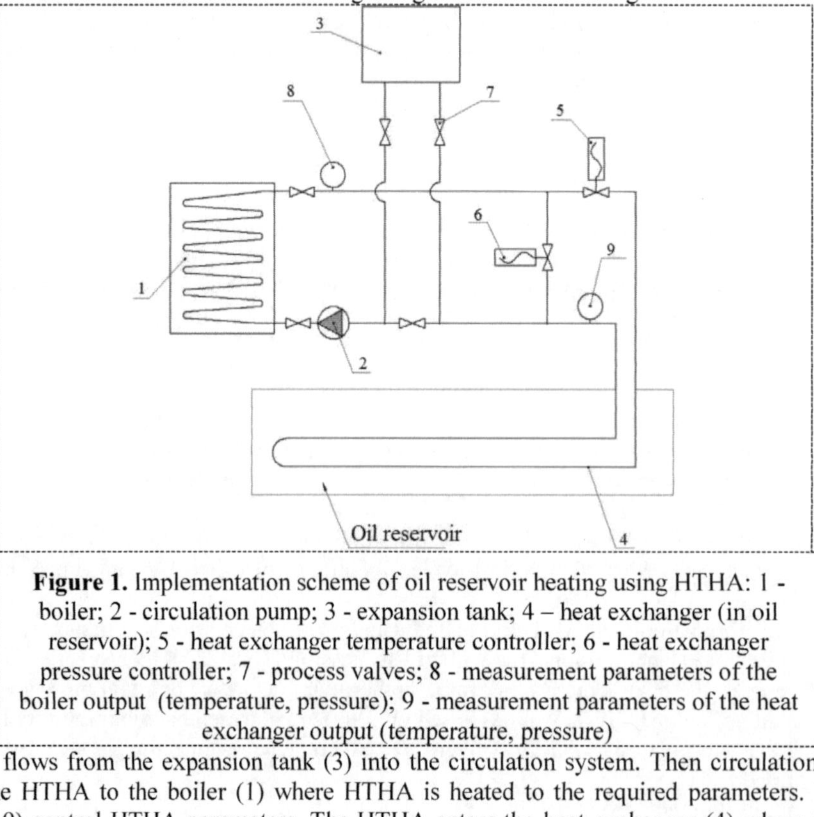

Figure 1. Implementation scheme of oil reservoir heating using HTHA: 1 - boiler; 2 - circulation pump; 3 - expansion tank; 4 – heat exchanger (in oil reservoir); 5 - heat exchanger temperature controller; 6 - heat exchanger pressure controller; 7 - process valves; 8 - measurement parameters of the boiler output (temperature, pressure); 9 - measurement parameters of the heat exchanger output (temperature, pressure)

HTHA flows from the expansion tank (3) into the circulation system. Then circulation pump (2) delivers the HTHA to the boiler (1) where HTHA is heated to the required parameters. Measuring devices (8,9) control HTHA parameters. The HTHA enters the heat exchanger (4) where it transfers heat to oil reservoir. After this HTHA through the circulation pump returns back to the boiler. Controllers (5,6) ensure the maintenance of the HTHA required parameters at the boiler outlet.

To preserve the physicochemical properties of HTHA and to preserve the possibility of its long-term recycling it is necessary to maintain the optimum temperature of coolant. Decreasing the maximum operating temperature of 10 °C increases the continuous operation time of the coolant about twice, therefore it is necessary to use HTHA with a maximum operating temperature of 20-30 °C above desired at the inlet of the heat exchanger.

To prevent the formation of scale on the inside of the pipes and for the enhancement of heat transfer it's necessary to maintain high Reynolds number of HTHA flow. This will turbulize boundary layer and reduce its thickness. It is proposed to use a boiler that ensure required parameters of the HTHA at its high speeds and minimum hydraulic resistance. The main objective of the creation high-speed boiler it is to provide HTHA high speed in the heating surfaces in any capacity. Moreover, increasing capacity and increasing the thermal stress of heating surfaces the speed of HTHA in them increases and increases resistance to overheating and boiling. The task of ensuring HTHA high speed in the heating surfaces can also be solved due to circulation circuit consisting of circulation pump constant performance that connected to the supply line. The circulation pump provides increasing and maintaining the speed of HTHA in a closed circuit "boiler-pump-oil reservoir". In addition, it is necessary to optimize the combustion process in the boiler furnace when fuel will be use associated petroleum gas [5-7]. This will greatly improve the energy efficiency of a heat transfer process.

3. Conclusion

The technological scheme of heating of an oil reservoir was developed. It is a closed circuit, which consists of a HTHA heating boiler, circulation pump providing recirculation of the coolant and a heat exchanger that transmits heat to the oil reservoir.

The recommendations for choice type of HTHA and the conditions of its movement were given. Recommendations will help to maintain the physicochemical properties of the coolant during prolonged recycling.

The work is performed according to the Russian Government Program of Competitive Growth of Kazan Federal University.

References

[1] Takhautdinov SH F, KHisamov R S, Ibatullin R R 2009 *Neftyanoe khozyajstvo* **7** pp 34-37
[2] Ibatullin R R 2011 *Russia B. Tekhnologicheskie protsessy razrabotki neftyanykh mestorozhdenij* 304
[3] Maksutov R A, Orlov G I, Osipov A V 2007 *Neftyanoe khozyajstvo* **2** pp 34-37
[4] Maksutov R A, Orlov G I, Osipov A V 2005 *Tekhnologii TEHK* **6** pp 36-40
[5] Saifullin E R, Vankov Yu V 2015 *IOP Conf. Ser.: Materials Science and Engineering* **86** 012006
[6] Saifullin, E R, Larionov V M, Busarov A V, Busarov V V 2016 *Journal of Physics: Conference Series* **669 (1)** 012037
[7] Saifullin, E R, Larionov V M, Busarov A V, Busarov V V 2016 *Journal of Physics: Conference Series* **669 (1)** 012043

IOP Conf. Series: Journal of Physics: Conf. Series **789** (2017) 012039 doi:10.1088/1742-6596/789/1/012039

Analysis of the heavy oil production technology effectiveness using natural thermal convection with heat agent recirculation method in reservoirs with varying initial water saturation

V B Osnos[1], V V Kuneevsky[1], V M Larionov[1], E R Saifullin[1], A V Gainetdinov[1], Yu V Vankov[2], I V Larionova[3]

[1]Kazan Federal University, 18 Kremlyovskaya str., Kazan 420008, Russian Federation
[2]Kazan State Power Engineering University, 51 Krasnoselskaya str., Kazan 420066, Russian Federation
[3]Kazan National Research Technical University named after A.N. Tupolev, 10, K.Marx str., Kazan 420111, Russian Federation

E-mail: mr.emilsr@gmail.com

Abstract. The method of natural thermal convection with heat agent recirculation (NTC HAR) in oil reservoirs is described. The analysis of the effectiveness of this method for oil reservoir heating with the values of water saturation from 0 to 0.5 units is conducted. As the test element Ashalchinskoye oil field is taken. CMG STARS software was used for calculations. Dynamics of cumulative production, recovery factor and specific energy consumption per 1 m^3 of crude oil produced in the application of the heat exchanger with heat agent in cases of different initial water saturation are defined and presented as graphs.

1. Introduction

In works [1-4] the following technology that uses thermal and gravitational impact on oil-containing layers was proposed.

Heat exchanger in the form of coaxially arranged pipes is entered in oil reservoir. High temperature heat agent moves downhole through a narrow pipe and returns to the surface through a space between the heat exchanger pipes, then the heat agent is heated to initial temperature and again returns to the heat exchanger. It implemented a closed circuit of heat agent flow. It is proposed to use high temperature heat agent which has a temperature range from - 115 °C to + 450 °C. As a result, the heat agent does not penetrate into the formation, and does not change its structure. Closed circuit scheme can significantly reduce the amount of heat agent consumption and avoid additional water saturation of the formation.

Due to natural thermal convection, heat exchanger walls are heated to high temperature. Water and light oil fractions, surrounding the heat exchanger, evaporate. The flow of hot vapor moves up and heats the high-viscosity oil. Over time oil moves to the lower part of the well due to gravity and retrieved to the surface. A mode of natural thermal convection that directs the upward flow of hot steam and the downward flow of heated oil is implemented. Using of one well allows to reach the zone of maximum heating at the oil production area.

These factors suggests that the proposed technology will be more effective than the widely used technology based on steam assisted gravity method of heavy oil extraction.

The aim of work is assessment of the effectiveness of the proposed technologies for reservoirs with different initial water saturation.

2. Research method

Evaluating the effectiveness of the new technology using in reservoirs with different water saturation was also carried out in the software product CMG STARS by calculation of hydrodynamic and thermal processes occurring in the reservoir.

As the test element site Ashalchinskoye oil field size – 300x100x10 m is taken [5]. The initial water saturation of the productive formation was varied and amounted to 0.5; 0.4; 0.3; 0.2 and 0 units, that is 50%, 40%, 30%, 20% and 0% water production, respectively.

A heat source for heating the reservoir was a heat-exchanger with high temperature heat agent. The heat exchanger 300 m length, an inner diameter of 0.15 m is placed horizontally in the bottom part of the productive formation.

The selection of oil at the level of the reservoir pressure maintenance (0.4 - 1 MPa) started after the preheat phase formation within three months heater capacity of 7 Gcal/h (coolant temperature is 330-335 °C).

3. Calculation results

The obtained results are summarized in graphs with dynamic data (Figures 1-3).

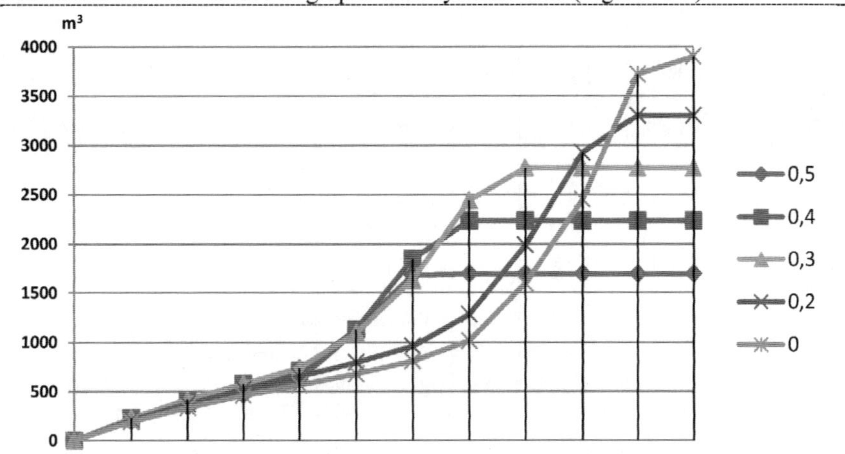

Figure 1. Dynamics of cumulative oil production in the application of the heat exchanger with HTC based on different initial water saturation

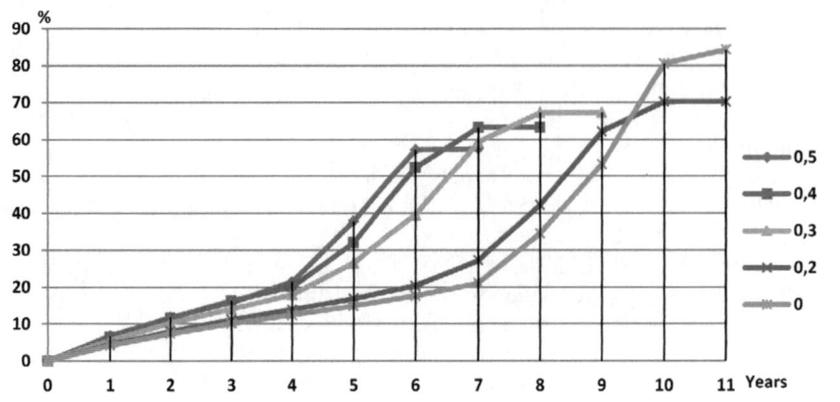

Figure 2. The dynamics of the oil recovery factor in the application of the heat exchanger with HTC based on different initial water saturation

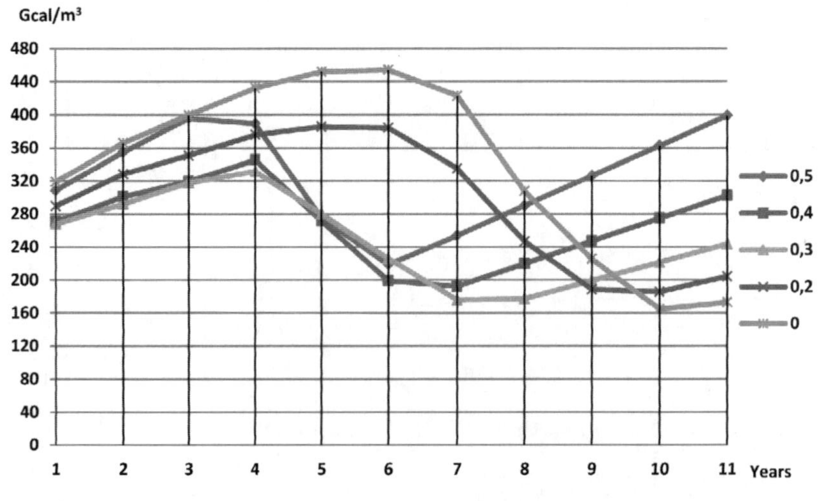

Figure 3. The dynamics of the specific energy consumption per m^3 of produced products in the application of the heat exchanger with HTC based on different initial water saturation

The figures 1-3 show that with the increase in the value of the initial water saturation reservoir is produced for a short period of time. Figure 3 shows the energy consumption needed for production of 1 m^3 of oil offered by technology. From figure 3 it is clear that at the initial stage in the first three years oil extraction is the most efficient from the formation water saturation 0.4 and 0.3, and is less efficient - 0.5 units, since it takes a lot of energy to warm more water (specific heat of water is about 4.5 times more oil heat capacity), the same high cost oil without water, because a lot of energy spent on reducing the viscosity of crude oil. In further operation from 4 to 11 years the most effective are the options with initial water saturation of 0.2 and 0.3, because they have low water content and viscosity of the oil the temperature above 130 °C, become approximately equal to the viscosity of water.

4. Conclusion

The efficiency of heavy oil production depends on water saturation reservoir. According to the calculations for Ashalchinskoye oil field water saturation optimal values were from 0.2 to 0.3 units. This is due to the fact that under these conditions less heat energy is required for heating the reservoir because the heat capacity of water is about 4.5 times more oil heat capacity. However, oil production without the water saturation is also high-cost, because a lot of energy spent on reducing the viscosity of crude oil.

The work is performed according to the Russian Government Program of Competitive Growth of Kazan Federal University.

References

[1] Ibatullin R R, Kuneevsky V V, Osnos V B, Zaripov A T, Absalyamov R Sh 2013 *Oil economy* **1** pp 62-64
[2] Kuneevsky V V, Vankov Yu V, Osnos V B, Zaripov A T, Absalyamov R Sh 2012 *Nefteservice* **4**(20) pp 54-56
[3] Kuneevsky V V, Vankov Yu V, Osnos V B, Zaripov A T, Absalyamov 2012 *Power Problems* **5-6** pp 84-89
[4] Vankov Yu V, Kuneevskiyv V V, Osnos V B 2014 *Exposition Oil Gas* 7(39) pp 20-26
[5] Ibatullin R R, Kuneevskiyv V V, Osnos V B, Zaripov A T, Absalyamov R SH 2013 *Neftyanoe khozyajstvo* **1** pp 62-64

IOP Conf. Series: Journal of Physics: Conf. Series **789** (2017) 012040 doi:10.1088/1742-6596/789/1/012040

Analysis of the impact of informative heat treatment parameters on the properties of hardening of the surface layers

R R Rakhimov, Ruz R Saubanov, I H Israfilov

Kazan (Volga) Federal University (KFU), Naberezhnye Chelny, Russia
E-mail:irmaris@yandex.ru, rafisih88@mail.ru

Abstract. The change features of surface steel layer were investigated under influence of plasma source of concentrated energy. The results of research of the effect of plasma treatment process on the microstructure and the thickness distribution of microhardness of the product and the formation of heat affected zone were presented. The resulting information (informative) data and measured values of microhardness in the hardened layers, depending on the plasma processing modes allow you to manage the process of heat treatment of metals. Designed programming model that works on the basis of the solution of the heat equation, allows to predict and investigate the temperature field after exposure to the plasma stream.

1. Introduction
Improving the machines characterized by a constant increase in power and performance. The increase in performance can only be achieved with a corresponding improvement of the performance of their main components and parts. Condition of parts of the surface layer largely determines the technical and economic parameters of exploitation of machines. Therefore, improvement of surface hardening technology is an important and urgent task. One promising direction to solve this problem is the development and introduction of technologies of hardening produced by the action of concentrated streams of energy, such as laser, ion-plasma, plasma, electron beam and others. However, it is known that a laser, ion-plasma and electron beam thermostrengthening require more capital investment in equipment and operating costs of these processes is twice as high compared to the plasma thermostrengthening [1].

2. Research heat treatment processes of the steel 45
The use of concentrated energy flows improves productivity, accuracy and quality of processing, improved working conditions, increased crop production. However, chemical-thermal plasma treatment using a nitrogen or argon-nitrogen mixture is complex process. With this method of processing hardened surfaces are prone to cracking. Therefore, the lack of automatic control of technological processes (TP), no tracking information environment of informative parameters of heat treatment may reduce the quality of the machined parts, an increase in production costs, etc. [2].

Also, despite the positive results of the plasma heat treatment, its industrial application is difficult lack of information data suitable for controlling the development and heat treatment processes, and results of the study and use of electric surface heat treatment in the working gas atmosphere lacking.

Therefore, the aim of this research is to determine the influence of information (informative) parameters on the spread of micro-hardness values and the formation of the microstructure of metals in the environment at the working gas pulsed plasma flow.

Experimental pulse plasma installation (PI) has been developed for the experiments (Fig. 1), which is the installation of a plasma generator of alternating current [3]. PI developed based on a three-electrode AC plasma torch, allows to clean the surface of the metal heat treatment in the working gas atmosphere or in air, depending on the type of process [7-8].

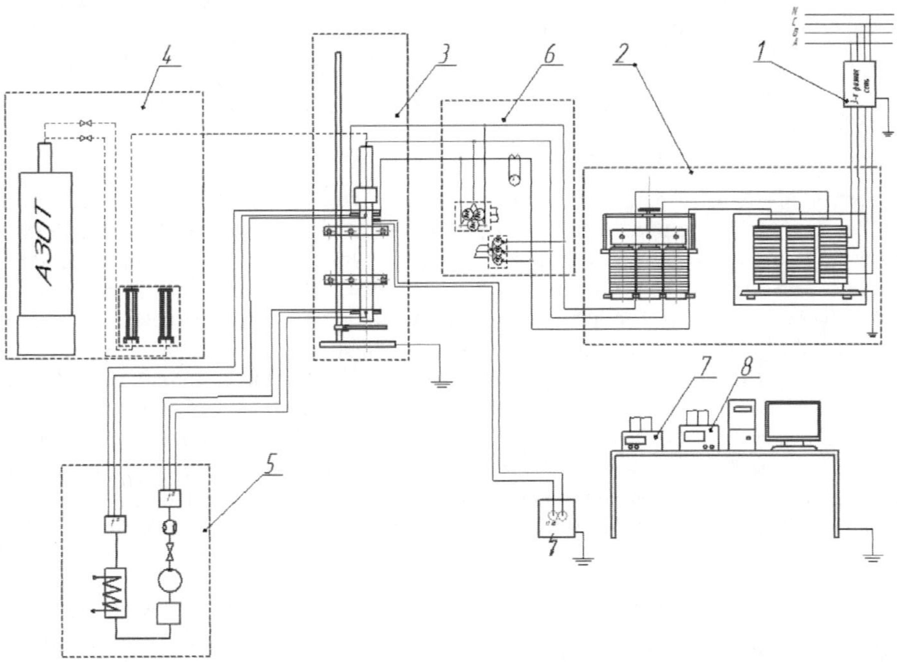

Figure 1. General view of the installation of plasma heat treatment:
1. Electromagnetic actuator. 2. Controllable power supply. 3. Three electrode plasma installation. 4. Induction system. 5. Cooling system 6. System of measurement 7.8 Area operator with monitoring and fixing of informative parameters of devices.

In the experiment, plasma flow diameter of 5 to 10 mm is directed at the sample in the working chamber with gaseous nitrogen. Heat flux on the surface of metal is 103 - 106 W / cm^2. The use of pulsed plasma jet allows precise control of the heat flux, with the possibility of reconfiguring the pulse time of 5 to 7 ms, and the delay time from 5 ms to continuous operation.

Structure and properties of the surface hardened layer is formed under the influence of temperature field arising in the metal when exposed to high-energy plasma jet [5]. The structure and properties of the treated surface determined by the nature of the thermal treatment cycle. Distribution of microhardness on the surface of the processed sample PI with optimized characteristics shown in Fig. 2.

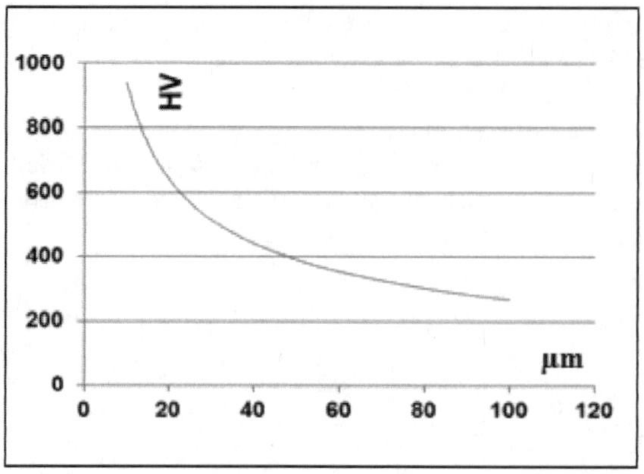

Figure 2. - Distribution microhardness depth of heat-treated zone sample steel 45

The microstructure of the treated steel is shown in Fig. 3. As can be seen from the figure the surface consists of two specific areas: the first is the hardening zone, which contains nitrogen nitrides, increase firmness; The second zone is the primary metal, the metal which has an initial hardness.

For reproducibility of quality indicators TP surface hardening of metal is necessary to control the characteristics of PI with informative parameters [6]. Informative parameters include power of the energy flow, its density on the surface of the workpiece, on the surface temperature field, energy absorption coefficient, etc. Information about the thermal state to determine the depth and width of the heat affected zone, to predict the final structure and properties of the reinforcing layer. Evaluating these parameters set, the direction and the degree of influence on them rationally processing mode using the parameters of the mathematical model steel heating surface concentrated heat source. For a preliminary assessment of technological modes of surface heat treatment process, it is advisable to use a model based on the solution of the boundary value problem of heat conduction.

Figure 3. The microstructure of treated steel (x500)

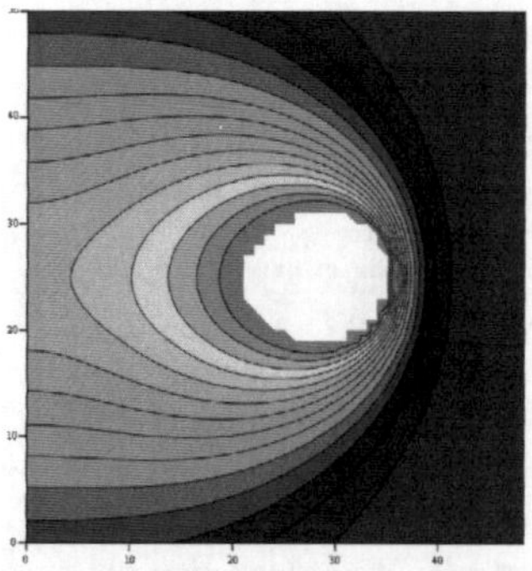

Figure 4. The model of the temperature field distribution of the ECP sources

The programming model is based on the heat equation, which allows you to: vary the heating points; calculate the temperature value at any time and anywhere in a given plane; investigate the temperature field, depending on the duration of exposure to a specific portion of the heating zone; change the speed of processing [1]. Thermogram calculation model is shown in Fig. 4.

For theoretical calculations of the temperature field can apply the equation (1) of heat propagation in a massive semi-infinite body by a powerful, fast-moving normally distributed source - torch.

$$T(x, y, t) = T_0 + \frac{q}{2 \cdot \pi \cdot \lambda \cdot v} \cdot \frac{\exp\left[-\left(\frac{y^2}{4 \cdot a \cdot t_v}\right)\right]}{\sqrt{t_v \cdot (t_0 + t_v)}}, \qquad (1)$$

where T – heating temperature, ^0C;x,y– width and the depth of heating patches, m; t– time, s; T_0 – be the initial body temperature, ^0C; q – effective power of the plasma jet, W; λ – coefficient of thermal conductivity, W/m·K; ν – processing speed, m/s; a – thermal diffusivity coefficient, m^2/s.

3. Conclusion

Metallurgical research after plasma hardening steels show improvement of the quality indicators of stability. The effectiveness of the experimental setup for thermal hardening of the surface layer of parts in a gaseous environment clear. Due to the fact that, in contrast to conventional methods of chemical-thermal treatment, not necessary to use costly vacuum chambers, reduced time and costs. However, to get indicators of quality are comparable to the methods of ion-plasma nitriding is necessary to observe the process conditions. These calculations can determine the necessary parameters of the pulsed plasma generator for the given parameters of quality of the hardened zone.

References

[1] Saubanov R R, Rakhimov R R, Zvezdin V V, Israfilov I Kh 2011 A method for nanostructured surface modification layer parts concentrated streams of energy *Proceedings of the 13th Int. Scient.* (Conf Section 3 St-Petersburg Univ. of the Polytechnic University) -pp.389-394.

[2] Khisamutdinov R M, Zvezdin VV, Saubanov Ruz R, Israfilov I H, Rakhimov R R, Spirin A A 2016 Study of processes of steels surfaces modification with highly concentrated energy flows *IOP Conf. Ser* (Materials science and engineering) Volume 669, Issue 1, Article number 012024.

[3] Saubanov Ruz R, Zvezdin V V, Israfilov I Kh, Rakhimov R R, Saubanov Rusl R Invention patent №2558713, Russia, Россия, IASC H05H1/24; Application of 11.03.2014; Published: 10.08.2015; Priority 11.03.2014. Bul. №22. Impulse alternating current plasma generator device.

[4] Saubanov R R, Rakhimov R R 2014 Investigation of metal nitriding at atmospheric pressure / *Modern instrumented systems, information technology and innovation* (a collection of scientific works of XI-th International scientific-practical conference) in 4 volumes. Editor Gorokhov AA 2014. pp 331-334.

[5] Gabdrakhmanov Az T, Israphilov I H, Galiakbarov A T, Bashmakov D A, Samigullin A D 2014 Pulse plasma surface thermostrengthening of machine parts *IOP Conf. Ser .:* (Materials science and engineering) Volume 69, Issue 1, Article number 012037.

[6] Israfilov I H, Galiakbarov A T, Gabdrakhmanov A T, Simonov L A, Bashmakov D A, Samigullin A D 2014 Automatic control system of plasma quenching technology complex with desired levels of quality *Proceedings of the higher educational institutions "Physics"*, №3 / 3, Volume 57: Siberian Physico-Technical Institute,. - P. 152-155.

[7] Kashapov N F., Kashapov R N. 2014 A study of plasma-electrolytic process for different ratios of the anode space to the cathode *Proceedings of the higher educational institutions. Physics*.. T. 57. № 3-3. pp 168-170.

[8] Denisov D G, Kashapov N F and Kashapov R N 2015 *IOP Conference Series: Materials Science and Engineering* **86** 012005

Investigation of wood combustion in the high-enthalpy oxidizer flow

S M Reshetnikov, I A Zyryanov, A G Budin, A P Pozolotin

Department of Engineering Physics ,610000, Kirov, Moskowskaya Street, 36, Russia

E-mail: b185@mail.ru

Abstract. The experimental data of wood combustion in the high-enthalpy oxidizer flowresearch is presented. Combustion laws of two wood species (pine and birch) in a hybrid rocket engine (HRE) are obtained. Heat flows from the flame to the condensed phase surface are defined. The prospects of the wood use in the HRE (based on thrust characteristics) are shown.

Wood combustion research in a high-enthalpy flow is an important scientific and technical task.On the one hand, the wood can be used as fuel in a hybrid rocket engines whose development are actively conducted in recent years [1-2].On the other hand, investigation results are interesting to forecasting and fire prevention systems developers [3].

Based on the above, the task of the work is wood combustion in high-enthalpy flow research.

The experimental setup is a stand with a hybrid rocket engine, described in detail in [4].As two woodspecies birch and pine are selected.The wood sample is a cylinder with initialmass m_0, length of 0.2m and channel radius r_0= 0.01m.Before the testing started the samples are dried in an oven at a temperature of 600 °C for 3 days until the moisture content of 6%. Samples moisture is determined by electric hygrometers.Combustion is performed in the gaseous oxygen withstream density [10-40] kg/m^{2}s.The linear combustion rate is calculated according to the formula given in [5], thrust measurement is madeby the load cell[4]. Heat flow calculation from the flame to the wood surface produced by the method described in [6]. Carbon black investigation is produced on the microscopes JEM-2100 and JSM-6510 LV.

Fig. 1 shows a graph of linear wood combustion rate against oxygen flux.

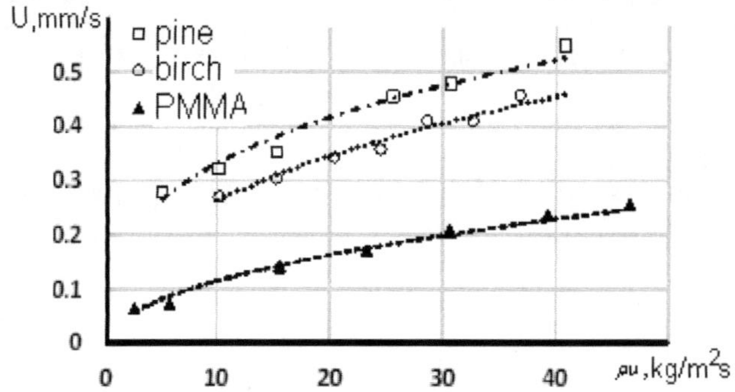

Figure 1. Dependence of linear wood and PMMA combustion rate against oxygen flux.

According to Fig. 1, combustion laws are the following:

$$U = 0.1(\rho \upsilon)^{0.39} \text{ for birch}$$

$$U = 0.16(\rho \upsilon)^{0.33} \text{ forpine}$$

$$U = 0.029(\rho \upsilon)^{0.57} \text{ forPMMA}$$

Combustion law for PMMA is in good agreement with the data presented in [7].The exponent in the combustion law refers to combustion mode, in our case for wood it is a mixed mode (diffusion-kinetic).

Fig. 2 shows a graph of thrust against oxygen flux.

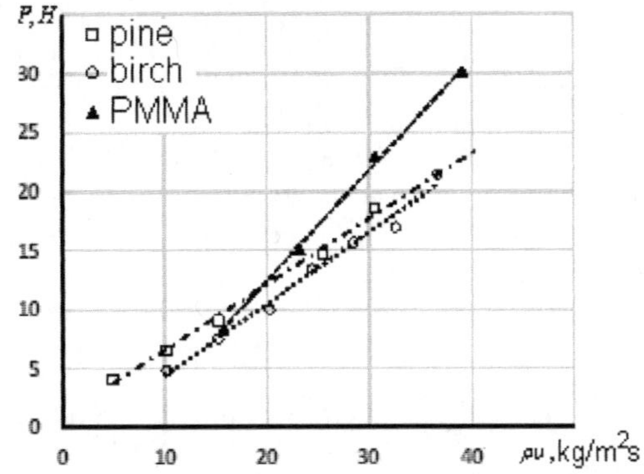

Figure 2. Dependence of the thrust against oxygen flux density.

According to Fig. 2, the thrust of HREwith wood almost as good asthrust of HREwith PMMA.This indicates the prospects of wood using as a propellant.

Fig. 3 shows agraph of the combustion rate against heat flux from the flame to the wood surface.

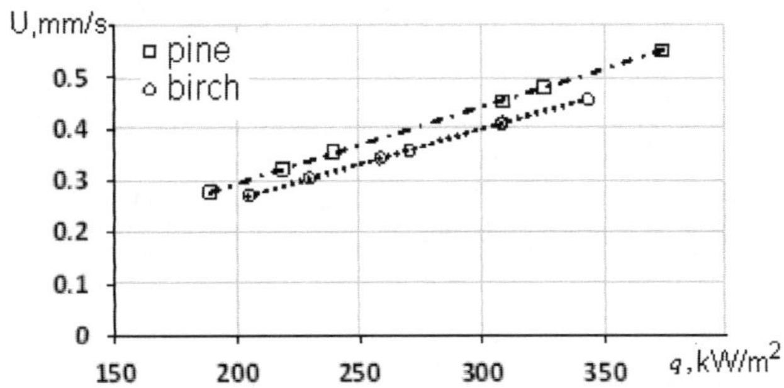

Figure 3. Dependence of linear wood combustion rate against heat flux density.

The value of the heat flux density from the flame to the wood surface is in the range [180-360] $\frac{kW}{m^2}$ (Fig. 3).However, combustibility tests carried out when the samples are exposed bythe heat flux value to $50\frac{kW}{m^2}$ [3]. Such data may not be sufficient to predict the development of a forest fire with a heat flow of about $200\frac{kW}{m^2}$, given in [8].

The thickness of the carbon layer formed during wood combustion in the HRE lies within [200-400] micrometers.In tests [3] carbon layer thickness reaches 5000 micrometers at the heat flux density $50\frac{kW}{m^2}$.

Thus, a comprehensive study of wood combustion in high-enthalpy flow is conducted, resulting in:
1. Combustion laws for birch and pine wood is defined, combustion mode is diffusion-kinetic;
2. Thrust of HRE with wood compared to thrust of HRE with PMMA, prospects of using of wood as a new type of fuel are shown;
3. It has been determined that the heat flux from the flame to the wood surface with the high-enthalpy combustion value in 3-7 times higher than the indicator used for assessing the flammability in case of fire.
4. It is shown that the thickness of the carbon layer after wood combustionin high-enthalpy flow is 12 times lower than thickness obtained by standard fire tests.

References

[1] Ivanov NN, Ivanov A N, The use of hybrid rocket engines in space vehicles, *Journal of FGUP NPO named afterLavochkin,*2010, V.3, P. 50-55.
[2] Gubertov A M, Mironov V V, Gollender R G and others, The processes in hybrid rocket engines, ed. ASKoroteeva,*Moscow: Nauka,* 2008, 405 P.
[3] AssevaR M, Serkov B B, Sivenkov A B, Wood combustion and its fire behavior, *Moscow: Academy of GPS MCHS Russia,* 2010, 262 P.
[4] ReshetnikovS M, ZyryanovIA, PozolotinA P, BudinA G, Electrostatic field influence on the combustion rate in hybrid rocket. *Kazan, Vestnik of the Kazan State Technical University,* 2015, V.71, P. 52-57.
[5] Hudson M K et al. Guanidiniumazo-tetrazolate (GAT) as a high performance hybrid rocket fuel additive, *J. Pyrotechnics*, 2004. V.19, P. 37-42.
[6] PozolotinAP, ReshetnikovS M, ZyryanovI A,BudinA G, ShaurkoP N, Methods of experimental study of wood combustion in a high-enthalpy flow, *Kirov, Society, science, innovation (SPC-2016): Coll. materials*, 2016.P. 1746-1751.
[7] Yuasa S, Yamamoto K, Hachiya H, Kitagawa K, Development of a Small Sounding Hybrid Rocket with a Swirling-Oxidizer-Type Engine *JAPAN – AIAA,* 2001, 353.
[8] ValendikE N, KosovIV, Thermal radiation of forest fires and its possible effects on the forest stand, 2008, V.25, № 1-2, P. 88-92.

Heterogeneous mass transfer in HRE in the presence of electrostatic field research

S M Reshetnikov, I A Zyryanov, A G Budin, A P Pozolotin

Department of Engineering Physics ,610000, Kirov, Moskowskaya Street, 36, Russia

E-mail: b185@mail.ru

Abstract. The paper presents research results of polymethylmethacrylate (PMMA) combustion in a hybrid rocket engine (HRE) under the influence of an electrostatic field. It is shown that the main mechanism of electrostatic field influence on the combustion rate is process changes in the condensed phase.

Hybrid rocket engine (HRE) is a chemical rocket engine, in which fuel and oxidant are in various aggregate states.In these engines, the combustion process is realized in the channel of the solid component through which a liquid or gaseous componentflows.The channel profile, as well as the fuel couple is selected based on the assigned task.

Burning rate of condensed substances is limited by the heterogeneous heat and mass transfer processes. Combustion control can be achieved by impact on flame and fuel.

During combustion the main source of heat is a flame.The flame is a dusty plasma, in which the charged particles are formed mainly as a result of chemi-ionization process. The excess charges in the flame accumulates due to the interaction of the charges with the dispersed particles (soot particles, droplets and particles of fuel, incombustible impurities) [1,2].The presence of dusty plasma in the combustion zone makes it possible to control the heat transfer by means of an electrostatic field.As noted in the papers [3-4] the impact of the field on the flame leads to changes in the position of the flame, turbulence in the combustion zone, the combustion temperature, etc.On the other hand, electric field can affect the properties of condensed fuel [5].The presence of electrostatic field in the HRE combustion chamber leads to combustion rateand thrust increase [6], however, processes responsible for this increaseremain unclear.

Proceeding from the above, to determine the cause of combustion rate changes, the task of the present paper isresearching of the HRE solid component combustion rate in the presence of an electrostatic field.

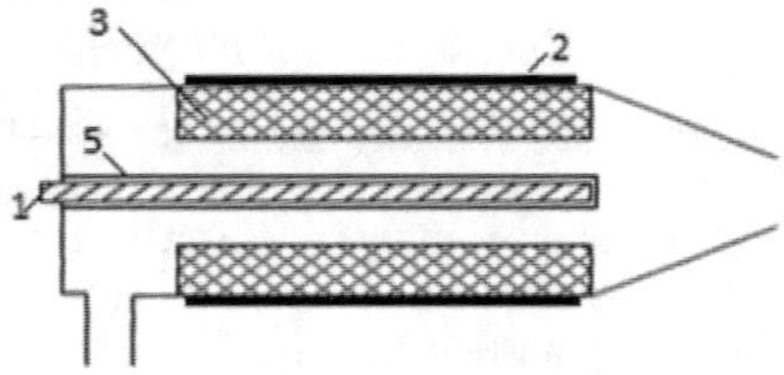

a)

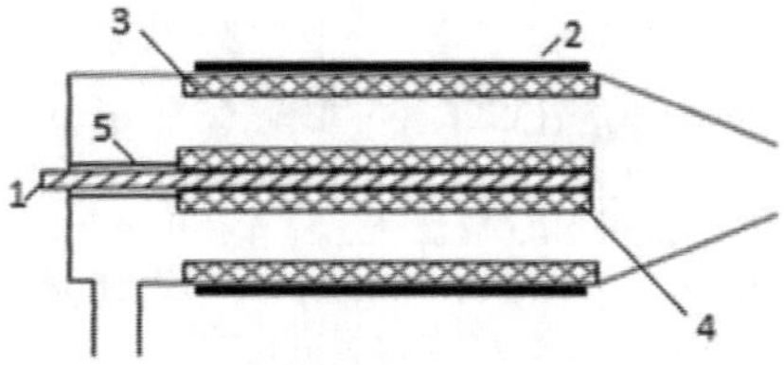

b)

Figure 1.HRE schemes: a) the center electrode in a quartz insulation; b) the central electrode is covered with fuel.1 - the central electrode, 2 - external electrode 3 - the fuel block 4 - fuel block covering the central electrode, 5 - quartz insulation.

Experiments were carried out with two types of HRE, combustion chambers schematic diagrams of which are shown in Fig. 1. Combustion chamber of the first engine (Fig.1a) is a cylindrical channel extending along the fuel a block of polymethylmethacrylate (PMMA) with initial diameter of 20 mm and length of 200 mm.The electrostatic field is generated between the central electrode located on the channel axis and a grid wrapped around the fuel block. In the second engine (Fig. 1b) the combustion chamber is formed by a PMMA tube with inner diameter of 22 mm and coaxially disposed PMMA rod with diameter of 12 mm. Combustion occur in the gap between the inner rod and the tube wall.The length of the combustion chamber is 50 mm. The first electrode is located inside the central axial fuel block and the second electrode is disposed around external fuel layer.As an oxidizer in the experimentsgaseous oxygen is used.Detailed description of the installation and the experimental procedure is given in [6].

Author's research conducted previously [6] shows that the combustion laws in the engine shown in (Fig. 1a) with a potential difference between the electrodes of 0, 3 and 5 kV are the following: $U = 0.029(\rho\upsilon)^{0,57}$ for 0 kV, $U = 0.033(\rho\upsilon)^{0,56}$ for 3 kV, and $U = 0.038(\rho\upsilon)^{0,56}$ for 5 kV. Here U is a linear combustion rate of the polymer,mm/s, ρ is an oxidant density, kg/m^3, υ is an oxidant velocity, m/s.

The exponent in the combustion law remains constant within the error. The relative change in the linear combustion rate under the influence of an electrostatic field will be determined by the coefficients before $\rho\upsilon$ ratio.

Fig. 2 shows the dependence of the relative combustion rate U/U_0 on the potential difference between the electrodes, where U and U_0 - linear fuel combustion rate in the field and without the field, respectively.The measurements were performed with a fixed oxidizer flow rate.According to the presented results it can be seen that with the potential difference increasing a linear combustion rate increasing is observed.

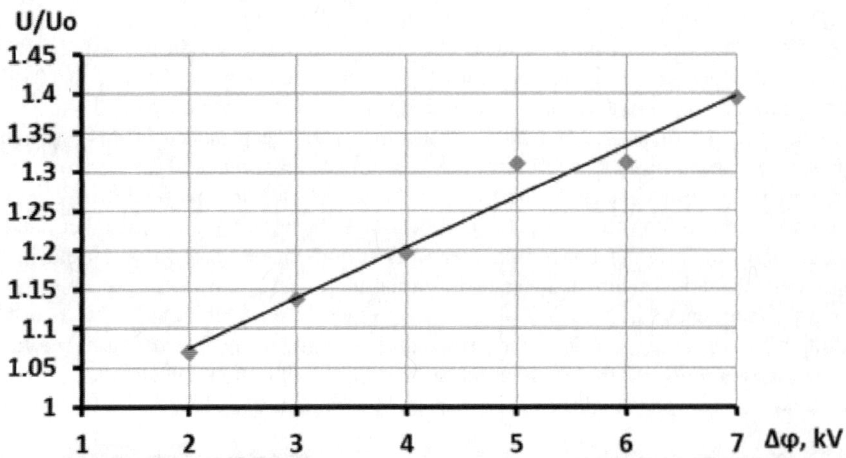

Figure 2. Dependence of the relative combustion rate against potential difference.

To detect the dominant mechanism of the field influence on the solid fuel block combustion rate experiments with the coaxial scheme engine were performed.In this experiments the electrode polarity is changed.The results are shown in Table 1, where $\rho\upsilon$ is an oxidant flow, kg/m^{2}s, U is a linear combustion rate, mm/s, $\Delta\varphi$is a potential difference, kV.

Table 1. Experimental results.

ρv, kg/m^{2}s	Δφ, kV	the central electrode sign	*U*, mm/s, inner block	*U*, mm/s, external block
8	0		0,11	0,106
	5	+	0,189	0,132
	5	-	0,185	0,138
10	0		0,142	0,137
	5	+	0,199	0,158
	5	-	0,205	0,161

The measurement results show that in the electrostatic field absence linear combustion rates of the central and external fuelblock are the same within error.The field presence leads to the linear combustion rateincreasing.A significant difference in the inner andexternalblock combustion rate is caused by the inhomogeneity of the electrostatic field.It was found that the direction of the field does not matter.

These results indicate that changes in the combustion rate are determined by the potential difference between the electrodes (Fig. 2). This may be caused by the influence on the gas or condensed phase.On the other hand, the polarity of the electrodes has no effect on the resulting combustion rate changing: for the external and inner fuel blocks inpresence of the field combustion rate increase remains unchanged with the polarity of both variants (Table 1).This allows to conclude that the effect of "ionic wind", which is usually seen as the main mechanism of combustion parameters changing in the electric field, is not dominant.Therefore, combustion rate increasing occurs due to the processes in the condensed phase.

Thus, in the paper the results of PMMA combustion rate in the HRE experimental study are presented.It is shown that the change of combustion rate in the electrostatic field is determined by the potential difference between the electrodes. It is concluded that the result of the field influence on the combustion ratechanges is determined by the processes in the condensed phase.

References:

[1] Reshetnikov SM, Zyryanov IA, Pozolotin A P, Budin AG, The distribution of excess charges in the diffusion flame of hydrocarbons,*Journal of Physics: Conference Series.* 2016. V. 669. P. 4.

[2] Dautov GU, SabitovShR, Fairushin II, Research Potential Distribution And Concentration Of Electrons In Dusty Plasma, *Kazan, Messenger of Kazan National ResearchTechnikal University named after A N Tupolev,* 2007, V.1, P.29-32.

[3] VolkovEN, SepmanAV, KornilovV N, KonnovAA, ShoshinY S, de GoeyLP H, *In Proceedings of the European Combustion Meeting*, 2009, P.1-6.

[4] Tretyakov P K, Tupikin A V, Zudov V N, Effect of laser radiation and electric field on combustion of hydrocarbon-air mixtures,*Combustion, Explosion, and Shock Waves*, 2009, V.45, № 4, P. 413-420.

[5] Tao R.,HuangK, TangH, BellD,Electrorheologyleads of efficient combustion, *Energy & Fuels*, 2008, V.22, P.3785-3788.

[6] ReshetnikovS M, ZyryanovIA, PozolotinA P, BudinA G, Electrostatic field influence on the combustion rate in hybrid rocket. *Kazan, Vestnik of the Kazan State Technical University,* 2015, V.71, P. 52-57.

Functional plasma sprayed coatings on magnesium ceramic substrates.

SadikovK.G.[1], SofronitskiyA.O.[1], Dautov I.G.[2]

[1] Kazan Federal University, Institute of Engineering, 420008, Kremlevskaya St, 18, Kazan, Russian Federation
[2] Kazan National Research Technical University named by A.N.Tupolev, 420111, K.Marx str., 10, Kazan, Russian Federation

E-mail: artempic8@mail.com

Abstract. The possibility of improving the quality of the titanium articles using ceramic molds with improved surface layer. The improvement is achieved by reducing a porosity of by plasma sprayed Al2O3.

1. Introduction

In the aviation and marine industries widely used products from titanium alloys. These products meet high requirements on quality and reliability, so the improvement of technologies for their production is an urgent task.

For the manufacture of titanium alloy parts using molds. They can be made of various materials such as graphite or ceramic-based MgO. Graphite dust that is formed during the manufacture and operation by graphite forms causes irreparable damage to human health, it is one of the most dangerous by the degree of its impact on the respiratory organs and provokes a many diseases. An alternative to graphite casting forms are ceramic molds, based on MgO, however, such molds have a porous surface, which contributes to preservation a residual gas on stage of evacuating.

Element	Weight %	Atomic %
C K	20.28	36.28
O K	28.83	38.73
Na K	0.53	0.49
Mg K	1.77	1.56
Al K	1.83	1.46
Si K	1.19	0.91
Cl K	0.35	0.21
K K	0.29	0.16
Ca K	0.34	0.18
Ti K	44.60	20.01

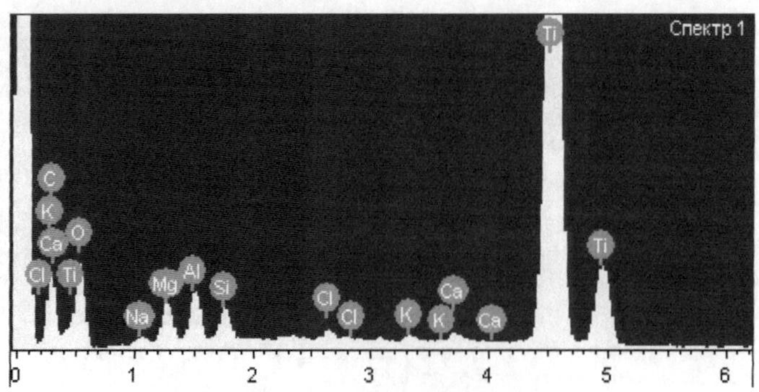

Figure 1. The spectra of the surface of titanium samples manufactured in ceramic forms.

2. Analysis

As shown at the spectral analysis of composition of the surface layer of a casting (Figure 1), an amount titanium and oxygen atoms in the layer approximately rated as 1:2. It is formed on the surface the layer, composed mainly of TiO^2 with high values of hardness and microhardness, which leads to disruption of strength characteristics of products. One of reducing methods of penetrating gases during casting processes is an application of plasma coatings [1-12] on molds (Figure 2).

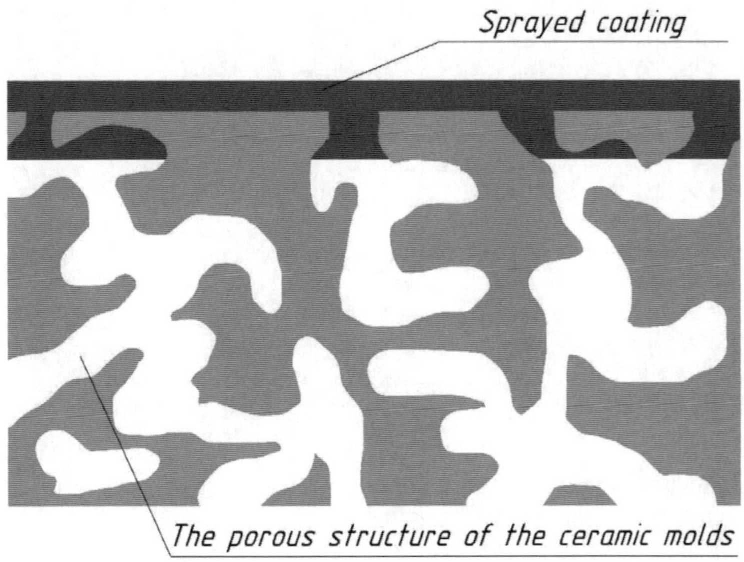

Figure 2. Spray coatings on porous ceramic molds.

3. Experiment and results

Series of experiments of Al_2O_3 deposition was carried out. Spraying was performed by arc plasma torch. The sprayed layer has formed a thermal barrier coating. The melting temperature of Al_2O_3 is 2345 K, and the dissociation temperature about 4000 K, which is significantly higher than the melting temperature of the titanium. Sputtering was carried out using natural gas, and as a result, particles of Al_2O_3 reached the surface of the casting mold in a molten state.

The composition of the casting, produced in a ceramic mold with Al_2O_3-coating, was investigated by X-ray spectral analysis, which showed decreased number of oxygen atoms and other impurities in the casting surface. The content data of TiO_2 are consistent with the values of hardness and micro-hardness, which allowed carrying out a comparative analysis of the properties of titanium casting for selected sections.

The experiments were performed on ceramic samples, which are constituent elements of complex casting molds. They were exposed at temperature 600 K within an hour, and then cooled off 2-3 degrees per minute speed, to ambient temperature under the conditions excluding H_2O ingress of the environment. At first the sample was exposed in a plasma flow for processing after which immediately was produced spraying. Upon receipt of casting the camera volume was pre- evacuated to a pressure of 1-2 Pa, then filled with technically pure argon.

Figure 3 shows a comparative analysis results of changes in the relative hardness of castings produced in the same conditions in magnesite forms, curve 1- with Al_2O_3 coating and curve 2 - without, where R - section radius, r - distance from the casting center, Hs - surface hardness, H- material hardness.

In the region I, closer to the center of a sample, values of the curve - 2 lies below than values of the curve - 1, that indicating higher value Hsu of castings made in a traditional way. In region II is seen decreased value of the curve 1 due to the lower content of TiO_2. In both cases, content of TiO_2 in the surface layers increased, so the area III is characterized by a sharp increase of the relative hardness.

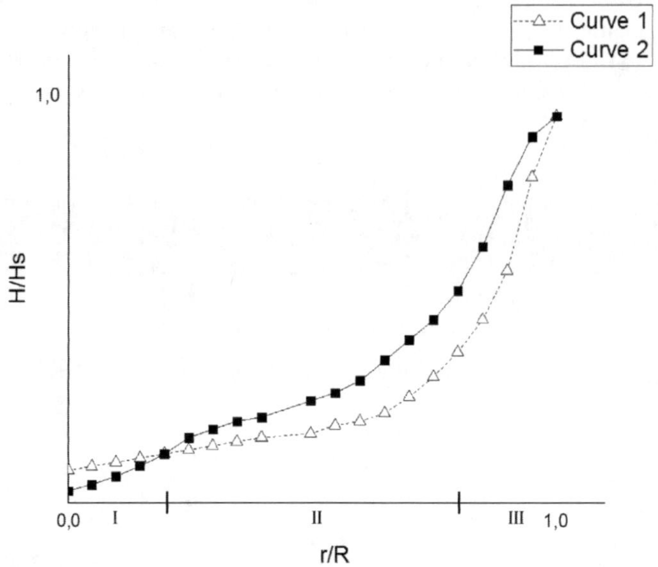

Figure 3. The results of the comparative analysis of changes in the relative hardness of the castings .

4. Conclusion

The proposed method of plasma spraying to magnesite forms can improve the strength properties of titanium articles by reducing of TiO_2 content in the casting volume. It is also possible to reduce the wear of the cutting and milling instruments on stage of preparation of the titanium articles.

References

[1] Fayrushin I, Kashapov N and Dautov I 2014 *J. Phys.: Conf. Ser.* **567** 012009
[2] Kashapov L N, Kashapov N F and Kashapov R N 2013 *J. Phys.: Conf. Ser.* **479** 012011
[3] Kashapov L N, Kashapov N F and Kashapov R N 2014 *J. Phys.: Conf. Ser.* **567** 012025
[4] Zaripov R G, Kashapov N F, Tkachenko L A and Shaydullin L R 2016 *J. Phys.: Conf. Ser.* **669** 012053
[5] Denisov D G, Kashapov N F and Kashapov R N 2015 *IOP Conference Series: Materials Science and Engineering* **86** 012005
[6] Gavrilova V A, Kashapov N F and Kashapov R N 2011 *Biomedical Engineering* **45** 198–200
[7] Saifutdinov A I, Fairushin I I and Kashapov N F 2016 *JETP Lett.* **104** 180–185
[8] Galyautdinov R T and Kashapov N F 2003 *Svarochnoe Proizvodstvo* **(3)** 27–31
[9] Azarov A I, Kashapov N F and Osipova O P 2003 Litejnoe Proizvodstvo **(11)** 30–31
[10] Abdullin I Sh, Galyautdinov R T and Kashapov N F 2001 *Inzhenerno-Fizicheskii Zhurnal* **74** 104–107
[11] Luchkin A G, Kashapov N F and Luchkin G S 2013 *J. Phys.: Conf. Series* **479** 012019
[12] Kornienko E E, Lapushkina E J, Kuzmin, V I, Vaschenko S P, Gulyaev I P, Kartaev E V, Sergachev D S, Kashapov N, Sharifullin S, Fayrushin I 2014 *Journal of Physics: Conf. Series* **567** 012010

High particle and heat load material testing device

Ya Sadovskiy, L Begrambekov, P Shigin, A Ayrapetov, O Bidlevich, A Grunin, N Utkov

National Research Nuclear University MEPhI, Moscow, Russian Federation

E-mail: yaroslav@plasma.mephi.ru

Abstract. A brief overview of the high heat flux material testing facilities is presented which forms background for a new small-scale laboratory device capable of high heat flux material testing. The scheme of the device is presented and its working principles are described. High heat and particle loads are achieved by focusing of either electron or ion beam on the sample being tested. The sample is fixed to an actively cooled copper table. Experiments with hydrogen ion beam irradiation of an MPG-8 graphite sample are presented, in which heat loads of 270 MW/m^2 were achieved. SEM images of the sample surface after testing are presented.

1. Introduction

Ion-surface interactions are well studied, though the flux range used in most papers is 10^{19}-10^{22} ion/m^{2}s. It is well known that some processes strongly depend on irradiating flux intensity. For example, in paper [1] it is shown that chemical erosion of graphite decreases with increasing ion flux. This tendency is measured up to 10^{24} ion/m^{2}s flux, but not further. In paper [2] blisters were formed on tungsten under high flux hydrogen ion irradiation. The temperature of the sample was 480-520 K which is favorable for blister formation on copper, as blisters are formed at 0.3-0.4 melting temperature. Same comparison with copper can be made for flake formation and flake melting which also occurred at copper-related temperatures. This clearly indicates that during high flux irradiation properties of target material can change thus leading to effects previously unexpected.

These and other unforeseen effects would take place during the operation of the ITER first wall, and especially divertor. At the same time, there are quite few experimental devices capable of continuous irradiation of samples with fluxes higher than 10^{25} ion/m^{2}s. Some of these devices are either limited in variation of parameters or are narrow aimed in their purpose. Most of these devices are used for modeling of the conditions of ITER first wall and divertor, where heat fluxes of tens to hundreds of MW/m^2 are expected.

Quasi-stationary plasma accelerators, such as QSPA-T [3] or QSPA-Kh50 [4] are capable of very high pulse loads (in the order of GW/m^2), though the pulse duration is less than 1 ms and after each pulse the device needs “recharging” for several minutes. Thus, a usual material testing study involves up to 100 pulses. Contrary, typical ELM frequencies in ITER lie in the range of tens of Hz. High heat loads can be achieved with electron beams, such as utilized in JUDITH II [5] device. In this device a highly focused electron beam rapidly scans the sample area thus leading to a desired mean heat flux. In this case considering the heat load being stationary indeed is arguable. High plasma density sources are used in devices like Magnum-PSI [6], MARION [7] and PISCES [8]. These devices allow stationary testing high heat loads of up to 20 MW/m^2. Operation time is limited to several tens or hundreds of seconds. Moreover, in Magnum-PSI a combined stationary and pulse load operation regime is realized. All these devices are highly complex installations requiring a dedicated engineering infrastructure and staff to maintain and operate them.

At the same time, if one does not want to test a full scale ITER divertor tungsten mock-ups, a rather compact laboratory device, working with small samples is enough for researching surface processes occurring at high pulse and stationary heat loads. High heat load density can be achieved by focusing of an ion beam on the sample. Independent of the sample size, an active cooling system is required when working with high heat loads.

A device capable of stationary and pulse irradiation of samples by either ion or electron beam with heat flux density of up to 200 MW/m^2 is presented in the paper.

2. Experimental device

The scheme of the experimental device is presented in Figure 1. The device has a cylindrical vacuum chamber (1) with a cylindrical plasma chamber (2) inside. Basically the whole system has axial symmetry. A DC plasma discharge is initiated between a heated tungsten wire cathode (3) and ring-shaped anode (4). By changing of the cathode heating current the plasma discharge current can be varied easily. Typical discharge currents vary from 2 to 20 A. Cathode sputtering is minimized by using low discharge voltage in the range of 50-80 V. The anode is in fact two rings with a gap between them. Ions are drawn to the sample through this gap. The outer anode ring can be moved vertically by means of a motion feedthrough (5). Thus plasma edge geometry and ion beam shape can be varied. Residual vacuum in the device is better than 10^{-6} Pa. Basic components of residual vacuum are $H_2O \approx$ 75% and $H_2 \approx$ 25%. A differentially pumped quadrupole mass spectrometer is used for residual vacuum control, working gas purity control and leak finding. A lock chamber (7) is positioned at the bottom of the vacuum chamber allows quick sample exchange without the loss of the vacuum in the device. A gate valve (8) is closed when the lock chamber is at atmospheric pressure. A motion feedthrough with a PTFE sealing (9) is used to move the sample holder.

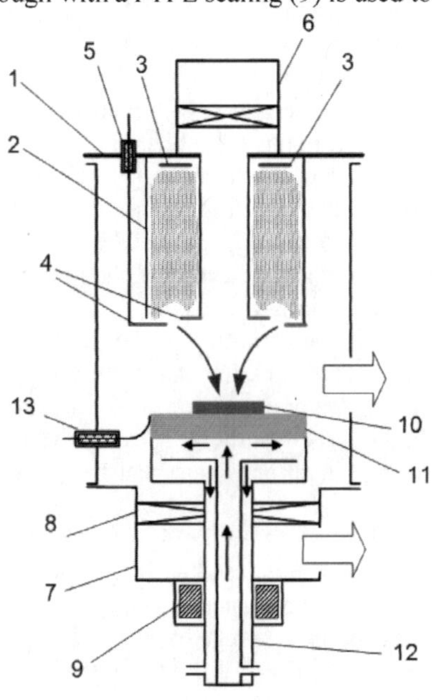

Figure 1. The scheme of the experimental device.
1 – vacuum chamber
2 – plasma chamber
3 – cathode
4 – anode
5 – motion feedthrough
6 – top lock chamber
7 – bottom lock chamber
8 – gate valve
9 – sealing
10 – sample for testing
11 – copper table
12 – stainless steel tube
13 – electric feedthrough with a slip contact

The sample (10) being tested is fixed to a sample holder (11) which moves on a polished stainless steel tube (12). The sample holder is basically a flat copper table which is actively cooled from the downside. To bias the sample an electric feedthrough (13) is installed in the vacuum chamber. When the sample holder moves into the vacuum chamber it contacts the feedthrough thus allowing its biasing. The copper table is connected to the stainless steel tube through a cylindrical ceramic insulator, capable of >30 kV insulation. Dielectric silicon-organic fluid with low viscosity is used for cooling. This fluid circulates in a closed loop with a fluid to water heat exchanger in it. The copper table has radial fins on the downside to increase its surface area contacting the fluid. Copper table thickness, fin height and number of fins were optimized using numerical modeling. The whole system was designed to be able removing of the maximum available beam power. Sample temperature could be varied by putting special inserts between the sample and the copper table. Sample temperature is monitored by an optical pyrometer through the window on the top of the vacuum chamber.

All valves in the vacuum system of the device are pneumatic and thus can be activated remotely. Every operation on the device is automated and controlled through a LabView application specially designed for this device.

High voltage power supply was also specially designed for this project. It allows biasing the sample either positively for electron irradiation or negatively for ion irradiation. Maximum bias voltage is 25 kV. The power supply has a maximum power of 4 kW and can output this power at a wide range of voltages. For example: 200 mA at 20 kV, 400 mA at 10 kV or 800 mA at 5 kV. Continuous feedback for controlling of the voltage and current is realized by a dedicated PC with LabView RealTime software allowing up to 500 measurements per second. The voltage change over time can be pre-programmed thus allowing pulse irradiation with desired frequency and duty cycle. Pulse frequencies of up to 100 Hz are available. Pulse irradiation combined with stationary irradiation regimes are also possible.

The beam shape is governed by a number of factors: bias voltage, plasma density, relative position of the sample table and anode rings. By varying these factors the beam spot size on the sample can be changed. Thus, desired heat flux can be achieved independently of bias voltage. Maximum heat fluxed are realized when the beam is highly focused. One can estimate that in the case of a 5 mm diameter spot and 4 kW beam the average heat density would be 200 MW/m^2.

Secondary electrons are emitted from the sample as the ion beam strikes it. These electrons form a secondary electron beam going from the sample upwards. Accelerating these electrons draws power from the same high voltage power supply that biases the sample and thus decreases the ion beam power. Moreover, the secondary electron beam is so focused that it can damage elements of the vacuum chamber. Therefore, a ring shaped anti-dynatron electrode is installed above the sample holder. This electrode is biased so that it blocks the electrons emitted from the sample. The bias voltage of this electrode is higher in value than that of the sample. Shape and position of this electrode were numerically modeled to allow blockage of the secondary electrons with a least possible bias voltage.

3. Experimental results

The first primary experiments made on the device were performed with MPG-8 graphite samples. Hydrogen was used as a working gas. Various regimes of ion beam irradiation were conducted with continuously increasing heat flux. Heat flux was estimated by beam power and spot size on the sample. After exposing of the sample to a highly focused ion beam for 15 minutes a macroscopic crater was found on the sample. This crater was measured with a help of a stylus profilometer. Its diameter and maximum depth were found to be 3 mm and 50 μm. By knowing the shape of the crater conclusion on the beam shape could be made and flux density calculation could be refined. It was then calculated that at the centre of the beam the heat flux density reached 270 MW/m^2. The sample was not cooled during irradiation and its temperature was 1300 K. SEM images of the same spot on the sample surface after 5, 10 and 15 minutes of irradiation are presented in Figure 2 a,b,c.

Grain sizes of the MPG-8 graphite are in the range of 15-25 μm and they can be seen on the images. Also a more fine relief is seen on the surface of the grains. This relief has characteristic dimensions of about 1 μm. As the irradiation continues grains are sputtered and pores between the grains are opened. The fine relief on the grains is formed already after 5 minutes of irradiation and does not evolve over time. The sample was deliberately cracked in to pieces so that its cross section could be examined. Well pronounced 15-25 μm grains were observed on the cross section with no signs of the fine relief. Thus a conclusion is made that this fine relief is a result of sputtering and redeposition processes at the surface and does not represent any structure in the bulk of material.

It is worth to mention that this kind of surface modification was observed at lower heat flux densities as well. Thus no effects special to high heat flux irradiation were found in this case.

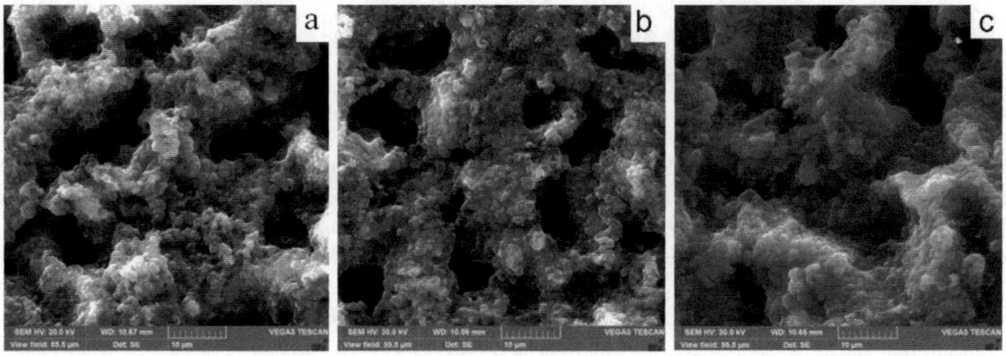

Figure 2. Same spot on the surface of the MPG-8 graphite sample after 5 (a), 10 (b) and 15 (c) minutes of highly focused hydrogen ion beam irradiation.

4. Conclusions.

A laboratory experimental device capable of high heat flux irradiation of materials with either ion or electron beams is presented. The device is easy to operate and is fully automated, and allows pulse irradiation, continuous irradiation or combined pulse and continuous irradiation. By varying of the beam size heat fluxes of up to 270 MW/m^2 could be achieved which is proved for a test experiment with an MPG-8 graphite sample. After the irradiation fine structure with a 1 μm grain size is observed on the surface of the 15-20 μm graphite grains.

Acknowledgements

This work was supported by the Ministry of Education and Science of the Russian Federation (agreement contract № 14.575.21.0049 of 27.06.14).

References

[1] J. Roth, R. Preuss, W. Bohmeyer, S. Brezinsek, et al. Nucl. Fusion 44 (2004) L21
[2] W.M.Snu, M Nakamichi, V.KH. Alimov, G.-N. Luo, K. Isobe, T. Yamanishi et al. J. of Nucl. Mater. 390-391(2009)1071
[3] Klimov N., et al. — J. Nucl. Mater., 2009, vol. 390—391, p. 721—726
[4] Garkusha I.E., et al., — J. of Nucl. Mater., 2007, vol. 363—365, p. 1021—1025
[5] Majerus P., et al.— Fusion Engineering and Design, 2005, vol. 75—79, p. 365—369
[6] De Temmerman G., et al. — Fusion Engineering and Design, 2013, vol. 88, p. 483—487
[7] Nicolai D., et al. — Fusion Engineering and Design, 2013, vol. 88, p. 2506—2509
[8] Goebel D.M., et al.. — J. Nucl. Mater., 1984, vol. 121, p. 277—282

IOP Conf. Series: Journal of Physics: Conf. Series **789** (2017) 012045 doi:10.1088/1742-6596/789/1/012045

The heat effect of combustion process depending on fuel composition fluctuations

E R Saifullin, S A Nazarychev, A O Malahov, V M Larionov, O V Iovleva

Kazan Federal University, 18 Kremlyovskaya str., Kazan 420008, Russian Federation

E-mail: mr.emilsr@gmail.com

Abstract. This article describes effects of changes in the fuel composition on the combustion process. Assessment of effect made on example of the steam boiler analysis.

1. Introduction

The problem of natural resources rational use, energy efficiency and energy saving is relevant at the moment. In this connection, the task of rational and efficient use of natural resources in their combustion is relevant, in particular in thermal power plants. One way of solving this problem is the use of alternative fuels, such as mine gas, landfill gas, biogas, gas of waste water and that is especially true - associated petroleum gas and waste oil refineries.

The problem of using such fuels is in the uncontrollable changes of their composition in time (Table 1) [1].

Table 1. The fluctuation ranges of fuel composition.

	CH_4	C_2H_6	C_3H_8	C_4H_{10}	C_5H_{12} and higher	CO_2	H_2	N_2
Natural gas	80-98	0-1	0-0.5	0-0.5	0-0.5	0-1	2-10	0
Associated petroleum gas	15-90	0-25	0-20	0-10	0-15	0-10	0-5	0-30
Waste of oil refineries	0-35	0-19	0-32	0-40	0-47	0	0-25	0

These changes negatively effect on the combustion process, reduce combustion process efficiency, causing damage to the environment [2-4].

The aim of this work was to assessment of the changes effect of fuel composition on the thermal effect of the combustion process.

2. Calculation results

The optimality of combustion process (stoichiometric ratio of "fuel-air") can quantify the excess air factor α. When $\alpha = 1$ the combustion optimal, combustion temperature maximum (Figure 1) [5].

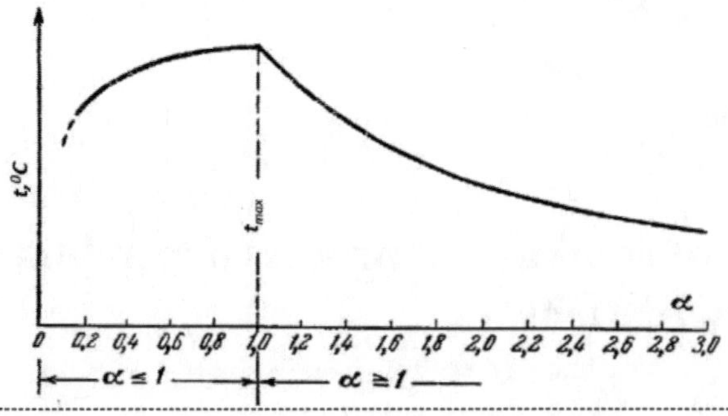

Figure 1. The nature of the theoretical fuel combustion temperature curve in case of lack and excess of air. The y-ordinate is fuel combustion temperature t, °C; x-axis – excess air factor α

The fuel composition affects the carbon number of fuel n, and hence the theoretically required amount of air V_0, which affects the factor α.

$$n = \frac{1CH_4 + 2C_2H_6 + 3C_3H_8 + ...}{(100 - B)},$$

where CH_4, C_2H_6, C_3H_8, etc. – the sum of the hydrocarbons components, %; B - the sum of the ballast components CO_2 and N_2, %.

$$V_0 = 7.13n + 2.28.$$

$$\alpha = V / V_0,$$

where V is the real amount of air.

Increasing the proportion of fuel heavy fractions with a large carbon number reduces α, reduction of carbon number - to increase α. In both cases, the combustion temperature is reduced, optimality of combustion process is impaired.

Assessment of the fuel composition impact on the thermal effect of the combustion process carried out by the example of analysis a steam boiler operating mode. The steam capacity D = 6097 kg/h; enthalpy of water at the inlet H_w = 0.508 MJ/kg; the required enthalpy of steam H_s = 2.77 MJ/kg; boiler efficiency η = 91.85 %. The optimal mode: the fuel composition CH_4=60 %, C_2H_6=4.5 %, C_3H_8=13 %, C_4H_{10}=7.5 %, C_5H_{12}=4.9 %, CO_2=0.1 %, N_2=10 %; fuel consumption G_f= 286.98 m^3/h.

The fuel composition affects specific heat of combustion (SHC) Q_m, that affects the enthalpy of coolant [6].

$$Q_m = 0.01(Q_{H_2S}H_2S + Q_{CO}CO + Q_{H_2}H_2 + \sum Q_{C_mH_n} C_mH_n),$$

where Q_{H_2S}, Q_{CO} etc. – combustion heat of gas components.

$$H_s = \frac{G_f Q_m \eta}{100 \cdot D} + H_w.$$

The graph (Figure 2) shows the change in enthalpy of the coolant in case of SHC deviation from the optimum values.

As can be seen from the graphs, in the case of reducing the fuel SHC enthalpy of the coolant begins to decline sharply. This is due to a decreasing in heat release of fuel combustion and the increasing of air excess. In the case of SHC increasing enthalpy of the coolant rises slightly or remains constant. This is because the increase of combustion heat was compensated by combustion incompleteness.

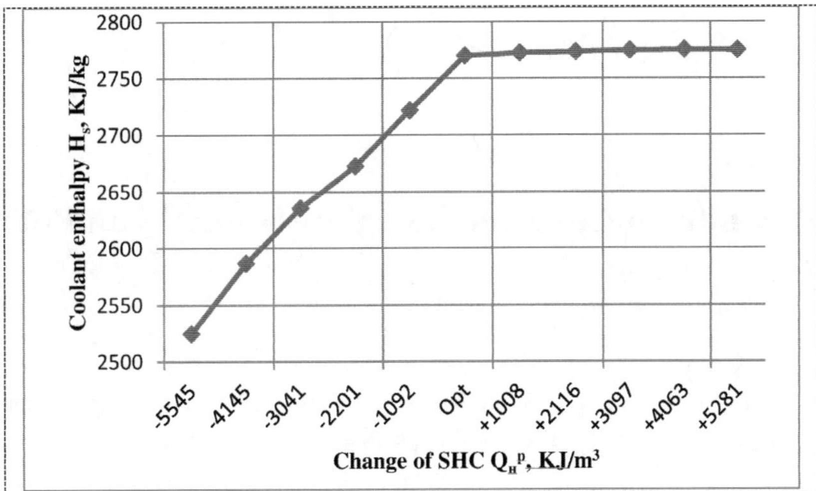

Figure 2. The effect of SHC change on the coolant enthalpy

3. Conclusion

Uncontrolled changes in fuel composition lead to instability of the thermal effect of the combustion process, which adversely affects the thermal performance of power plants.

Reducing the SHC leads to a sharp decrease in the enthalpy of the coolant at the outlet because of the excess air coefficient increasing. With increasing fuel SHC the coolant enthalpy is not substantially changed due to the growing underburning value.

The work is performed according to the Russian Government Program of Competitive Growth of Kazan Federal University.

References

[1] Lisienko V G, SHHelokov Ya M, Ladygichev M G 2004 *Russia B. Toplivo. Ratsional'noe szhiganie, upravlenie i tekhnologicheskoe ispol'zovanie* 608.

[2] Saifullin E R, Vankov Yu V 2015 *IOP Conf. Ser.: Materials Science and Engineering* **86** 012006

[3] Saifullin, E R, Larionov V M, Busarov A V, Busarov V V 2016 *Journal of Physics: Conference Series* **669 (1)** 012037

[4] Saifullin, E R, Larionov V M, Busarov A V, Busarov V V 2016 *Journal of Physics: Conference Series* **669 (1)** 012043

[5] Knorre G F, Aref'ev K M, Blokh A G *1966 Russia B. Theory of combustor processes* 1042

[6] Izdatel'stvo NPO TSKTI 1998 *Russia B. Teplovoj raschet kotlov(Normativnyj metod)* 256

Ion-plasma nitriding as a method of instruments and parts durability

A.D. Samigullin, A.T. Galiakbarov, R.T. Galiakbarov, A.R. Samigullina

Kazan Federal University, 423800, 68/19 Mira avenue, Naberezhnye Chelny, Russia

E-mail: samigullin86@mail.ru

Abstract: Improvement of the machines, parts, devices reliability as well as improvement of their quality and operation are topics of interest at the present time. Solution to these problems is related to hardening of the product surface layers in the first place. This article deals with parameters of nitriding process using the example of 38XM steel which is applied in essential parts of turbine installations and compressors operating at temperatures up to 400°C. The article also provides the results of nitriding at different modes.

1. Introduction:

Physical entity of ion-plasma nitriding processes in gas environment is identical. However quality and quantity characteristics of particular phases of these processes have their differences which affect the nitrided case development structure and kinetics. Besides the nitrided machine parts obtain increased corrosion resistance in damp atmosphere and other atmospheres with low corrosive power. Their ability to maintain high hardness and durability at high operating temperatures are also significant. Main technological advantage of nitriding is low temperature of process (400-500 degrees) which is the reason of insignificant volumetric changes and deformation of machined parts and instruments. [2]

2. Nitriding stages:

The process of ion-plasma nitriding includes 4 stages: cleaning, heating, aeration and cooling.

After vacuum degassing of the chamber and filling it with working gas to 0,5-0,8 mbar cathode cleaning of parts begins during which discharge rate is permanent and makes 10-20% of its maximum value, coefficient of electrical pulses charge is 50% and pressure is permanent within 0,5-0,8 mbar. This stage is used for initial heating and additional cleaning of parts. Low current and low filling ratio prevent the appearance of electric arcs at cold and contaminated parts. Cleaning stage takes 5-20 minutes.

Cleaning leads to heating which consists of 6 steps. Heating rate, target temperature, pressure, frequency, filling ratio and composition of working gas are specified for each step. The currency is specified by temperature regulator. Heating process of parts leads to step-by-step (graded) pressure and filling ratio increase. During each step they may be persistent or may increase with temperature raise. At the last step of this stage the parts are heated to the defined nitriding temperature and its final parameters – temperature, pressure, filling ratio and gas ratio usually correspond to operating parameters of saturation mode. When nitrogen and hydrogen are used as working gas mixture their proportion may be changed from 1:1 to 1:10. When specified nitriding temperature is reached the isothermic saturation process begins. Temperature, filling ratio, pressure and gas ratio are permanent. Their values are 480-500°C, 80%, 2,0-8,0 mbar and 3:1 – 6:1 accordingly.

Nitriding mode takes 5-20 hours depending on design-engineering requirements and performance characteristics of hardened products.

At the end of isothermic saturation parts may be cooled down in discharged atmosphere of working and inert gas. If necessary slow cooling may be used under conditions of plasma burning with decrease of discharge power as well as atmosphere pressure of working gases. When temperature 150°C is reached the chamber is opened and the parts are cooled down.

3. Nitriding process:

Composition and properties of hardened layer of nitrided parts depend on the following technological factors: current between electrodes, composition of gas atmosphere, degree of its exhaustion, operating temperature, process duration, relative position of parts and electrodes.

Nitriding temperature is usually 470-580°C, current 400-1000 V, exhaustion 1-10 mm of mercury. Operating pressure is limited by properties of glow discharge. At the pressure lower than 1 mm of mercury the energy of ions is insufficient for heating of machined part to operating temperature, at the pressure higher that 10 mm of mercury the discharge consistency is interrupted, glow discharge develops into arc discharge accompanied by development of melted microcraters on the surface. At the pressure 2 mm of mercury and temperature 520 °C and 6 mm of mercury and 620°C accordingly the maximum depth of diffused layer is reached. Increase of current also affects the depth of layer.

Change of current intensity within a broad range (0,5-20 mA/cm2) does not affect the nitriding process.

Ammonia, nitrogen and mixture of nitrogen and hydrogen are used as nitrogen containing gases. Nitrogen plasma does not allow the presence of oxygen because it may decrease activity of operating atmosphere, hydrogen insignificantly affects the growth of layer.

4. Impact of nitriding parameters:

Besides typical factors for all nitriding methods such as duration of treatment and temperature other factors may affect the nitrided layer, for example presence of gas in plasma, width of glowing edge etc.

Data provided about reactions of plasma nitriding shows that temperature does not affect the diffusion mechanism. Nitrogen content at the temperature of 400°C depends on the same parameters as at the temperature of 500 or 600°C. This shows that the temperature range is very large and limited only by physicochemical processes within the metal. Thus, for example, although nitrogen diffusion at the temperature lower than 350°C is very weak it allows to obtain the nitrided layers of shallow thickness.

Plasma nitriding at low temperatures. Nitriding in plasma of glow discharge within the range of temperatures 400-500°C corresponds to modern engineering development providing maintenance of high strength of machined parts core [4].

Figure 1 using 38XM steel as an example shows comparison of test piece strength changing curves after plasma nitriding at the temperature of 450°C and 570°C. After 46 hrs of processing at the temperature of 450°C (curve 5) the depth of nitriding is almost same as after 24 hrs of processing at the temperature of 570°C although the character of their hardness change is different.

In case of processing at temperature of 450°C the hardness of surface is almost 150 HV higher. If for example the hardness of surface HV2 and plasma nitriding temperature relation is analyzed (Figure 2) it shows that (under condition of equal thickness nitrided layer) if the temperature of nitriding decreases the surface layer hardness increases.

 IOP Publishing
doi:10.1088/1742-6596/789/1/012046

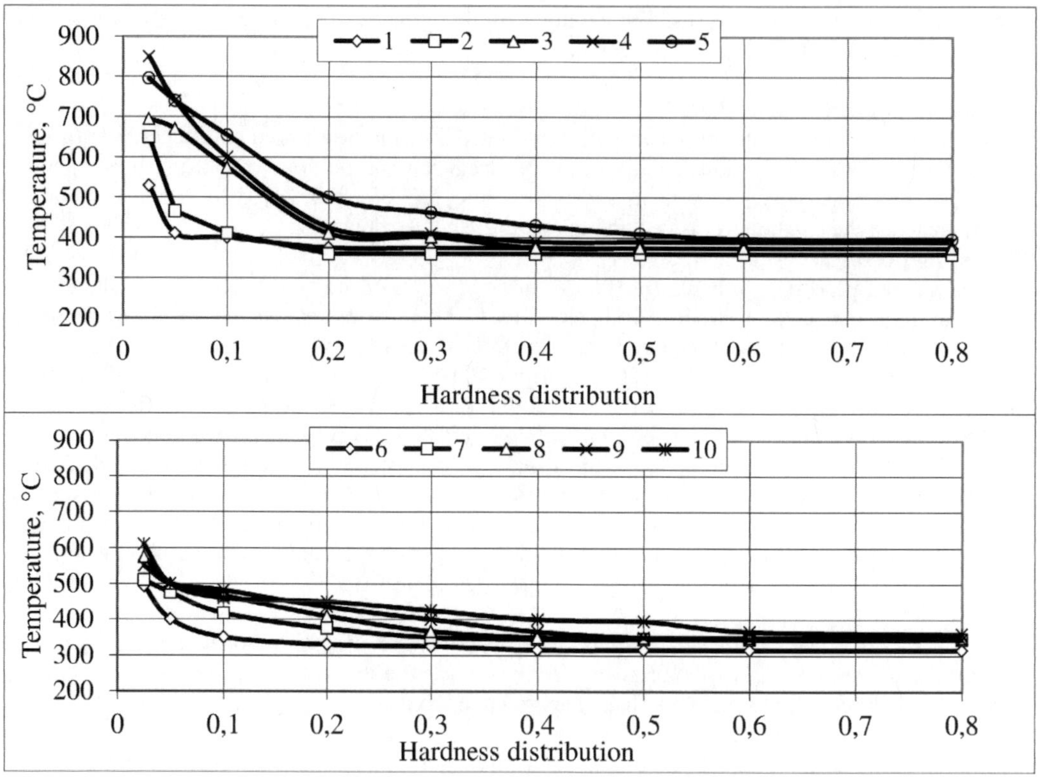

Figure 1. Influence of nitriding temperature on the value and distribution of hardness (steel 38XM). Plasma nitriding at the temperature of 450°C (1-5) and 570°C (6-10) within: 1 – 1 hr 20 min. 2 – 2 hrs., 3 – 9 hrs, 4 – 22 hrs. 5 – 46 hrs, 6 – 5 min, 7 – 30 min, 8 – 2 hrs 15 min, 9 – 8 hrs, 10 – 24 hrs.

The reason of the higher hardness is fine distributed complex nitrides appearing in perlite at low temperatures at a great amount.[5]

Durability of parts nitrided this way in the result of sliding friction, especially dry sliding, is shown in Figure 3. Degree of wear was defined as relation of mass loss and number of rotations.

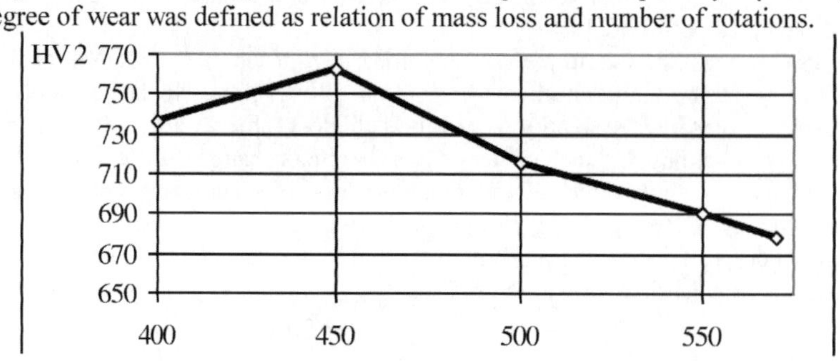

Figure 2. Hardness and temperature relation at equal nitriding depth (method of treatment – plasma nitriding, material steel 38XM).

Nitrided layers obtained at the temperatures of 450, 530 and 580 °C in the result of nitriding length change had the diffusion layer with thickness ~ 0,2 mm and nitride γ-layer with thickness 4-6 micron. Results show that lower wear of test pieces corresponds to lower temperatures of nitriding.

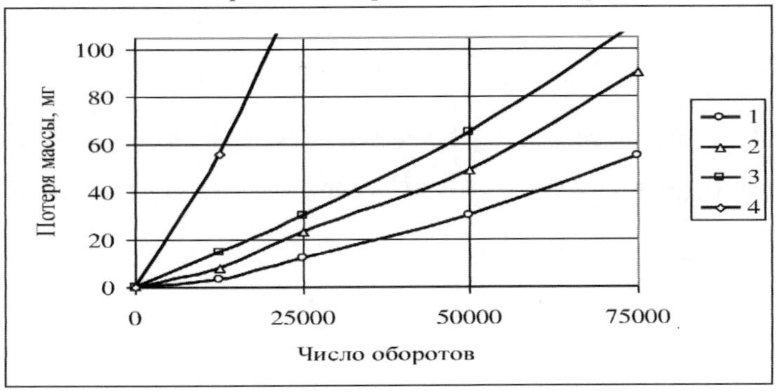

Figure. 3. Wear of steel 38XM after plasma nitriding at different temperatures (F = 196,2 H, vrel= 1,43 m/s, emergency operation test):1, 2, 3 – plasma nitriding at 450°C, 30 hrs; 530°C, 9 hrs and 580°C, 1 hr; 4 – improvement to σв = 1200 N/mm2.

5. Conclusion:

The research shows that nitriding temperature does not affect the depth of appearing hard layer but at the decrease in processing temperature the growth of surface layer hardness is observed. Structural analysis shows than file distributed complex nitrides having higher hardness appear in perlite at lower nitriding temperature. This in its turn affects the part durability which is increased.

References

[1] Arzamasov BN Bratukhin AG Eliseev Yu, Panayotov TA Ionic chemical heat treatment of alloys in gas media. - M: *Izd.MGTU them. NE Bauman*, 1999. - 400 p.

[2] Research on ion-plasma nitriding of parts. Israfilov IH, Zvezdin VV, Israfilov DI, Chernova MA *The low-temperature plasma in the deposition of functional coatings processes. 2015.* T. 1. C. 6. number 211-216.

[3] Research on ion-plasma nitriding of tool steels. Israfilov IH, Bashmakov DA Zvezdin VV, Fairuz RS, Nugumanova AI In: *Design and study of technical systems interuniversity scientific collection. The Ministry of Education and Science of the Russian Federation, the GOU VPO "Kama State Engineering and Economic Academy".* Naberezhnye Chelny, 2007, pp 100-106.

[4] Impulse plasma surface thermohardening of manufacturing engineering parts / Gabdrakhmanov A T, Israphilov I H, Galiakbarov A T, Bashmakov D A and Samigullin A D // *International scientific and technical conference proceedings «Innovative engineering technologies, equipment and materials - 2013» (Interbranch scientific and technical complex «IMTOM-2013») and «Improving competitiveness and energy efficiency of machine builders under conditions of WTO» forum.* Part 2. - Kazan, 2013. - Pages 44-48

[5] Pulse plasma surface thermostrengthening of machine parts /Israfilov, I.H., Galiakbarov, A.T., Bashmakov, D.A., Gabdrakhmanov, A.T., Samigullin, A.D. // *IOP Conf. Ser.: Mater. Sci. Eng.*, Volume 69, Issue 1, 2014, Article number 012037

[6] Lahtin JM, Cohen JD, D. Spies, H. Boehmer Theory and nitriding technology. - M .: Engineering 1991.

Localized thermal cleaning method for pumping and compression pipes from asphalt, resin and paraffin deposits using plasma

A.D. Samigullin, A.R. Samigullina, Al. D. Samigullin, A.T. Gabdrakhmanov

Kazan Federal University, 423800, 68/19 Mira avenue, Naberezhnye Chelny, Russia

samigullin86@mail.ru

Abstract: The article deals with temperature calculation results at the inner surface of oil-well tubing and side-wall melting of asphalt, resin and paraffin deposits.

1. Introduction:

Cleaning of pipes from dismounted main oil pipeline is not a simple task however reuse of such pipes after thorough cleaning of surface and application of insulating coating is a promising direction.

It is a known fact that during operation the hard asphalt, resin and paraffin deposits are built on the steel pipes surfaces.

To clean these pipes from oil residues the abrasive-spray, solvent washout method or treatment by overheated steam with following mechanical cleaning are applied as a rule. Other unconventional methods may be used as well. All methods are quite effective but energy and time-consuming.

To reduce the energy and time consumption of the process the localized thermal method of cleaning of pumping and compression pipes from asphalt, resin and paraffin deposits was suggested.

Method. The localized thermal method of cleaning of pumping and compression pipes from asphalt, resin and paraffin deposits is based not on the complete melting of asphalt, resin and paraffin deposits but only of the side-wall layer. This is achieved by heating of the outer surface of the cleaned pipe using the circular plasmatron; due to high-thermal conductivity of pumping and compression pipes material the heat flow leads to softening and melting of the side-wall layer adjunct to heated area of the cleaned pipe.

Gradual heating of the cleaned pipe starting from the lower part leads to consequent melting of the side-wall layer of asphalt, resin and paraffin deposits along the cleaned pipe; as a result of that free outflow of built gases and melting products of asphalt, resin and paraffin deposits is provided without causing explosion during gas formation and temperature expansion of side-wall layer of asphalt, resin and paraffin deposits in closed space.

Main mass of asphalt, resin and paraffin deposits in the center of heated area of the cleaned pipe does not melt due to low conductivity of asphalt, resin and paraffin deposits.

Melted side-wall layer of asphalt, resin and paraffin deposits can fulfill a function of lubricant at the outflow of hard mass of asphalt, resin and paraffin deposits from cleaned pipe under gravity.

After outflow of hard mass of asphalt, resin and paraffin deposits from cleaned pipe the heating of the cleaned pipe by several actuations of circular plasmatron is continued until all melted residues of asphalt, resin and paraffin deposits flow down to receiving container.

2. Calculation:

Important factor for removal process of asphalt, resin and paraffin deposits is melting temperature which allows to estimate the mobility of asphalt, resin and paraffin deposits and defined in the first place by chemical structure of asphalt, resin and paraffin deposits. The higher the melting temperature of asphalt, resin and paraffin deposits the higher the content of high-molecular and high-melting compounds in it, n-paraffines in a first place, the more difficult these residues can be removed. Chart 1 shows the physicochemical characteristics of the researched asphalt, resin and paraffin deposits.

Chart 1. The physicochemical characteristics of the researched asphalt, resin and paraffin deposits (ARPD)

№	Key figures	ARPD 1	ARPD 2	ARPD 3	ARPD 4	ARPD 5	ARPD 6
		Refinery ARPD			Field ARPD		
1	Density, g/cm3	0,94	1,0	1,3	1,0	0,9	0,81
2	Mechanical impurities content, mass %	6,97	49,9	44,5	37,34	12,06	18,53
3	Water content, %	5	2,6	1,4	-	0,3	4
4	Melting temperature, °C	44	57	53,4	75	70	66
5	Sulfur content, mass %	0,3	1,7	1,5	0,4	0,2 0,1	0,1
6	Boiling temperature of ARPD, °C	≈550	≈550	≈550	≈550	≈550	≈550

As Chart 1 shows the melting temperature depending on the type of asphalt, resin and paraffin deposits varies. Based on these data the decision was taken to calculate the temperature at the inner surface of the cleaned pipe.

The calculation was carried out using Star CCM+ software package. The calculation requires the development of the precise detailed computer model of the research object. Computer model of the research object was created in COMPAC 3D program. Then the computer model was imported to Star CCM+ program, volume mesh of finite elements was created. The further action was correct definition of initial and design parameters. The accuracy of such calculations depends on parameters of finite element mesh parameters. The volume mesh of finite elements is shown in Figure 1.

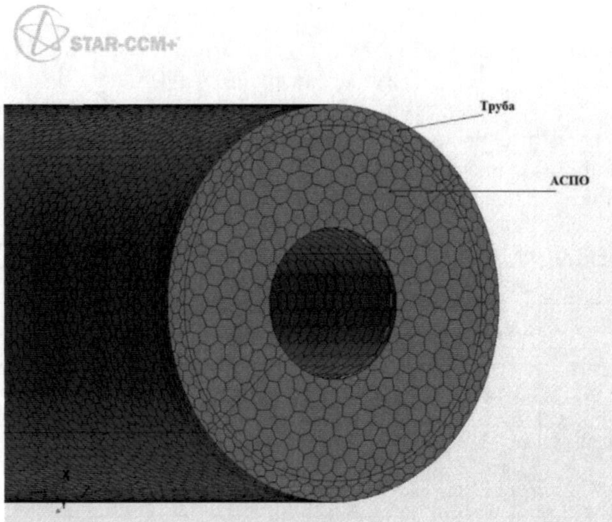

Figure 1. Computer model of research object in volume mesh of finite elements.

For computer modelling the steel 20 pipe with diameter 70 mm and wall thickness 4 mm was used. Physicomechanical properties of asphalt, resin and paraffin deposits in the computer model: density 1000.0 kg/m3, heat conductivity – 0,26 W/m*K, specific heat capacity – 2200.0 J/kg*K.

After computer modeling the temperature curve was obtained according to time in the beginning of pipe at the inner surface after thermal source removal i.e. circular plasmatron 2. Temperature at the inner surface of the pipe after removal of the thermal source will increase due to temperature conductivity of the pipe material before the temperature equalization throughout the pipe.

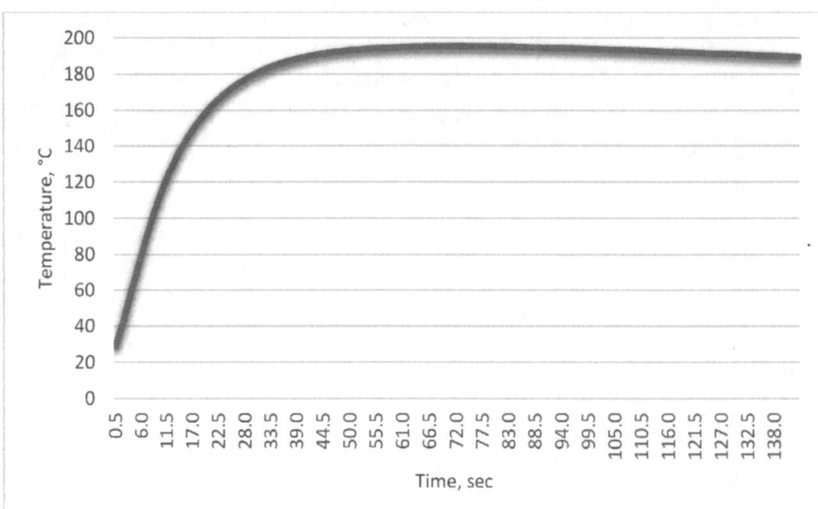

Figure 2. Temperature variation according the time in the beginning of pipe at the inner surface.

Melting temperature of asphalt, resin and paraffin deposits according to Chart 1 is not higher than 75°C and the boiling temperature ≈550°C. Figure 2 shows that temperature of the inner pipe surface does not exceed 200°C which is acceptable for this process of oil-well tubes cleaning from asphalt, resin and paraffin deposits [7].

3. Conclusion

Results of conducted research show that the temperature at the pipe inner surface is enough to melt the thin side-wall layer of asphalt, resin and paraffin deposits. This method of thermal cleaning allows to effectively clean pumping and compression pipes from asphalt, resin and paraffin deposits. Also this method is more energy-efficient due to application of low power plasmatron and low cost equipment.

References

[1] Samigullin A. D., Galiakbarov A. T., Galiakbarov R. T., Study of the petroleum schedules thermal cleaning process from asphalt, ressin and paraffin deposits using low-temperature plasma//*VII Conference on low temperature plasma in the processes of functional coating preparation*. - 2016. - Vol.669, Is. - Art. №12017.

[2] Ivanova LV The study of the asphaltene deposits of different nature and the way of their use / Ivanova LV1, V.N.2 Koshelev, OA Stokolos // *Electronic scientific journal "Oil and Gas Business"*, 2011, №2 http://www.ogbus.ru

[3] Russian Federation Patent № 2437726, IPC (2006.01) B08B9/00. Priority from April, 14 2010. Published on December, 27 2011. Patent description

[4] Impulse plasma surface thermohardening of manufacturing engineering parts / Gabdrakhmanov A T, Israphilov I H, Galiakbarov A T, Bashmakov D A and Samigullin A D // *International scientific and technical conference proceedings «Innovative engineering technologies, equipment and materials - 2013» (Interbranch scientific and technical complex «IMTOM-2013») and «Improving competitiveness and energy efficiency of machine builders under conditions of WTO» forum.* Part 2. - Kazan, 2013. - Pages 44-48

[5] Pulse plasma surface thermostrengthening of machine parts /Israfilov, I.H., Galiakbarov, A.T., Bashmakov, D.A., Gabdrakhmanov, A.T., Samigullin, A.D. // *IOP Conf. Ser.: Mater. Sci. Eng.*, Volume 69, Issue 1, 2014, Article number 012037

[6] Russian Federation Patent № 161034, IPC (2006.01) МПК B08B 9/02. Priority from August, 17 2015. Published on April, 10 2016. Patent description

[7] Samigullin AD. Improving the thermal treatment process of OCTG from asfaltosmolo-paraffin deposits / AD Samigullin, AT Galiakbarov. *International scientific-practical conference "Information technologies. Automation. Actualization and problem-solving training of highly qualified personnel (PAI-2015) ", 2015 .: Proceedings / red.kol. Simon LA [and etc.]; under. Ed. Simonova LA, Savitsky SK* - Naberezhnye Chelny: Publishing house Naberezhnochelninsky Institute (branch) FGAOU VPO "Kazan (Volga) Federal University", 2015. - S.54-59.

Hybrid welding of dissimilar metals

A D Samigullin[1], D A Bashmakov[1], I Kh Israphilov[1], G A Turichin[2]

[1]Kazan Federal University Naberezhnye Chelny Institute,
[2]Peter the Great St. Petersburg Polytechnic University

E-mail: amigullin86@mail.ru

Abstract: The article addresses issues laser - plasma welding (LPW) dissimilar metals and the results of metallographic studies of the microstructure of welds ferrite - 40 steel and molybdenum - steel 40. Increasing potential opportunities the high-energy processing is carried out by integration the laser radiation (LR) and plasma, which allows you to create the desired spatial distribution of the energy flow for technological processes (TP) of laser-plasma heat treatment (LPT) of metals. The distribution of the thermal field is determined by the density distribution of energy flow LR and plasma exposure time, and the thermal characteristics of the treated metal. The most interesting is the treatment of details with ring flow of plasma and LR axial impact.

1 Introduction:

In order to ensure a high quality of performance of the welded joint in the manufacture of critical parts is used LPW. Application of LPW allows obtaining permanent connections in inert gas at atmospheric pressure. In contrast to laser welding, LPW can significantly increase the energy characteristics of technological installation by increasing the absorption coefficient of LR metals by preheating the details by plasma.

Another problem of laser welding is the residual stresses in the weld zone. Therefore, to reduce the residual stresses use thermal tempering of details. This requires additional time and energy expenditure.

This problem is solved by the use of LR and ring plasma torch. Linear annular plasmatron generates a ring-shaped plasma on the surface of the details, where the leading edge provides a temperature rise in the target area LR on the metal, and the rear zone heats the weld to a tempering temperature.

The most perspective way to improve the laser-plasma technological complex (LPTK) is an integrated approach that includes the creation of a high-energy heat treatment of CAD technology; synthesis of automatic control system (ACS) LPTK with feedback on the process parameters in the zone of interaction of plasma and LR with metals and the development of new processing methods with information parameters, allowing the creation of databases, in order to optimize the structure LPTK.

The urgency of the task of joining dissimilar materials explained by the difference the functional purpose of separate parts of details. For example, the gyroscope consists of a rotor core and shaft. axis function is to provide mechanical strength of the structure when the rotor rotates. Function of rotor Core is to provide high-permeability magnetic fields. Combination of this functions in homogenous materials increases the dimensions and weight of the rotor performance at a given mechanical and electromagnetic characteristics.

An important element of the TP manufacture is to ensure that the rotor the mechanical permanently attached connection with the magnetic core axis. This problem can be solved by applying laser welding to the steel ferrite core axis with specified mechanical properties.
It should be noted that the welding process in the manufacture of precision parts is responsible for the geometry and prevents the formation of defects in the structure of the weld seams (shells, poor fusion, burning, etc.), while required to provide minimum mechanical residual stress in the weld zone, that are may lead to its destruction [1]
Fig. 1 shows a linear plasmatron to generate thermal field at the weld. This plasmatron consists of two electrodes 1 and 2 formed in a ring shape with nozzles 3 and 3 ', 4 and 4', which are designed for supplying and removing coolant. They are also the current-supplying ends of the electrodes. The electrode spacing 5 is supplied plasma gas that directed to welded items 6 (Pat Russia 59931). The speed of the arc at a direct current of 100 - 600 A reaches 10 - 60 m / s (determined by means of high-speed motion picture).

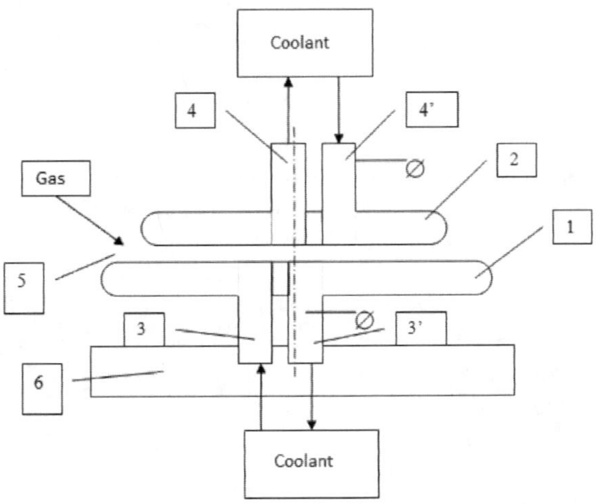

Fig. 1. The annular plasmatron.

LR is served in the center of the ring electrode linear plasmatron. Temperature field distribution is shown in Fig. 2.

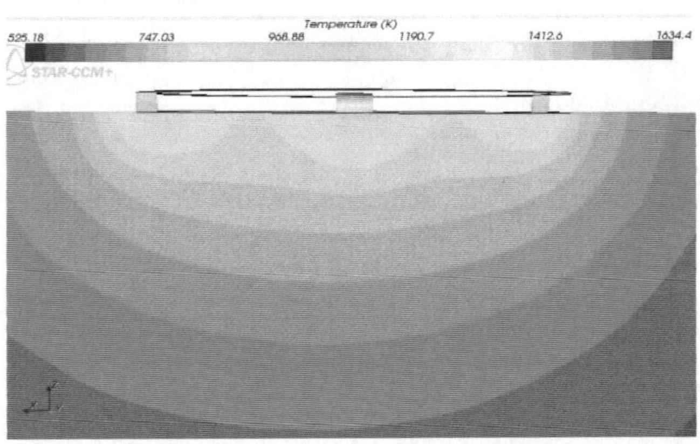

Fig 2. The distribution of the temperature field along the weld at the impact of LR and the annular plasma of plasma torch.

One of the essential steps of the laser welding process is the pre-treatment of surfaces to be welded. The surface of the metal in the weld zone cleaned of scale, rust and other contaminants as well as moisture. Dirt and moisture conditions are created for the formation of porosity, oxide inclusions, and in some cases, and cold cracking in the weld metal and heat affected zone due to saturation with hydrogen. After cleaning the surfaces of parts degreased.
Because thermal effect zone of LR in the metal has segment shape whose dimensions depend on the parameters of the energy source and the physical properties of metals [2], it is appropriate weld plane tilted at an angle, tangentially to this segment. In this case, by melting the refractory metal is melted the fusible metal by heat transfer. Thus, there is an interaction between the active metals with the surrounding gas medium, which is present in the junction area of the two metals due to surface roughness. When crystallization of metal occurs formation of voids in the metal, which impairs the quality of the weld. It is therefore necessary to minimize the roughness of the mating surfaces by mechanical lapping.
In order to assess the depth of penetration can be used formulas [1]:

$$h = \frac{P}{2\pi\lambda_{т} T_{к}} \ln \frac{r_{л} + a/\upsilon_{св}}{r_{л}},$$

Where: λт – material thermal conductivity; Tк – boiling temperature; a – thermal coefficient material.
Zone of thermal effects is formed by LPW is composed of several sections (fig. 3). At the bottom of the picture there is a ferrite microstructure, at the top - a modified steel structure. Between them is visible the zone of interaction of two dissimilar metals, characterized by the diffusion processes taking place in the transition zone. LR direction on.

Fig. 3. A micrograph of ferrite Welding & Steel 40, x500

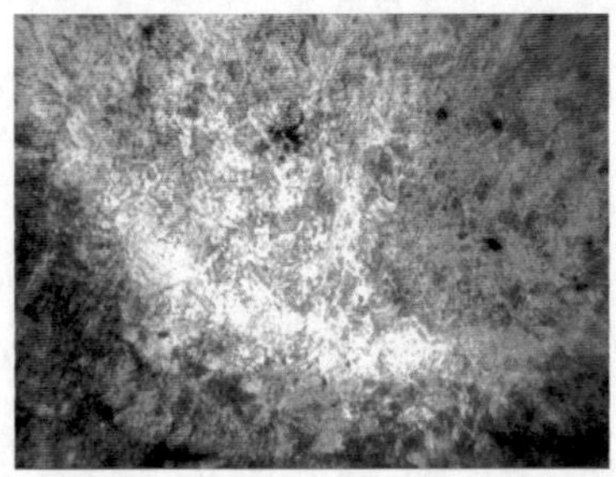

Fig. 4. A micrograph of a weld molybdenum and steel 40.; x500

Another pattern was observed in the analysis of photomicrographs of molybdenum welding (Tm. = 2620 ° C) and steel (Tm. = 1510 ° C). At the upper portion (Fig. 4, 5) molybdenum is visible region with a uniform structure throughout the thickness [6].

Fig. 5. Photo weldment of molybdenum and steel 40.

2. Conclusion:

The main influence on the process of welding of dissimilar materials have physical and chemical properties of both the material and the environment in which the TP. This is due to the distribution of thermal field in a variety of materials characterized by varying reactivity and thermal properties (thermal conductivity, heat capacity, melting point, crystallization temperature-tion, etc.). To achieve the required quality of the TP welding dissimilar materials is need to provide welding in neutral gas atmosphere (e.g., argon), and, the geometric shape of interface on the weld should be determined by the shape of the temperature distribution in the boundaries of the refractory material, and its value at the junction with the low-melting equal to the melting point of the fusible.

References

[1] Pat. Russia № 2415739. Method of laser welding of components made of dissimilar metals / Zvezdin V.V., Israphilov I.Kh., Veliev D.E. 06.16.2009.

[2] Pat. Russia № 59931. Plasmotron / Israphilov I.Kh., Israphilov Z.Kh., Israphilov D.I., galiakbarov A.T., 27.12.2006.

[3] Grigor'yants A.G., Shiganov I.N., Misyurov A.I. Technological laser machining pro cesses: Proc. The manual for high schools / Ed. A.G. Grigoryants. - M.: *Publishing House of the MSTU. Bauman*, 2006. - 664 p.

[4] Improving the quality of welded joints of knots and car of details in laser welding. Zvezdin V.V., Zamorskiy V.V., Karimov V.S., Sabirov I.S. *"Problems of quality and operation of vehicles." Materials III International scientific and technical confer ence. Part 1.* - Penza: PGUAS, 2004. pp. 54-58.

[5] Zvezdin V.V., Khamadeev A.V., Portnov S.M. Investigation of laser welding process, molybdenum and steel 45 / *Information and socio-economic aspects of modern tech nology: Online Electronic Science and Technology Journal,* 2008. - № 4.

[6] Pulse plasma surface thermostrengthening of machine parts /Israfilov, I.H., Galiakba rov, A.T., Bashmakov, D.A., Gabdrakhmanov, A.T., Samigullin, A.D. // *IOP Conf. Ser.: Mater. Sci. Eng.*, Volume 69, Issue 1, 2014, Article number 012037

IOP Conf. Series: Journal of Physics: Conf. Series **789** (2017) 012049 doi:10.1088/1742-6596/789/1/012049

Acoustic gas oscillations in coaxial tubes

E V Semenova[1], V M Larionov[1], O V Iovleva[1]
[1]Kazan Federal University, 18 Kremlyovskaya str., Kazan 420008, Russian Federation

E-mail: evgeniya.yallina@gmail.com

Abstract. Pulsating combustion is one of the possible solutions to improve energy efficiency of combustors that use hydrocarbon fuels. In this paper analysis of gas oscillations in coaxial tubes is provided. An influence of geometric parameters of the combustion chamber and the resonance tube on the installation frequency is considered.

1. Introduction

It is known [1-4] that the variation of gas significantly accelerate the process of burning fuels in power plants, for example, in the combustors of Rijke type tube. In the works [5-8] developed a mathematical model of pulsating combustion of solid fuel and method of calculation of acoustic vibrations in installations of the type under consideration. One of these installations is shown in Fig.1. Its peculiarity is that it consists of two resonators that have their own spectrum of frequencies of oscillations.

The purpose of this study is to analyze the influence of the natural frequencies of the resonators and the extent of their interaction on the frequency of oscillations of gas in the system of coaxially located tubes.

2. Frequency equation

In the studied setting are considered four sites. The first part of the combustion chamber befor the lattice that holds the fuel. The second section – the area of heat dissipation, which are located between the grill and the section of the combustion chamber, which introduced the organ pipe. The third part – annular channel between the tubes, which is fed secondary air. The fourth phase resonance pipe that moves the hot gas. In the acoustic analysis it was obtained the equation of frequency of oscillations of gas in the considered setting, which has the following form:

$$\frac{S_c}{Y_1}+\frac{S_3}{Y_3}=\frac{S_r}{Y_r} \qquad (1)$$

The imaginary parts of the acoustic impedances of the selected areas defined by the relations:

$$Y_1=\rho_{1,0}c_1 tg(\omega x^*/c_1+\varphi_1),\quad Y_3=\rho_{3,0}c_3 tg(\omega l_3/c_3+\varphi_3),$$

$$Y_r=\rho_{4,0}c_{4,0} tg\left(\frac{b_4}{2\omega}+\beta_4 tg\varphi_4\right) \qquad (2)$$

in which $\varphi_1=\frac{\omega\Delta l_c}{c_1}$, $l_c=\frac{\omega\Delta l_3}{c_1}$, $\Delta l_c=0.61R_c$, $\varphi_3=\frac{\omega\Delta l_3}{c_1}$, $\Delta l_3=0.61R_3$, $R_3=\sqrt{R_c^2-R_r^2}$,

IOP Conf. Series: Journal of Physics: Conf. Series **789** (2017) 012049 doi:10.1088/1742-6596/789/1/012049

$$l_r^* = l_r + 0.61R_r, \tag{3}$$

where ω - the cyclic frequency ρ_0 - the average density of the gas, c_1 - the velocity of sound in air, $\rho_{4,0}$ and $c_{4,0}$ - average gas density and velocity of sound at the inlet to the resonance tube, a_4 and b_4 - distribution coefficients of sound velocity in the resonance tube: $c_4 = a_4 - b_4 x_4$, $S = \pi R^2$, subscript «*c* corresponds to the combustion chamber, the index «*r*» - resonance tube.

Equation (1) and the relation (2), (3) make it possible to calculate the frequency of oscillations of gas in the test installation with arbitrary geometrical parameters.

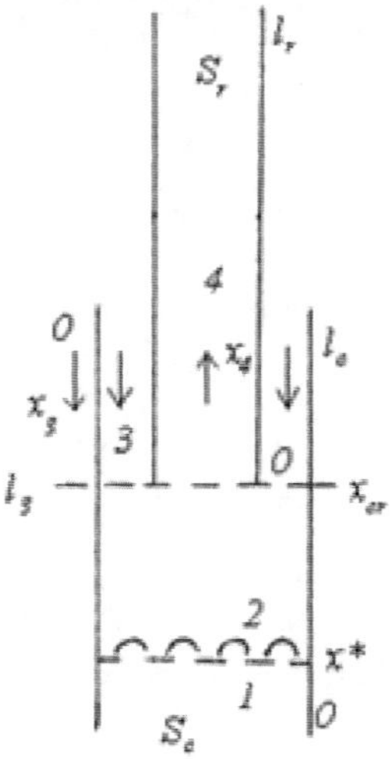

Figure1. Scheme study unit.
1, 2 - cold and hot part of the combustion chamber; 3 - an annular channel, 4 - resonance tube.

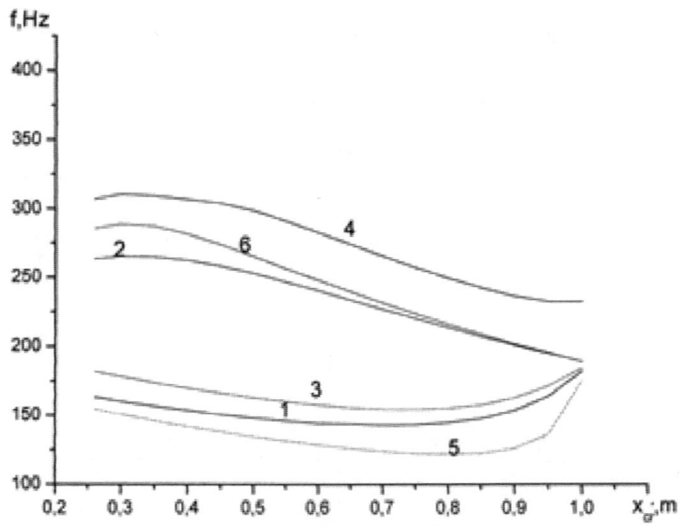

Figure 2. The frequencies of oscillations of the gas in the resonance tube, depending on the position and the length. 1,2 - l_r=0.95m; 3, 4 - l_r=0.75m; 5, 6 - l_r=1.15м

Figure 2 shows the results of the calculation of the frequency oscillations of the gas pipe - Combustion chamber length of 1 m and an inner diameter of 55mm. These values and the calculation of the conditions were the same as in [1, 5], which investigated pulsating combustion pellet in the tube, open at both ends. Resonance tube had an inner diameter of 40 mm. Calculations have shown that in the resonance tube of length 0.95m, open at both ends, the frequency of oscillation is the same gas as in the pipe - the combustion chamber when the resonance tube is fully withdrawn and the fuel combustion takes place at a distance $x^*= l/4$ from lower end. This value is equal to 186 Hz.

Calculations were performed for the pipe - the combustion chamber of the same length - 1 m, but larger, having an inner diameter - 0.1m. The resonance tube has a diameter of 0.07m. Calculations have shown that in this case the natural frequencies of the pipe - Combustion chamber and the resonance tube - 175 Hz when lr = 1.01m.

3. Conclusion

It were obtained the following main results. One of the frequencies in the system is always below the gas oscillation, and the other is higher than oscillation frequency in the tubes when its interaction is absent. Moving of the resonance tube within combustion chamber increases their acoustic interaction and extends the interval between the frequencies of the system.

The work is performed according to the Russian Government Program of Competitive Growth of Kazan Federal University.

References

[1] Ermakov R A, Sadykov A F, Galiullin R G, Larionov V M 2011 *Chemical and Petroleum Engineering* **46** pp 749-751

[2] Ermakov R A, Galiullin R G, Larionov V M, Nikolaev A N 2008 *Chemical and Petroleum Engineering* **44** pp 387-392

[3] Larionov V M, Saifullin E R, Semenova E V 2016 *Journal of Physics: Conference Series* **669** (1) 012047

[4] Larionov V M, Sadikov K G, Mitrofanov G A 2016 *Journal of Physics: Conference Series* **669** (1) 012042

[5] Iovleva O V, Larionov V M, Semenova E V 2016 *Journal of Physics: Conference Series* **669** (1) 012023

[6] Larionov V M, Nazarenko T I 2000 *Izvestiya Vysshikh Uchebnykh Zavedenij. Aviatsionnaya Tekhnika* (4) 69

[7] Iovleva O V, Larionov V M 2012 *Russian Aeronautics* **55** pp 388–391

[8] Yallina E V, Larionov V M, Iovleva O V 2013 *Journal of Physics: Conference Series* **479** (1) 012017

Determination of steel bar dispersed mass in electric discharge with alternative electrode

G Yu Shakirova[1], Yu I Shakirov, V I Ilyin, R A Valiev and L N Drogaylova

Kazan Federal University Naberezhnye Chelny Institute, 68/19 Mira Street, Naberezhnye Chelny, 423810, Russian Federation

E-mail: guzya81@mail.ru

Abstract. The mathematical model of plane problem of metal bar dispersion in electric discharge with liquid electrolyte is suggested in this research. The analogy with the plane problem of the theory of jets in an ideal fluid is used to solve the task. It actually means determination of analytic function in the field with one area of unknown boundary. The formula for determination of dispersed metal powder mass assuming the bar axial symmetry has been calculated.

1. Introduction

Powder developers have been more and more successfully used for the production of heavy duty details over recent years. Composite powders are produced by synthesis under the conditions of technological burning followed by pulverization and classification due to the fractions. Along with the traditional methods of powder production, electric discharge dispersion of metals and alloy materials with the usage of liquids is being applied more and more [1-5].

The mathematical model suggested in this work allows to evaluate the influence of the task parameters on the eventual result using at least qualitative and partially quantitative data.

2. Experiment

Figure 1 shows the scheme of a powder developer synthesis unit that consists of a constant power supply, an electrolytic bath with an electrolyte and a steel conductive electrode with different carbon content (0.2 % - 1.0 %).

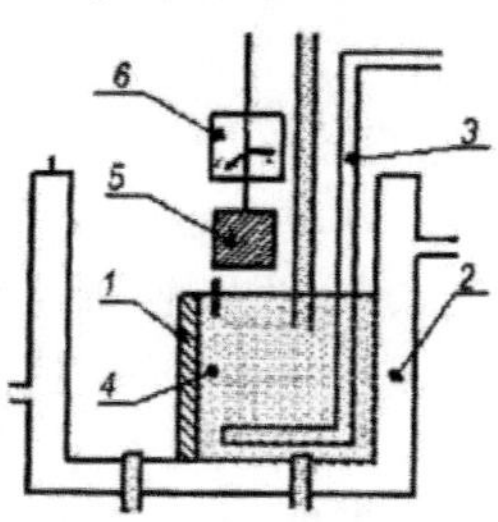

Figure 1. Scheme of a powder developer synthesis unit: 1 - electrolytic bath, 2 - water-cooling jacket, 3 – bubbler, 4 – electrolyte, 5 – steel electrode, 6-co-ordination unit.

[1] To whom any correspondence should be addressed.

The results of the trials performed have shown big dependency of powder quantity and quality (the size of granules) on many geometrical and physical parameters [1, 2].

3. Results and their discussion

To solve the task first of all it should be mentioned that in the process of powder production the dependency of η current efficiency on anode current density j_a (figure 2) is identical to the analogical dependency $\eta(j_a)$ characteristic of electrolytic processing of metals [6, 7]. Therefore, we consider that the speed of metal removal V_m from the anode surface per unit weight is calculated similarly to Faraday's law $V_m = j_a\eta\varepsilon$, where $\eta = \eta(j_a)$ - current efficiency, equal to the amount of energy spent for the powder production in the electric discharge, j_a- current density, ε- electrochemical equivalent of metal. We consider that the anode surface is moving with the constant speed v and the linear speed of the points on the anode surface is calculated like the following:

$$V_a = v \cdot \cos\theta, \tag{1}$$

where θ — angle between the vector of anode delivery speed and the unit vector $\overrightarrow{n_a}$ of outward normal to the surface of the anode.

In this case the general scheme of the process is not changing within time and the process can be considered steady or stationary.

Due to formula (1) steady state current density distribution j_a on the stationary anode boundary is determined by the following one:

$$\eta(j_a)j_a = \frac{\rho v}{\varepsilon}\cos\theta$$

where ρ – anode material density.

As it is seen in figure 2, dependency of $\eta(j)_a$ current efficiency can be approximated by the hyperbolic equitation and it can be considered that the following relation is fulfilled on the anode boundary [8]

$$j_a = \frac{\rho v}{\varepsilon}(a_0+b_0\cos\theta) \tag{2}$$

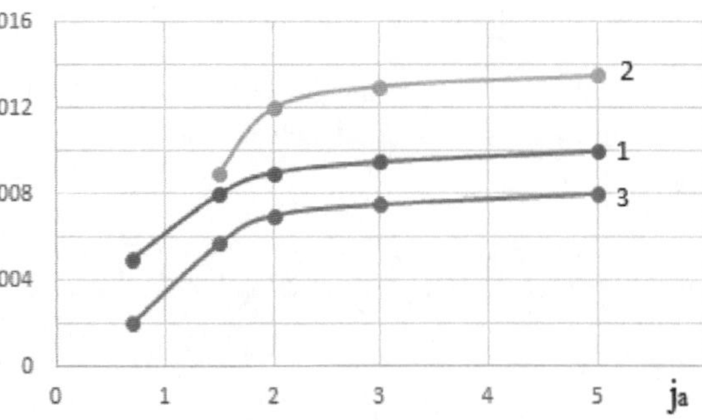

Figure 2. Dependency of current efficiency on anode current density while producing Fe_3O_4 powder of У10 steel under different current intensity.

Let's take a closer look at the two-dimensional model of the process (figure 3). We introduce the Cartesian coordinate systemx_1, y_1 connected with the anode moving towards the coordinate axis.

Ignoring the electrode phenomena, we consider that there exists electric field potential ψ_1 in the interelectrode space (IES) consistent with Laplace's equation

$$\Delta\psi_1 = 0 \tag{3}$$

and the condition of potentials' constancy $\psi_{1a} = u_a$, $\psi_{1k} = u_k$ is fulfilled on the electrodes' boundaries. Due to equation (3), there exists function $\varphi_1$the harmonicψ_1 conjugate and it is possible to allow the complex potential of electrostatic field W (x_1,y_1) = $\varphi_1(x_1,y_1)$ + $i\psi_1(x_1,y_1)$ that is the analytical function in the field $z_1=x_1+iy_1$.

Let's allow typical values of current density $j_a = \frac{\rho v}{\varepsilon}$, length $H = \kappa(u_a - u_k)/j_0$ (κ – electrical conductivity of the circumference) and move on to the dimensionless variables

$$x = \frac{x_1}{H}, \quad y = \frac{y_1}{H}, \quad z = x + iy, \quad W = \varphi + i\psi = W_1 - \frac{iu_k}{u_a - u_k}$$

Then, taking into account (2) that function ψ in the interelectrode field is consistent with Laplace's equation and the boundary conditions on the electrodes' surface

$$\psi_a = 1, \quad \psi_k = 0, \quad \frac{\partial\psi_a}{\partial n} = \frac{j_a}{j_0} = a_0 + b_0 \cos\theta$$

where a_0, b_0 – constants sensitive to the dependency of current efficiency on its density.

Therefore the initial task is modeled by potential plane-parallel flow of ideal incompressible liquid. Moreover, the analogue of $\vec{E}$ electric field intensity is the speed of dummy flow $\vec{V}$ while vectors $\vec{E}$ and $\vec{V}$ are orthogonal. The following equation $\partial\psi/\partial n$= V, V =$|\vec{V}|$ is fulfilled along the line ψ =const. V =$a_0 + b_0 \cos\theta$ on the anode surface.

A two-dimensional problem is studied in this research. The IES field in planez $= x + i$, determined by symmetry lines A_1C and B_1D is represented in figure 3, where CD – cathode boundary, A_1ABB_1 – anode boundary, A_1B_1, - points at infinity. The anode boundary is divided into two fields: straight-line segmentsA_1A, B_1B, where V_a=0 and segment AB where condition (2) is fulfilled. The origin of coordinates is C.

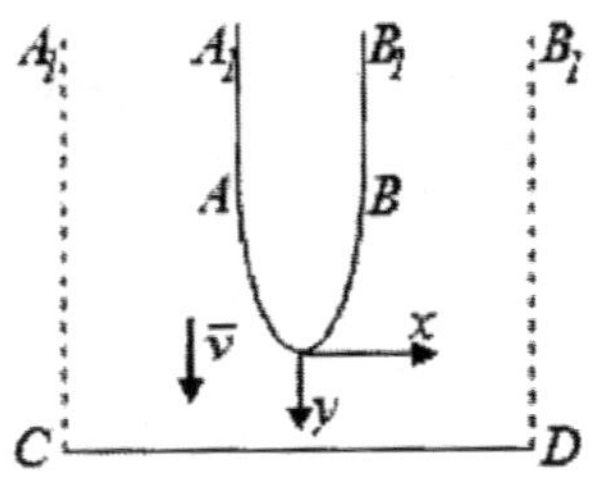

Figure 3. IES field.

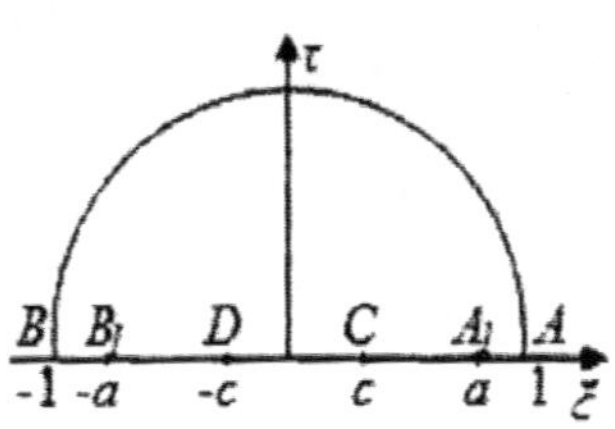

Figure 4. D_u field.

The task at issue in dimensionless form reduces itself to the search of boundary *AB* on the anode surface in the following task. Function ψ is consistent with Laplace's equation $\Delta\psi = 0$, on the electrodes' boundaries the following conditions are fulfilled: $\psi|_{CD} = 0, \psi|_{A_1BBB_1} = 1, \frac{\partial\psi}{\partial n|_{AB}} = a + b\cos\theta$, on symmetry lines A_1C *and* B_1D – conditions $\frac{\partial\psi}{\partial n|_{A_1C}} = 0, \frac{\partial\psi}{\partial n|_{B_1D}} = 0$.

For the task at issue the flow of dummy stream is created by the system of continuously distributed sources and sinks: sources – along lineA_1C, sinks – along lineB_1D. Condition $V = a_0 + b_0 \cos\theta$ is fulfilled on the unknown boundary AB.

Let's suppose that in the plane of additional complex variable $u = \xi + i\tau$ flow range D_z is equivalent to (figure 4)

$$D_u = \{|u| \le 1, \tau \ge 0\}$$

and function z (u) conformally reflects field D_u to field D_z c with the points' correspondence as shown in figures 3,4.

Let's calculate two functions: complex potential of dummy flow [3] W $(u) = \varphi(u) + i\psi(u)$ and Joukowski function [9]

$$x(u) = ln\left(\frac{V_0 dz}{dW}\right) = r(u) + i\theta(u) \quad (4)$$

where $r(u) = \ln V_0 / V$, V- dummy flow speed module, Θ - inclination angle of speed vector to axis x, $V_0 = a_0$- speed value at point A.

Complex potential W (u) is consistent with the boundary conditions

$$\psi(u)=\begin{cases} 0, u = \xi, |\xi| \leq c \\ 1, u = \xi, d \leq |\xi| \leq 1; u = e^{i\sigma}, 0 \leq \sigma \leq \pi \end{cases}$$

$$\varphi(u)=\begin{cases} 0, u = \xi, c \leq \xi \leq d \\ \varphi_0, u = \xi, d \leq |\xi| \leq 1, -d \leq \xi \leq -c \end{cases}$$

W (u) function range – rectangular (figure 5)

$$D_w = \{\varphi + i\psi | 0 \leq \varphi \leq \varphi_0, \quad 0 \leq \psi \leq 1\}$$

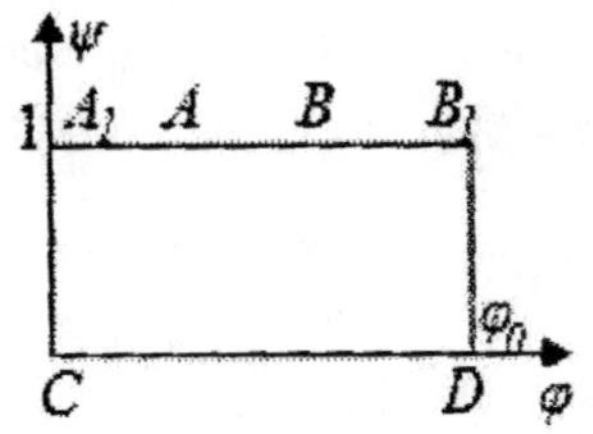

Figure 2. D_w field.

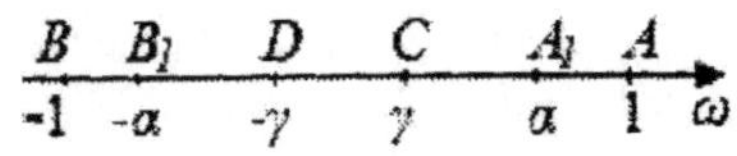

Figure 3. D_w field.

To determine derivative *dW/du* of complex potential we reflect D_u field to upper semiplane D_a with the points' correspondence as shown in figures 5, 6, using the transformation

$$x\omega(u) = \frac{2u}{1+u^2} \tag{4}$$

then, in accordance with Schwarz-Christoffel formula [7] we find the function reflecting D_W field to the $W(u)$ function range:

$$W(\omega) = N \int_\gamma^\omega \frac{d\omega}{\sqrt{(\omega^2-\alpha^2)(\omega^2-\gamma^2)}},$$

$$N = \left(\frac{1}{\gamma}K\left(\frac{\alpha}{\gamma}\right) - \frac{1}{\alpha}K\left(\frac{\gamma}{\alpha}\right)\right)^{-1}, \ \varphi_0 = \frac{2K(\gamma/\alpha)}{\gamma K\left(\frac{\gamma}{\alpha}\right) - \alpha K(\alpha/y)},$$

$$K(k) = \int_0^1 \frac{d\omega}{\sqrt{(\omega^2 - 1)(k^2\omega^2 - 1)}}$$

where $K(k)$ – complete elliptic integral of the first kind [10]. Differentiating $W(\omega)$ on variable ω and taking into account the dependency (5), we see:

$$\frac{dW}{du} = \frac{2N(1-u^2)}{\sqrt{4u^2-a^2(1+u^2)^2}\sqrt{4u^2-c^2(1+u^2)^2}}$$

Let's take a closer look at Joukowski function for the dummy flow. On the cathode boundary imaginary part $\lambda(u)$ is equal to zero. The stationary condition of forming (2) is fulfilled on the anode boundary AB. There are the following boundary conditions for Joukowski function $\lambda(u)$

$$\begin{cases} \frac{\pi}{2}, u = \xi, -1 \le \xi \le -\alpha \\ 0, u = \xi, |\xi| < \alpha \\ -\frac{\pi}{2}, u = \xi, \alpha \le \xi \le 1 \end{cases}$$

$$r(u) = -\ln\left(1 + \frac{b_0}{a_0} cos\theta(u)\right), u = e^{i\sigma}, 0 \le \sigma \le \pi.$$

We are going to find Joukowski function $x(u)$ in the following way

$$x(u)x = x_0(u) + f(u),$$

where function $x_0(u) = r_0(u) + i\theta_0(u)$ is consistent with the boundary conditions

$$\theta_0(u)=\theta_0(u), u = \xi, |\xi| \le 1,$$

$$r_0(u) = 0, u = e^{i\sigma}, 0 \le \sigma \le \pi$$

and has the same peculiarities in field D_u that $\lambda(u)$has, function $\lambda(u)$- analytical in D_u and continuous in $\overline{D_u}$.

The boundary conditions for $\lambda_0(u)$ allow to construct function $d\lambda_0/du$, which is real when $u = \xi, |\xi| \le 1$ and has a simple pole with residue ½ at points $A_1(u = a), B_1 = (u = -\alpha)$. Let's extend function $\frac{\lambda(u)}{du}$ to the whole plane in accordance with the principle of symmetry and construct it using the singularity method [9]. Then, integrating we find out:

$$\lambda_0(u) = \frac{1}{2}\ln\left(\frac{u^2 - a^2}{1 - a^2u^2}\right) + A_0u + B_0$$

Using the boundary conditions at points C and D, we find constants $A_0 = 0, B_0 = 0$.

Comparing the boundary conditions for functions $x(u)$ and $x_0(u)$, we get the boundary conditions for the unknown function $f(u) = \lambda(u) + i\mu(u)$:

$$\mu(u) = 0, u = \xi, |\xi| \le 1,$$

$$\lambda(u) = -\ln\left(1 + \frac{b_0}{a_0}\cos\theta(u)\right), u = e^{i\sigma}, 0 \le \sigma \le \pi.$$

Taking into account the boundary conditions (6), (7) and the symmetry of the dummy flow, function $f(u)$ can be analytically extended to the whole circle and presented in the form of Laurent expansion:

$$f(u) = \sum_{n=0}^{\infty} c_{2n}\, u^{2n},$$

where c_{2n} – actual coefficients.

Based on the condition $f(1) = 0$ we find $c_0 = -\sum_{n=0}^{\infty} c_{2n}$,

Condition (7), as function $f(u)$ has been presented in the form of the expansion, looks like

$$c_0 + \sum_{n=1}^{\infty} c_{2n}\cos(2n\sigma) = -\ln\left(1 + \frac{b_0}{a_0}\cos\theta(\sigma)\right)$$

Multiplying both parts of equation (8) by$\cos(2n\sigma)$, integrating over σ within $0, \pi/2$, we get infinite system of equations for defining coefficients c_{2n}:

$$c_{2n} = -\frac{4}{\pi}\int_0^{\frac{\pi}{2}} \ln(1+\frac{b_0}{a_0}cos\theta_0(\sigma)+\mu(\sigma))\cos(2n\sigma)d\sigma, n=\overline{1,\infty}.$$

We calculate the dimensionless coordinates of AB anode boundary points based on (4) using the formula

$$z(u) = \frac{1}{a_0}\int_i^u \frac{dW}{du}\exp\big(x(u)\big)\,du = e^{i\sigma}, 0 \le \sigma \le \pi$$

To solve the task it is necessary to find mathematical parameters c and a. We can do it specifying intervals |CD| and $|A_1B_1|$:

$$z(-c)\text{-}z(c)=|CD|,\ \ z(-a)\text{-}z(a)=|A_1B_1|$$

It is possible to determine the mass of dispersed powder considering the bar axially symmetric and using the formula for calculation of square of the rotary surface.

$$m = 4\pi\alpha_0\rho H^2\int_0^{\pi/2} x(\sigma)\exp\left(-r(\sigma)\right)|\frac{dz}{d\sigma}|d\sigma$$

where ρ – metal density.

The results of calculations of the boundary shape of anode surface for different special cases are shown in figures 7, 8.

0

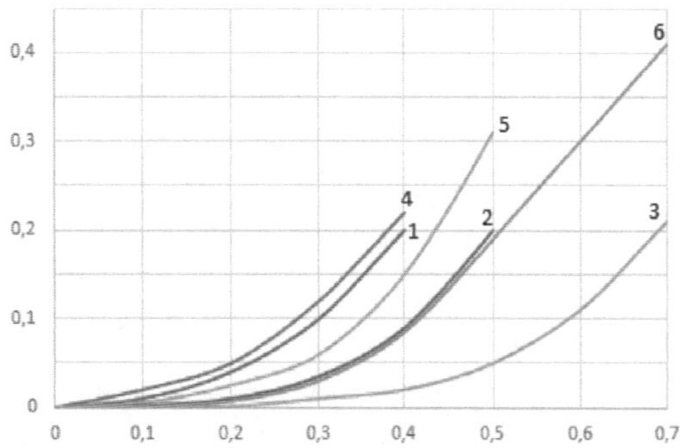

Figure 7. The results are calculated for α=0.1; b=0.9; |CD|=2.0 and different $|A_1B_1|$. Curves 1, 4 – $|A_1B_1|$=0.6; 2, 5 – $|A_1B_1|$=1.0; 3, 6 – $|A_1B_1|$|=1.4. Curves 4, 5, 6 – unlimited cathode.

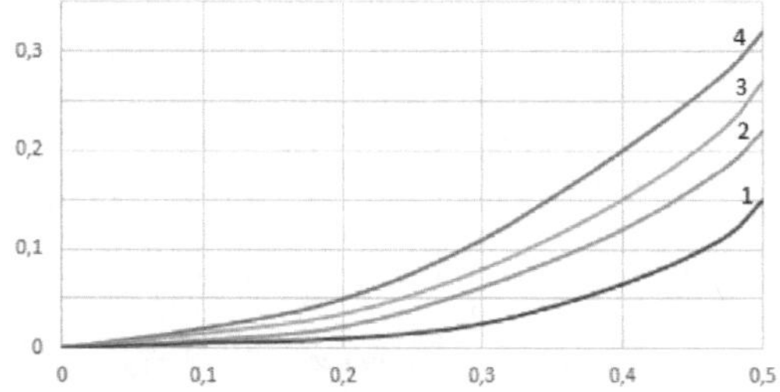

Figure 8. The results are calculated for α=0.1; b=0.9; $|A_1B_1|$=1.0 and different |CD|. Curves 1 – |CD|=1.5; 2 – |CD|=2.0; 3 – |CD||=3.0; 4 – unlimited cathode.

Curves 4, 5, 6 in figure 7 and also curve 4 in figure 8 fit the case when the cathode width tends to infinity in figure 3 |CD|→ ∞, what corresponds with c→a in figure 4.

4. Conclusion

The mathematical model qualitatively enough describes the removal of metal from the anode surface process, and also allows you to determine the approximate weight of the dispersed powder of a steel electrode in the discharge of liquid electrodes.

References

[1] Valiev R A, Gaisin F M and Shakirov Yu I 1991 Special traits of powder obtained in discharge between steel electrode and electrolyte *Soviet Powder Metallurgy and Metal Ceramics* **30** p 448-50

[2] Valiev R A, Gaisin F M, Romanov E S and Shakirov Yu I 1991 Synthesis of iron oxide powders in a discharge with a liquid electrode *Physics and chemistry of materials treatment* **25** p 654-8
[3] Sadikov K, Fayrushin I, Shamsutdinov A and Kashapov N 2016 Experimental setup for plasma treatment of disperse materials in the arc plasma jet *J. of Phys.: Conf. Ser.* **669** 012048
[4] Shakirov Y I, Valiev R I, Khafizov A A, Valiev R A and Khakimov R G 2016 Erosion of electrode metal in the electric discharge under the exposure of the electrolyte stream *J. of Phys.: Conf. Ser.* **669** 012064
[5] Khafizov A A, Shakirov Yu I, Valiev R A, Valiev R I and Khafizova G M 2016 Study of thermal and electrical parameters of workpieces during spray coating by electrolytic plasma jet *J. of Phys.: Conf. Ser.* **669** 012030
[6] Davydov A D and Kozak E 1990 *Vysokoskorostnoe jelektrohimicheskoe formoobrazovanie* (Moscow: Nauka)
[7] Lavrentyev M A and Shabat B V 1987 *Metody teorii funkcij kompleksnogo peremennogo* (Moscow: Nauka)
[8] Kotljar L M and Minazetdinov N M 2003 *Prikladnaja mehanika i tehnicheskaja fizika* **44** p 179-84
[9] Gurevich M I 1979 *Teorija struj ideal'noj zhidkosti* (Moscow: Nauka)
[10] Janke E, Jemde F and Ljosh F 1977 *Special'nye funkcii* (Moscow: Nauka)

Improving the quality indicators fuel pump of plasma technology

S N Sharifullin, A S Pirogova
Kazan Federal University, 18 Kremlyovskaya Street, Kazan, 420008, Russian Federation

Saidchist@mail.ru

Abstract. It is shown that the product quality control is not only to control the process of its manufacture. To the product was in demand and the competitive need to search for new promising technologies for its production. One such technology is the plasma technology.

1. Introduction.

According to GOST R ISO 9000-2015 2.2.1 The quality of the products and services of the organization depends on the ability to meet consumers' intentional or unintentional influence on relevant stakeholders. Indeed, the quality is largely determined by the consumer and, therefore, the manufacturer must monitor the quality of work - namely, to follow the development of modern science. It is no secret that today, technologies such as plasma technology - a progressive trend in the field of engineering, which allows repeatedly and effectively improve the reliability and durability of machine parts. What does "progressive trend"? This means that there will always be those who are interested in the development of issues related to this field. So here rightly be competition. It was she who pushed the organization to maintain and enhance its position in the market. Usually this makes quality management.

Quality management is defined by a coordinated activity on all matters of organization management manual. Recall quality management principles: Customer focus, defining the interests and desires of buyers; Leadership that provides unity of purpose and direction of the state; The interaction of employees, ensuring the involvement and expertise in manufacturing processes; Process approach, implying the achievement of the goal through the relationship of all the elements; Continuous improvement, suggesting success; Decision making based on facts, which help to achieve the desired result; Management relations, which is the link between suppliers and the organization. All these factors must be met to obtain the expected, and most importantly - quality product (service) at the output. Multifaceted concept of quality, and therefore the characteristics of a particular object, it will be different. We say that the quality characteristics of the product parameter. In this aspect, usually takes into account the functional characteristics, reliability, durability, defect-free (free from defects), safety, design, sustainability, etc.

This material is considered a quality indicator as longevity or would be correct to say, life of the high pressure fuel pump (fuel pump) diesel internal combustion engines of transport and technological machines. Presumably, the more the life of the product, so it is more popular and competitive.

2. Materials and methods of research.

This article considers the possibility of increasing the service life of the high pressure fuel pump by creating a plasma coating of titanium nitride on the working surface of the plunger. For the base model of the fuel pump fuel pump is taken 33-02 model car engine KamAZ 5320 with factory plunger pairs. Application of titanium nitride on the plunger to produce magnetron installation UVN-4-2. Analysis of the results of research carried out with a theoretical model for the construction characteristics of the work pump. Construction characteristics of pump operation is based on the Lagrange polynomials.

IOP Conf. Series: Journal of Physics: Conf. Series **789** (2017) 012051 doi:10.1088/1742-6596/789/1/012051

3. Results and discussion.

In [1] it was proved that reliability and durability (high pressure pump) diesel engines, primarily depends on their resource plunger. Unfortunately, the quality of domestic manufacturing plunger such that their life is not more than 3 thousand operating hours, while that of foreign analogues it reaches 12 thousand or more. As was shown in [2], a resource plunger is determined by the pressure and wear of the developing mating parts. A pair of pressure developed by the plunger is directly dependent on the clearance between the sleeve and the plunger. The smaller the gap, the higher the pressure developed by plunger pair hence the greater its working life. Backlash in domestic plunger pairs is about 1.5 - 2.0 microns and a foreign - one micron or less. In addition, the wear resistance of domestic plunger is 1.5 - 2.0 times lower than the wear resistance of the foreign. Studies have shown [2-18] that the achievement of their foreign counterparts resource plunger possibly producing ion-plasma deposition of titanium nitride on the working surface of the plunger.

In [2] we developed a theoretical model for the construction of resource characteristics of the high-pressure fuel pump works, or briefly pump. Fig. 1 is a block diagram of the developed model. The essence of this model is as follows: The circuit has two branches: the left and right. The left branch of the pump is seen with known parameters based on experimental data obtained in the course of the performance test. The right branch of the block diagram of the theoretical model refers to the pump with unknown parameters. It requires build characteristic of his work and to determine resource. Pump with known parameters is taken as the base fuel pump. For the base model pump define the cyclic fuel delivery q, developed by a pair of plunger pressure p and the operating time t since the start of the pump up to the limit of his condition. Removing the data produce a certain period of duration of pump operation. Using one of the mathematical programs, using the Lagrange polynomial characteristic of building a working base model pump.

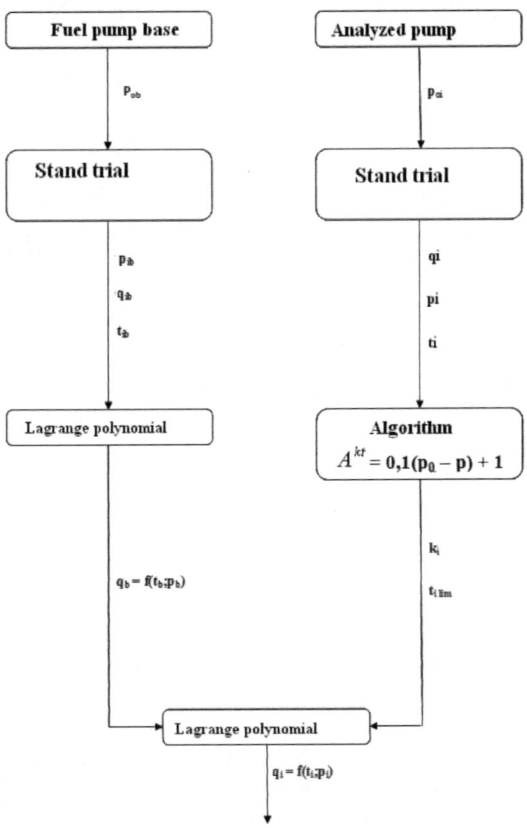

Figure 1 - Block diagram of a theoretical model for the construction characteristics of the work pump

A theoretical model for the construction of resource characteristics of the work can not only build a characteristic pump works, but also to determine its service life. This was used by the resulting changes in the law developed by a pair of plunger pressure from developments [2]:

$$A^{kt} = 0,1(p_o - p) + 1 \quad (1)$$

where A - constant characterizing the influence of the initial developed by a pair of plunger ro pressure; k - coefficient taking into account the intensity of wear plunger; t - duration of the operation; p - to develop a pair of plunger pressure at operating time t.

Going from the plunger assembly to the fuel pump and the fact that its operating time is determined by the service life plunger according to the formula (1) can be found and the service life of the p

$$t_r = \frac{\log\{0.1(p_0 - p_{\lim}) + 1\}}{k \cdot \log A} \quad (2)$$

where t_r - the life of the fuel pump; $p_{\lim}$ - the limit value developed by a pair of plunger pressure equal to the minimum pressure of the fuel injection through the nozzle.

Take for the base model fuel pump 33-02 model car engine KamAZ 5320 with factory plunger pairs. According to the experimental data (Fig. 2) of this resource injection pump 3000 hours at the initial developing a pair of plunger pressure of 70 MPa. Let the base model pump coefficient taking into account the wear plunger is unity, k = 1. If the resource limit injection pump 3000 hours developed by a pair of plunger pressure is the operating pressure of the injector. In this case $P_{\lim}$ = 20 MPa. Then, from the formula (2) we find the constant A: lq A = 2,6 • 10^{-4} or A = $10^{0.00026}$.

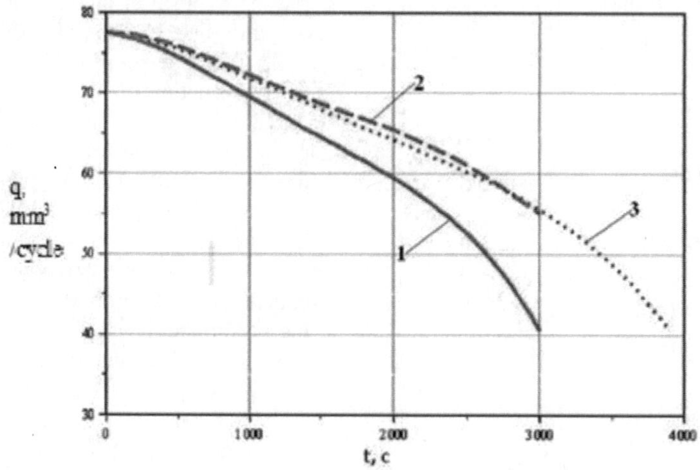

Figure 2 - Schedule features laws pump model 33-02 with factory pair of plunger. 1 - the experimental curve (base) at P_0 = 70 MPa; 2 - for polynomial curve at P_0 = 110 MPa; 3 - curve for the model when P_0 = 110 MPa

This constant A will not change its value, since the initial value of the injection pump setting does not depend on the state of the plunger assembly. Original developed by a pair of plunger pressure will influence only the operating time. This pattern of change of the characteristics of the work of the pump operating time is also saved.

Determine the probability of life of the fuel pump at an initial base model to develop a pair of plunger pressure of 110 MPa. From the formula (2) that the resource will then be equal to 3900 hours. To validate the calculation in the polynomial matrix for y = 70 MPa to change the ratio of fixed values q of x for the resource 3900 hours and plotted the cycle of fuel supply by operating time (curve 2 in Fig. 2). Fig. 2 curve 3 obtained according to the base curve at 1 time between the plunger assembly with a source to develop a pressure P_0 = 110 MPa. As can be seen, the curve 2 by the polynomial at P_0 = 110 MPa is fully integrated into the resulting curve. That means having the basic experimental data pump and conducting additional experiments can plot its basic parameters from the achievements and define the life of any source to a pair of pressure developed by plunger. It also follows from the graphs that the pattern of changes of the parameters of the pump operating time is stored in any source developed by a pair of plunger pressures. The difference will be only in the operating time.

Similarly, calculations are made for the injection pump having a plunger covering the plunger assembly of titanium nitride. Determine the probability of life of the fuel pump plunger with said cover when the original pair to develop plunger pressure of 110 MPa. According to the experimental data of the wear resistance of the titanium nitride coating is 2.4 times higher than the wear resistance of the material of the plunger the plunger assembly factory. Based on the above calculations for the value of the coefficient k assumed to be 1/2.4. Then the formula (2) determine the probable life of the pump plunger plunger coated pair of titanium nitride in the initial developing a pair of plunger pressure of 110 MPa: t = 9230 hours. To validate the calculation in the polynomial matrix for pump base model at y = 70 MPa to change the ratio of fixed values of x for the resource 9230 hours and plotted the cycle of use of the fuel supply (Fig. 3).

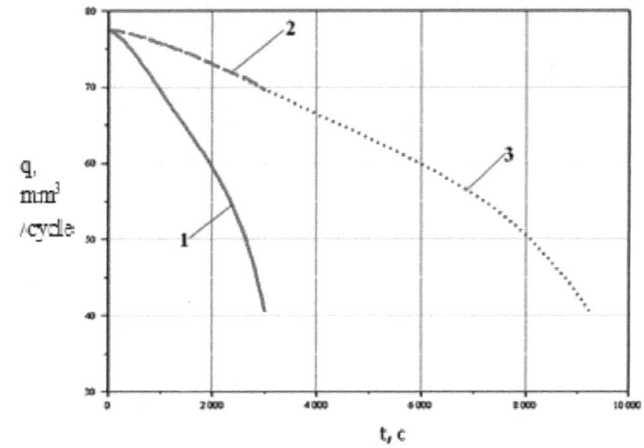

Figure 3 - Charts laws work performance pump models 33-02 coated plunger plunger pair of titanium nitride. 1- The experimental curve (base) at P0 = 70 MPa; 2 for polynomial curve at P0 = 110 MPa; 3- curve for the model when F0 = 110 MPa.

Analysis of the curves in Fig. 3 gives the same results as the analysis from the curves in Fig. 2. Then, we can conclude that the model works for the high pressure pump with a different state of plunger. As can be seen from the graphs in Fig. 1, a coating of titanium nitride on the plunger allows a three-fold increase in life of the diesel injection pump.

4. Conclusions

1. Plasma thin-film coatings can increase service life of the pump. For coating of titanium nitride may increase the service life of the fuel pump in 2 - 3 times.

References

[1] Adigamov N R, Sharifullin S N 2009 The complete solution recovery problems fuel equipment of diesel engines *Tractors and farm machinery* **(3)** 38 - 40

[2] Sharifullin S N 2009 *Increase of operational reliability of automotive diesel engine high pressure fuel pumps. - Diss. Doctor. tehn. Sciences.* (Moscow: GOSNITI)

[3] Solovev R Y, Sharifullin S N, Adigamov N R 2016 Plasma technology for increase of operating high pressure fuel pump diesel engines *J. Phys.: Conf. Ser* **669** 012050

[4] Kashapov L, Kashapov N and Kashapov R 2013 *J. Phys.: Conf. Ser.* **479** 012005

[5] Gavrilova V A, Fazlyyyakhmatov M G and Kashapov N F 2013 *J. Phys.: Conf. Ser.* **479** 012010

[6] Fayrushin I, Kashapov N and Dautov I 2014 *J. Phys.: Conf. Ser.* **567** 012009

[7] Kashapov L N, Kashapov N F and Kashapov R N 2013 *J. Phys.: Conf. Ser.* **479** 012011

[8] Kashapov L N, Kashapov N F and Kashapov R N 2014 *J. Phys.: Conf. Ser.* **567** 012025

[9] Zaripov R G, Kashapov N F, Tkachenko L A and Shaydullin L R 2016 *J. Phys.: Conf. Ser.* **669** 012053

[10] Denisov D G, Kashapov N F and Kashapov R N 2015 *IOP Conference Series: Materials Science and Engineering* **86** 012005

[11] Israfilov Z K and Kashapov N F 1991 *Journal of Engineering Physics* **60** 364–368
[12] Galyautdinov R T, Kasparov N F and Luchkin G S 2002 *Inzhenerno-Fizicheskii Zhurnal* **75** 170–173
[13] Saifutdinov A I, Fairushin I I and Kashapov N F 2016 *JETP Lett.* **104** 180–185
[14] Galyautdinov R T and Kashapov N F 2003 *Svarochnoe Proizvodstvo* **(3)** 27–31
[15] Azarov A I, Kashapov N F and Osipova O P 2003 Litejnoe Proizvodstvo **(11)** 30–31
[16] Abdullin I Sh, Galyautdinov R T and Kashapov N F 2001 *Inzhenerno-Fizicheskii Zhurnal* **74** 104–107
[17] Luchkin A G, Kashapov N F and Luchkin G S 2013 *J. Phys.: Conf. Series* **479** 012019
[18] Kornienko E E, Lapushkina E J, Kuzmin, V I, Vaschenko S P, Gulyaev I P, Kartaev E V, Sergachev D S, Kashapov N, Sharifullin S, Fayrushin I 2014 *Journal of Physics: Conf. Series* **567** 012010

Electrochemical behavior of the graphene materials synthesized using low temperature plasma

M B Shavelkina[1], R H Amirov[1], A Y Richagov [2], T B Shatalova[3]

[1]Joint Institute for High Temperatures of Russian Academy of Sciences, Izhorskaya Street 13/2, 125412 Moscow, Russia
[2]Russian Academy of Sciences A.N. Frumkin Institute of Physical Chemistry and Electrochemistry RAS, Leninsky prospect 31, 199071 Moscow, Russia
[3] Lomonosov Moscow State University, Leninskie Gory, 119991 Moscow, Russia

E-mail: mshavelkina@gmail.com

Abstract. By means of DC plasma torch of up to 45 kW power, few-layered graphene sheets were obtained. Their properties and structure were characterized by using electron microscopy, thermal analysis, Raman and infrared (IR) spectroscopy. Boundary surface of samples have been investigated using the method of "limited evaporation" and BET method.
Electrochemical examination of their properties was conducted. Due to the activity and stability of synthesized materials the conclusion was made regarding the possibility of the use of them as catalysts carriers for fuel cells electrodes, electric current sources, conducting additives for electrodes in non-aqueous electrolytes.

1. Introduction

The most of the existing methods of graphene production are related to the mechanical [1-3] or chemical [4-6] decomposition of graphite or epitaxial growth of graphene films [7]. In the first instance obtaining the high quality graphene layers is based exclusively on the fundamental research. In the second instance the chemical methods possess low cost and high productivity rate but do not completely separate graphite placed in the solution and flaked graphene in the solution does not preserve the sheet shape. There exists a method of graphite transformation into graphite oxide. It employs an approach named "oxidization - exfoliation -reduction" in which course basic graphite planes are coated with covalent-linked functional groups of oxygen. That oxidized graphite becomes hydrophilic and separates into individual graphene sheets in aqueous solution under the impact of ultrasound. However, the obtained graphene would possess electric conductivity by a few orders of magnitude lower compared with the conductivity of graphene produced by the Scotch-tape method from highly oriented pyrolitic graphite. That method does not provide for the possibility to free from oxygen completely and leads to the formation of graphene with high concentration of defects of hybridization.

The use of plasma is considered as one of the promising methods for graphene production. In the study [8] the gas discharge was ignited in the mixture of helium and carbon dioxide. The optimal conditions for synthesis of graphene with few numbers of layers were discovered whereas the number of defections was lower than in the graphene obtained by the chemical methods. However, because of the low speed of carbon evaporation the mass production of high quality carbon nanostructures cannot be implemented. From the viewpoint of large scale production of graphene it is of interest to use a plasma jet.

Our approach is relatively simple and easy for the commercial-scale production for electrochemical applications compared with CVD or wet chemistry techniques. The essence of it is the synthesis of functional groups - free graphene materials with a large area in a plasma jet generated by DC plasma torch of up to 45 kW power.

2. Methods

The basis of the plasma jet synthesis method is the principle of simultaneous input of plasma forming gas with carbon containing precursor (soot, hydrocarbons, alcohols) and its evaporation in the discharge gap with formation of carbonic vapor and further condensation in the form of solid carbon. Figure1 presents the schematic setup of the facility which main component is a plasma torch of the original design. The choice of plasma torch with the expanding channel of the outlet electrode is grounded by the ability of such design to ensure burning of the arc in the laminar flow at the sufficiently high speed of gas flow at the inlet [9]. Specifically the expanding channel is capable to provide for distribution of the electric current over the outlet electrode surface and, respectively, the magnetic field, which promote the increase of the positive column stability in the discharge gap.

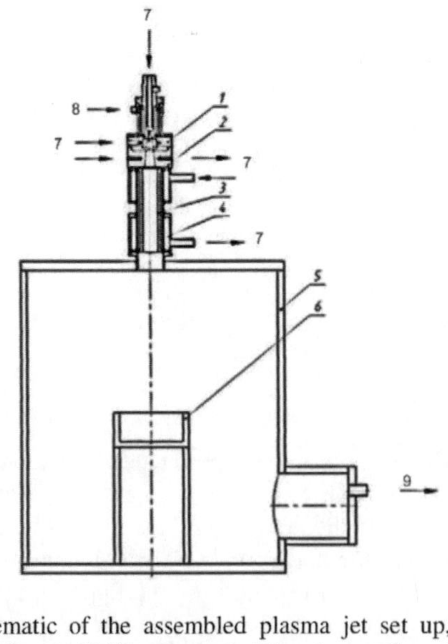

Figure 1. Schematic of the assembled plasma jet set up, used for the synthesis of grapheme. (1) Plasma torch, (2) water cooling of plasma torch, (3) graphite tube, (4) water cooling of graphite tube, (5) vacuum chamber, (6) collector, (7) water cooling, (8) plasma forming gas with carbon precursors, (9) vacuum pump.

Helium, argon (technical and high-purity) and nitrogen are used as plasma forming gases. Carbon sources are presented by hydrocarbons: propane-butane, methane and acetylene. Hydrocarbons flow rates were: propane-butane in the ratio 30:70 mass % - 0.11-0.30 g/s, methane - 0.15-0.37 g/s, acetylene - 0.05-0.16 g/s.

The optimal regime of functional groups - free graphene materials is achieved by widely ranged varying of pressure, energy input and flow rates proportion of plasma forming gas, initial substance and catalyst, formation of a flow with the using of the outlet electrode-anode expanding channel and by changing the geometric parameters. A detailed description of the experimental setup was given in the studies [10]. The synthesis parameters are presented in table 1.

For direct registration of the dimensional parameters and images the method of scanning electronic microscopy (SEM) was applied – microscope Hitachi S5500 with in-lens technique for obtaining a super-fine resolution as well as high sensitivity of EDS analysis.

Efficiency of the synthesis, phase composition and thermal stability of the carbon products were evaluated by the thermal methods: Differential Scanning Calorimetry (DSC) and TG (STA 449 platform F3 Jupiter (Netzsch)) with linear heating of sample in the air at the rate 10 K/min in the temperature range -- 20 -1000 °C.

The topology of formed graphene structure was estimated by Raman scattering (Ntegra Spectra). Excitation wavelength was 532 nm, output power was adjusted so as to prevent overheating of the material.

To determine the characteristics of the porous structure (pore volume, pore radius, surface area) we used a relatively new method of adsorption "Limited Evaporation" based on the analysis of the kinetics of evaporation of the adsorbate from the test material. To evaluate the specific surface area of materials the classic BET method was applied using a low-temperature nitrogen adsorption analyzer Quadrasorb SI.

IR spectroscopy (FT-IR spectrometer VERTEX 70 (Bruker, Germany) used in the mid-IR range (4000-400 cm^{-1}, or 2.5-25 μm) at a resolution of 2 cm^{-1} to determine the structure of samples on the presence of the or other functional groups.

Electrochemical investigation was conducted in a universal laboratory cell of filter-press structure [10], with counter-electrode made of high capacitance activated carbon (CH-900 produced in Japan) which mass exceeded the mass of the working electrode by more than 100 times. The electrolyte was 0.5 M sulfuric acid solution. The graphs and the text indicate the potentials relatively to the stationary potential of activated carbon in the given electrolyte (around 700 mV RHE). Cyclic voltammetry was measured by potentiostat Elins P-30J(M).

Table 1. Technological conditions.

Power (kW)	Current (A)	Voltage (V)	Gas pressure (Torr)	Helium flow rate (g/s)	Nitrogen flow rate (g/s)	Argon flow rate (g/s)	Carbon flow rate (g/s)	Times of work (min)
30-45	150 - 400	35-110	77-740	0.5-0.9	1.5-2.5	3.0-3.75	0.047-0.276	6-10

3. Results and discussions

The structures which morphologies are presented on figure 2 were obtained by plasma jet decomposition of propane-butane, acetylene and methane at varying the vacuum chamber volume parameters by shifting target perpendicularly to vapor-gas flow. The 2D particle structures are labeled as flakes and showed a stacking averaging around 2 graphene layers (range 5 atomic layers, up to 2 nm thickness) and in plane dimensions between 50 and 1.5 μm.

Raman spectroscopy is established as the most accurate and easy tool to characterize graphene [1], the number of layers and the presence of defects (quality) especially when the number of layers is less than 5 [12]. Figure 3 demonstrates a typical Raman spectrum of graphene materials produced by the decomposition of acetylene when argon plasma was used at pressure of 350 Torr. There are characteristic of single-layer graphene materials [1] - G mode (1587 cm^{-1}), 2D strip (G') - (2669 cm^{-1}) and the D breathing mode (1355 cm^{-1}) on the spectrum. These are identified as the Raman fingerprints of graphene, namely of D, G and 2D peaks. Oxidation of these samples in the air shows that pure carbon phase amounts to 78-86 mass%.

Electrochemical studies of graphene synthesized in the plasma jet show high hydrophobicity and resistance to oxidation. Their hydrophobic properties are preserved for a long time even in intense (greater than 1 V) cathode-anode polarization in sulfuric acid solution and in durable keeping in concentrated sulfuric acid in the absence of anode polarization.

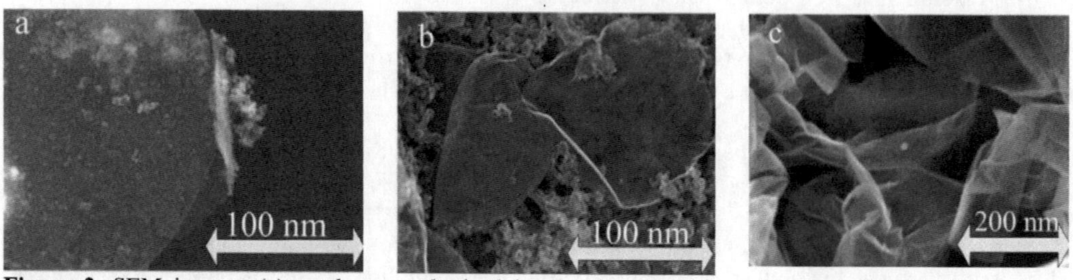

Figure 2. SEM images: (a) products synthesized by the decomposition of propane-butane, (b) products synthesized by the decomposition of acetylene, (c) products synthesized by the decomposition of methane at 710 Torr under a helium atmosphere.

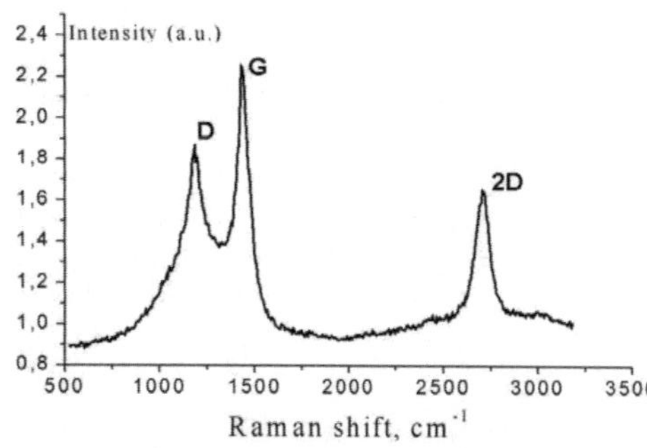

Figure 3. Raman spectra of synthesis products by the decomposition of acetylene using argon at a pressure of 350 Torr.

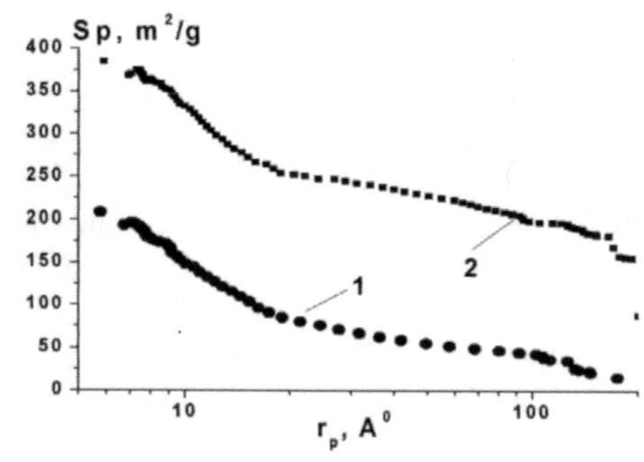

Figure 4. Specific surface area as a function of the pore radius for the samples synthesized by the decomposition of acetylene using helium (1) and argon (2) at a pressure of 350 Torr.

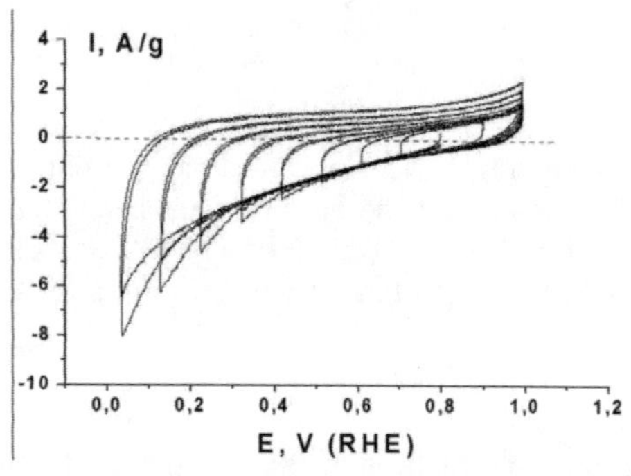

Figure 5. Cyclic voltammogram of graphene materials obtained by decomposition of acetylene at argon plasma at 500 Torr in 1M KOH at 50 mV/s.

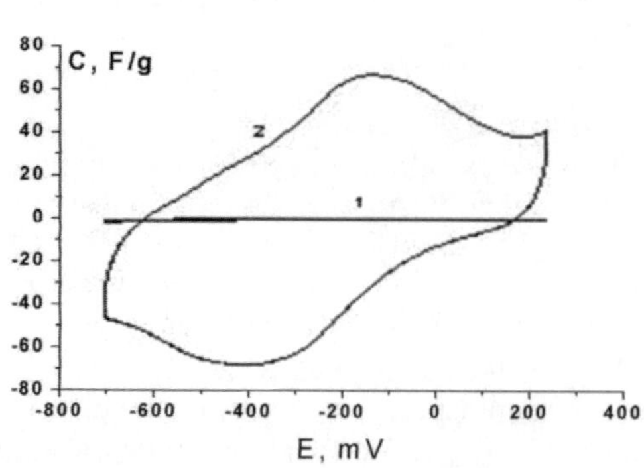

Figure 6. The VF curves of graphene materials in 0.5 M H_2SO_4 solution at 20 mV/s. (1) Initial graphene, (2) oxidized graphene.

Despite the large value of specific surface area (from 280 to 500 m^2/g) the samples demonstrate very low capacitance in aqueous electrolytes (around 0.1 F/g). Oxidation of graphene materials at potentials 1.6-1.7 V (versus RHE) leads to the capacitance growth by 2-3 orders of magnitude. Figure 4 presents the change of the value of specific surface area when different carbon precursors were used.

Figure 5 shows the typical cyclic voltammetry data for graphene materials in 1 M solution of potassium hydroxide (60 °C) at 50 mV/s. Extremely low capacitance of samples is reasoned by low wettability (hydrophobicity) of these materials. Figure 6 demonstrates the change of capacity of materials after oxidation in 0.5 M sulfuric acid solution. Analysis of the capacitance curves shape after oxidation shows that the capacitance of oxidized graphene materials is determined mainly by pseudo-capacitance of the surface functional groups formed in the result of electrochemical oxidation. According to the IR spectrum of the initial graphene materials obtained by the decomposition of hydrocarbons in a helium plasma do not contain functional groups. Capacitance of the double-electric layer does not exceed 10 F/g which may be related to the low electron-hole conductivity of the said materials.

Fuel cells consist of three key parts: cathode, anode and separation membrane. Among them, cathode materials are pivotal elements for the commercialization. Currently, expensive noble metals such as platinum, gold, ruthenium, and their alloys are best known cathode materials for oxygen reduction reaction. One is to minimize the use of expensive catalysts by increasing surface area via nanoparticle approach [13]. Due to the high specific surface area, good chemical stability and universal electrical properties the plasma jet graphene can be used as catalysts carriers for gas diffusion electrodes of fuel cells and air-metallic current sources. In addition, the low defectiveness and high electrical conductivity allow considering them as the promising conductive additives in a matrix of the composite electrode materials in non-aqueous electrolytes.

4. Conclusion

The present study has shown that it is possible to synthesis of few-layers graphene material at decomposition of hydrocarbons in plasma jet reactor based on DC plasma torch with expanding channel-anode, and with helium, argon or nitrogen used as plasma forming gas. The kind of plasma forming gas and the precursor influences the morphology of graphene materials, their amount and the value of specific surface area. When using propane-butane mixture as a precursor solid products with a high content of pure carbon are formed. The highest value of the specific surface area (500 m^2 / g) was found for the samples obtained by the decomposition of acetylene. The effect of the electrolytes on electrochemical properties for graphene materials as the electrode for supercapacitor and has been investigated by cyclic voltammetry, good chemical stability are established. A simple and cost-effective oxidation technique of graphene materials with good capacitive behavior encourages their possible use in commercial applications as catalysts carriers for fuel cells electrodes, electric current sources, conducting additives for electrodes in non-aqueous electrolytes.

5. Acknowledgements

The authors gratefully acknowledge Russian Foundation for Basic Research for the support by grants No 15-08-00165 and 16-08-00145.

6. References

[1] Novoselov K, Geim A, Morozov S, Jiang D, Zhang Y, Dubonos S, Grigorieva I and Firsov A 2004 *Science* **306** 666
[2] Novoselov K, Geim A, Morozov S, Jiang D, Katsnelson M, Dubonos S, Grigorieva I and Firsov A 2005 *Nature* **438** 197
[3] Zhang Y, Tan Y, Stormer H and Kim P 2005 *Nature* **438** 201
[4] Booth T, Blake P, Nair R, Jiang D, Hill E, Bangert U, Bleloch A, Gass M, Novoselov K, Katsnelson M and Geim A 2008 *Nano Lett.* **8** 244
[5] Eda G, Fanchini G and Chhowalla M 2008 *Nature Nanotech.* **3** 270
[6] Park S, Sungjin P and Ruoff R 2009 *Nature Nanotech.* **4** 217
[7] Berger C, Song Z, Li X, Wu X, Brown N, Naud C, Mayou D, Li T, Hass J, Marchenkov A, Conrad E, First P and de Heer W 2006 *Science* **312** 1191
[8] Wu Y, Wang B, Ma Y, Huang Y, Li N, Zhang F and Chen Y 2010 *Nano Res.* **3** 661

[9] Isakayev E, Sinkevich O, Tyuftyaev A and Chinnov V 2010 *High temperature* **48** 97
[10] Amirov R, Isakaev E, Shavelkina M and Shatalova T 2014 *J. of Phys.: Conf. Ser.* **550** 0120
[11] Rychagov A and Volfkovich Yu 2009 *Russian Journal of Electrochemistry* **45** 304
[12] Ferrari A, Meyer J, Scardaci V, Casiraghi C, Lazzeri M, Mauri F, Piscanec S, Jiang D, Novoselov K, Roth S and Geim A 2006 *Phys Rev Lett* **97** 187401
[13] Choi H, Jung S, Seo J, Chang D, Dai L and Baek J 2012 *Nano Energy* **1** 534

Formation of sensor array on the AFM chip surface by magnetron sputtering

I D Shumov[1], S L Kanashenko[1], V S Ziborov[2], Yu D Ivanov[1], A I Archakov[1], T O Pleshakova[1]

[1]Institute of Biomedical Chemistry, Moscow, Russia.
[2]Joint Institute for High Temperatures RAS, Moscow, Russia.

E-mail: shum230988@mail.ru

Abstract. Development of atomic force microscopy (AFM)-based nanotechnological approaches to highly sensitive detection of proteins is a perspective direction in biomedical research. These approaches use AFM chips to concentrate the target proteins from the test solution volume (buffer solution, diluted biological fluid) onto the chip surface for their subsequent registration on the chip surface by AFM. Atomic force microscope is a molecular detector that enables protein detection at ultra-low (subfemtomolar) concentrations in single-molecule counting mode. Due to extremely high sensitivity of AFM, its application for multiplexed protein detection is of great interest for use in proteomics and diagnostic applications. In this study, AFM chips containing an array of sensor areas have been fabricated. Magnetron sputtering of chromium and tungsten nanolayers has been used to form optically visible metallic marks on the AFM chip surface to provide necessary precision of AFM probe positioning against each sensor area for scanning. It has been demonstrated that the marks formed by magnetron sputtering of Cr and W are stable on the surface of the AFM chips during the following activation and intensive washing of this surface. The results obtained in our present study allow application of the developed chips for multiplexed protein analysis by AFM.

1. Introduction

Development of methods for the analysis of protein markers of human diseases at low and ultralow concentrations is a relevant area of research in modern biochemistry. In this regard, the use of nanotechnology-based approaches employing molecular detectors [1] including AFM is promising. Atomic force microscopy (AFM) enables to register single protein molecules and their complexes on the support (AFM chip) surface in counting mode [2-4], reaching subfemtomolar sensitivity of protein detetion [1, 3]. For protein analysis by AFM, preliminary concentrating of target protein molecules on the AFM chip surface at the expence of specific intermolecular interaction of target proteins with chip-immoblilzed probe molecules — so called biospecific fishing — is used [3]. The use of biospecific AFM-based fishing for highly sensitive analysis of viral hepatitis C [3] and HIV [5] oncomarkers is experimentally demonstrated.

Nevertheless, in the previous studies [3, 5, 6] one AFM chip was used to determine only one target protein, as only two sensor areas – an area with immobilized probe molecules against the target protein, and a control area were formed on its surface. The formation of an array containing multiple sensor areas which can contain different types of immobilized probe molecules, will allow multiplexed analysis of protein biomarkers by AFM with single-molecule sensitivity. It should be noted that due to small scanned area of AFM (of the order of several square microns), precise positioning of the AFM probe against each sensor area should be provided to obtain correct data. In the present study, the use of an array of marks formed on the AFM chip surface by magnetron sputtering is proposed to provide visual orientation upon precise positioning of the AFM probe.

The choice of method applicable for the formation of an array of marks and sensor areas is conditioned by the surface quality requirements upon the single-molecule experiments using AFM. As a rule, the size of protein molecules is below 5 nm [7]. The surface roughness therefore should not exceed 1 nm. Hence, no particulates causing excessive roughness of the AFM chip surface must be produced during the surface marking. The marks thus formed must also be stable on the chip surface, i.e. must have good adhesion to this surface, and must not disintegrate during the further use of the chip.

Magnetron sputtering allows obtaining coatings of high purity with good adhesion to the substrate surface and low porosity even at small thicknesses. This method also enables to vary deposition rate in a broad range by changing sputtering gas pressure and power applied to the target [8, 9]. In the case of high (of the order of 10^{-7} Torr) basic vacuum in the sputtering chamber, high quality coatings can be produced using not only heavy (Os, Pt) metals, but also the lighter ones (for instance, Cr). In the present study, DC magnetron sputtering deposition of chromium and tungsten nanolayers was used for the formation of an array of marks on the surface of AFM chips. After that, an array of sensor areas with immobilized probe molecules against hepatitis C protein biomarker (HCVcoreAg) was formed.

2. Methods

Rectangle (of 7×15 mm size) pieces of silanized muscovite mica (SPI, USA) treated as described in [7] were used for AFM chips fabrication. After metallic marks were formed on the chip surface, it was chemically activated to provide covalent attachment of probe molecules to the surface. Then, an array of sensor areas was formed on the activated surface by covalent immobilization of probe molecules from microdrops of their solutions dispensed on the surface. After that, the sensor areas were scanned using AFM to estimate the quality of the fabricated chips.

2.1. AFM chips marking

The chips were covered with copper shield with 16 round orifices. The diameter of each orifice was 450 μm, and the distance between the adjacent orifices was 1450 μm. The shield was fixed on the chip surface with a spring holder, and then the chip was placed into the sputtering chamber. Orion-3 magnetron sputtering system (AJA Inc., США) was used to form metallic marks (consisting of Cr and W nanolayers) on the AFM chip surface. The basic vacuum was 5.6×10^{-7} Torr. The sputtering was carried out using argon plasma at sputtering gas pressure 3×10^{-3} Torr. The distance between the metallic target and the specimen (AFM chip) was 15 cm. During sputtering, rotation of the substrate to be coated was provided with constant angle velocity 40 rpm. Sputtering was carried out in DC mode at constant power 70 W. In these conditions, the coating rate was 0.5 Å/s and 0.4 Å/s for Cr and W, respectively. At the first stage of coating process, Cr was deposited onto the substrate for 300 - 1200 s; At the second stage, W was deposited for 400 s. Both metallic targets were pre-sputtered for three minutes with the closed shutter prior to the deposition onto the substrate. It has to be noted that the marks formed on the AFM chip surface were insufficiently stable if Cr deposition time was less than 400 s, including the case of sputtering W directly on the chip surface without preliminary deposition of Cr (data not shown).

2.2. Formation of sensor areas

After the array of metallic marks was formed on the AFM chip surface, it was chemically activated using 3,3'-dithiobis(sulfosuccinimidyl propionate) (DTSSP) analogously to the procedure described in our previous studies [4, 5]. Briefly, 100 μL of 0.12 mM DTSSP solution in Dulbecco modified phosphate buffered saline (PBSD, Pierce, USA; pH 7.4, 10 mM phosphate, 150 mM NaCl) containing 30% v/v ethanol was dispenced onto the chip surface and incubated for 20 minutes in humid chamber, and then the chip was washed in EtOH : H_2O (1:1 v/v) at 650 rpm for 10 minutes and dried in nitrogen flow. Immediately after surface activation, the solutions of probe molecules — aptamers (Evrogen, Russia; 1 μM aqueous solution; nucleotide sequences are described in [10]) and antibodies (clone 1E5; Virogen, USA; 1 μM solution) against HCVcoreAg — were precisely dispensed onto the activated chip surface as an array of microdrops (of ~3 nL each) using Piezorray high accuracy system (Perkin Elmer Inc., USA). The microdrops were incubated on the activated AFM chip surface for 30 minutes in humid chamber, and after the incubation were intensively (650 rpm, 37°C) washed off with ultrapure water in a shaker for 45 minutes. In such a way, an array of sensor areas with immobilized probe molecules against HCVcoreAg was formed on the surface of AFM chips.

2.3. AFM scanning

NTEGRA atomic force microscope (NT-MDT, Zelenograd, Russia) was used to obtain AFM images of the chip surface. AFM scanning was performed in tapping mode in air using standard NSG10 AFM tips. The scan resolution was 256×256, and the scan size was 5×5μm^2. The number of scans collected for each sensor area was no less than 5.

3. Results obtained

Figure 1 displays optical images of AFM chip with metallic marks before (figure 1 (a)) and during (figure 1 (b)) the incubation of probe molecules solutions on the chemically activated surface of sensor areas. The optical image of AFM chip after the chip washing after the incubation is also shown (figure 1 (c)). As seen from Figure 1, the marks formed by magnetron sputtering did not undergo any visible changes during the activation and subsequent washing of the chip surface.

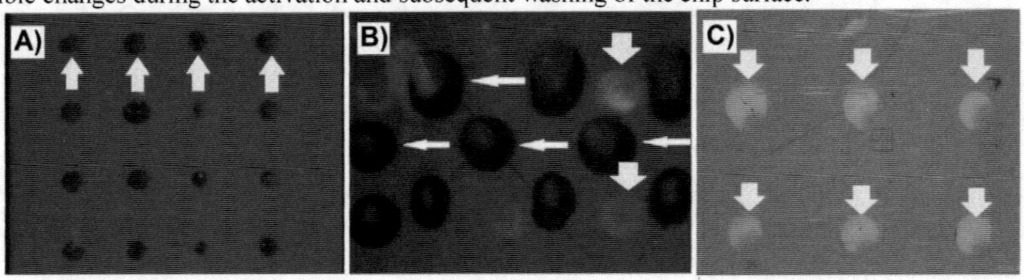

Figure 1. Optical images of AFM chip before (a), during (b, array fragment) and after (c, array fragment) the immobilization of probe molecules. Thick yellow arrows indicate the metallic marks, thin white arrows indicate microdrops (of ~3 nL each) of probe molecules solutions in sensor areas. The diameter of each mark is ~450 μm, the distance between the centers of the marks is ~1450 μm.

Figure 2 (a) displays typical AFM image of the surface of sensor area with immobilized aptamers against HCVcoreAg; the corresponding cross-section is also shown in figure 2 (B).

IOP Conf. Series: Journal of Physics: Conf. Series **789** (2017) 012053 doi:10.1088/1742-6596/789/1/012053

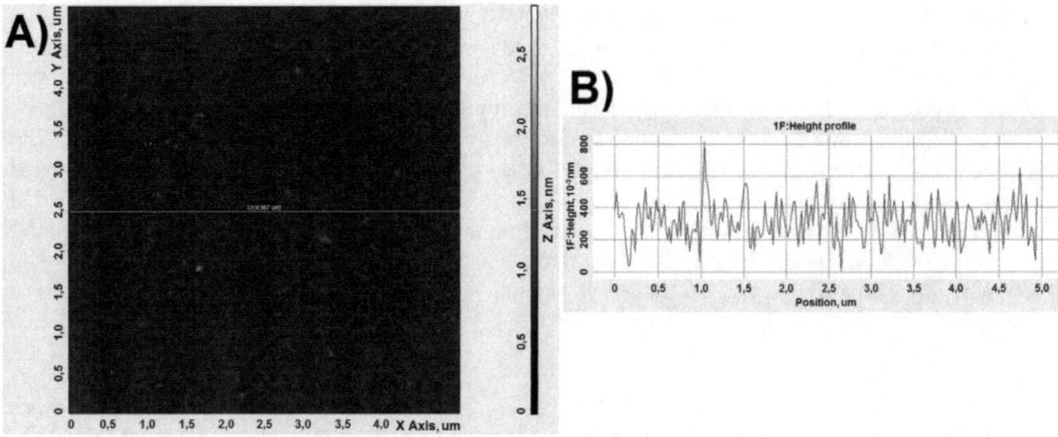

Figure 2. Typical AFM image of the sensor area surface with immobilized aptamers (a) and corresponding cross-section (b) indicated by a thin line in (a). AFM scan size 5×5 μm^2, Z scale 3 nm. The data were obtained with NTEGRA AFM (NT-MDT, Russia).

As seen from figure 2, no foreign objects with heights exceeding 1 nm is observed in the AFM image of the chip surface with immobilized aptamer probe molecules. The surface topography in the scanned sensor areas corresponds to that of silanized mica surface with immobilized aptamers [5]. This indicates that the process of AFM chip surface marking by magnetron sputtering does not cause any obstructive particualtes to appear on the chip surface, and the metallic marks thus formed are sufficiently stable on the chip surface during the chip activation with further formation of sensor areas and washing. The fabricated (magnetron sputtering)-marked chips with an array of sensor areas are therefore suitable for further use in highly sensitive detection of HCVcoreAg by AFM-based fishing.

4. Conclusions
Novel method of AFM chip surface marking has been developed. The results obtained allow application of the developed AFM-chips for multiplexed protein analysis. This is of interest for analytical proteomics and medical diagnostics.

5. Acknowledgements
The present study was performed in the framework of the Program for Basic Research of State Academies of Sciences for 2013-2020 and was granted with Russian Federation President scholarship for young scientists for 2016-2018 (project identificator SP-4280.2016.4).

References
[1] Pleshakova T O, Shumov I D, Ivanov Yu D, Malsagova K A, Kaysheva A L, Archakov A I 2015 *Biochemistry (Moscow) Suppl. Series B: Biomedical Chemistry* **9** 244-57
[2] Ramacviciene A, Snitka V, Mieliauskiene R, Ramanavicius A 2006 *Cent. Eur. J. Chem.* **4** 194-206
[3] Archakov A I, Ivanov Yu D, Lisitsa A V, Zgoda V G 2009 *Proteomics* **9** 1326-43
[4] Ivanov Yu D, Danichev V V, Pleshakova T O, Shumov I D, Ziborov V S, Krokhin N V, Zagumenniy M N, Ustinov V S, Smirnov L P, Archakov A I 2013 *Biochemistry (Moscow) Suppl. Series B: Biomedical Chemistry* **7** 46-61
[5] Ivanov Y D, Bukharina N S, Pleshakova T O, Frantsuzov P A, Andreeva E Yu, Kaysheva A L, Zgoda V G, Izotov A A, Pavlova T I, Ziborov V S, Radko S P, Moshkovskii S A, Archakov A I 2014 *Int. J. Nanomedicine* **9** 4659-70
[6] Ivanov Y D, Kaysheva A L, Frantsuzov P A, Pleshakova T O, Krohin N V, Izotov A A,

Shumov I D, Uchaikin V F, Konev V A, Ziborov V S, Archakov A I 2015 *Int. J. Nanomedicine* **10** 1597-1608
[7] Kuznetsov V Yu, Ivanov Yu D, Archakov A I 2004 *Proteomics* **4** 2390-6
[8] Kelly P J, Arnell R D 2000 *Vacuum* **56** 159-72
[9] Petroff P, Sheng T T, Sinha A K, Rozgonyi G A, Alexander F B 1973 *J. Appl. Phys.* **44** 2545-54
[10] Shi S, Yu X, Gao Y, Xue B, Wu X, Wang X, Yang D, Zhu H. 2014 *J. Virol.* **88** 1990-9

Effect of composition of electrolyte cathode on emission intensity of metal atoms in the discharge plasma at atmospheric pressure

N A Sirotkin, V A Titov

G.A. Krestov Institute of Solution Chemistry RAS, Ivanovo, 153045, Russia

E-mail: alexsad8@yandex.ru

Abstract. The emission spectra and the plasma parameters of discharges were investigated. Water solutions of copper and nickel chlorides with additions of sodium chloride were used as electrolyte cathodes. The change of the relative intensities of the emission lines of the elements (Cu and Ni) with increasing NaCl concentrations in the solution is shown. The gas temperature and the reduced electric field strength in the plasma were founded. The electron energy distribution function and the excitation rate coefficients for emitting states of copper and nickel atoms by the numerical solution of the Boltzmann equation were obtained. It was shown that in plasma emission spectroscopy measuring it is necessaryto considerthe correlation between rates of excitation and composition of the sample solution in order to determine the concentration of metal ions in the water solutions.

1. Introduction

Determination of metal ion concentrations in drinking water and wastewater industry is a crucial challenge associated with the problems of health and environmental protection. Several studies have shown that in order to control the composition of aqueous solutions the direct current discharge excited between the cathode electrolyte (analyzed solution) and metal anode can be used [1-5]. At that the lines of the metal atoms were registered in the emission spectra of the discharge. The method has a high sensitivity (up to 10^{-9} gl^{-1} [2]), low-power [3], does not require expensive equipment and allows analysis under atmospheric pressure without using an inert atmosphere. The experiments with solutions of salts showed that the intensity of the analyzed lines corresponding cations concentration in solutions is linear in a wide concentration range (0.1 to 100 mgl^{-1}) [4]. However, it is known that the emission intensity of atoms strongly depends on the pH of the solution [5]. Furthermoreit is established that various additives to the solutions have the great influence on the line intensity of atoms [2].

Reliable basis of the method requires to solving a number of questions. It is necessary to definethe mechanism of occurrence of analyzed atoms in the gas phase, mechanisms of the excitation and deactivation of the emitting states, the influence of the transfer of liquid cathode components to the gas phase on the physical parameters of the plasma. The concentration of the metal atoms in the plasma is determined by the transport processes of the liquid components of the cathode due to ion bombardment, as well as all the chemical processes that lead to the formation and expenditure of atoms [6]. If atoms go into emission state due to electron impact, it is necessary to know the excitation rate coefficients that depend on the electron energy distribution function (EEDF). In turn, EEDF depends on reduced electric field strength in the plasma (*E*/*N*) and a full set of electron collisions. The transfer processes alters the composition of thegas phase and they lead to a change of dischargecharacteristics.Thus, it can be expected that the composition of the cathode liquid electrolyte (solution) will affect the intensity of emission lines analyzed metals even at constant concentration of the corresponding ions in solution.

The aim of this work is experimental study of the effect of electrolyte composition of the cathode on the intensity of the emission lines of copper and nickel atoms and influence on the rates of excitation coefficients of the emission states.

2. Experimental part

Simplified scheme of experimental set-up is shown in Figure 1. Direct current (DC) discharge (i = 10 – 50 mA) was excited by applying high voltage between the surface of aqueous cathode and copper cathode which placed at the position 1–10 mm above the liquid surface. Water solutions of copper and nickel chlorides with concentrations of $(1–5)\times10^{-3}$ mol l^{-1} were used as cathodes. Solution of sodium chloride (c = 0.05 – 0.5 mol l^{-1}) is used as an additional electrolyte.

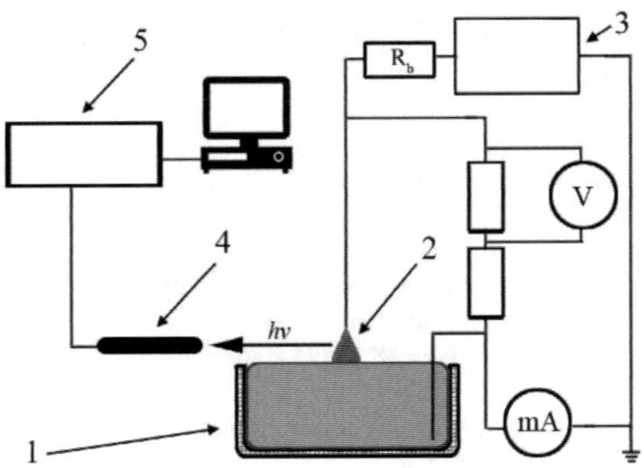

Figure 1. Experimental set-up. 1 – electrolyte cathode, 2 – zone of discharge, 3 – HV–source, 4 – optical fiber light guide, 5 – spectrometer AvaSpec-2048.

The electric field strength in plasma (E) and the cathode voltage drop (U_c) were calculated from dependences of the discharge voltage on the inter-electrode distance. These dependencies are approximated by a linear function with a correlation coefficient is not less than 0.99. The emission spectra of the positive column (λ = 200 – 950 nm) were recorded by the AvaSpec-2048 spectrometer (grating 600 lines mm $^{-1}$). Gas temperature was found fromthe intensity distribution in the radiation band of N_2 ($C^3\Pi_u \rightarrow B^3\Pi_g$, 0–2) transition as described in [7]. To evaluate the reproducibility of the results were performed five independent measurements. Limits of the confidence intervals were calculated with a confidence level 0.9.

3. Results and discussions

The dependences of the relative intensity of Cu atoms emission lines (λ = 324.7 nm, ε_{ex} = 3.82 eV) and Ni atoms emission lines (λ = 361.9 nm, ε_{ex} = 3.54 eV) on the concentration of the corresponding salts in solution at different concentrations of the additional electrolyte NaCl are shown in Figure 2. The emission intensity is directly proportional to the concentration of respective cations in the liquid cathode, but the angular dependency coefficients are significantly various for different concentrations of sodium chloride. Minimum slope of the straight line I = f(c) is observed without of NaCl in the solution, the maximum is observed in additional electrolyte concentration 0.05 mol l^{-1}. At least there are two reasons of influences of the sodium chloride additive in liquid cathode on the intensity of emission analyzed atoms.

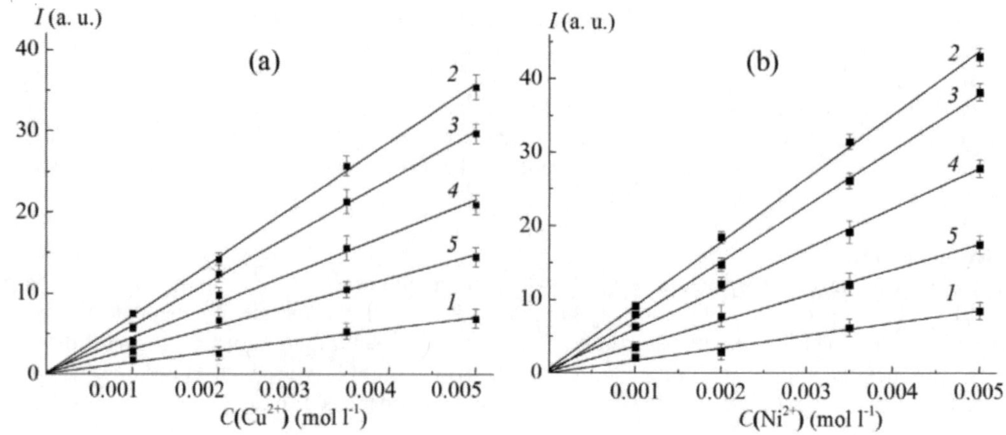

Figure 2. The intensities of the lines of copper (a) and nickel (b) atoms at various concentrations of sodium chloride in solution: 0 (*1*), *2* – 0.05 (*2*), 0.1 (*3*), 0.2 (*4*), 0.5 mol l^{-1} (*5*). $i = 50$ mA.

First, NaCl additive changes the value of the cathode voltage drop which in its turn determines the average energy of positive ions bombarding the surface of liquid cathode and the rate of cathode sputtering. Second, the electric field strength in plasma and character of the EEDF are changed and as a result the rates of excitation of the emitting states by electron impact are altered.

Experiments showed that the rotational temperature of N_2 ($C^3\Pi_u$), which can be identified with the temperature of the gas is (1600 ± 100) K and weakly depends on the discharge current and on the composition of the electrolyte cathode. At once, the reduced electric field strength in plasma is decreased with increasing concentration of sodium chloride in the solution (Figure 3). As a result, it changes behavior of EEDF and the rate coefficients of threshold processes including processes of electron impact excitation states Cu and Ni atoms. The rate coefficients of the electron impact excitation of Cu ($4p^2P^o_{3/2}$) and Ni ($4s^2D$) were calculated by using EEDF (Table 1). The electron energy distribution function was obtained by numerical solution of the Boltzmann equation (software package BOLSIG+) [8]. The collisions of the electrons with the molecules N_2, O_2 (at a ratio of mole fraction $y(N_2)$:$y(O_2)$ = 4:1), H_2O and NO were taken into account in our calculations with using cross-sections from [9, 10]. The mole fraction of water in plasma (~ 0.01) was evaluated by the rate of accumulation of condensate in a trap, and for NO molecules (~ 2.5×10^{-3}) by using data obtained by means of numerical modeling of processes in plasma [6].

The calculations showed the nearly twofold decrease in the rates of excitation coefficients of the emission states with the addition of sodium chloride in solution-cathode. However, it does not fully explain the changes observed in the emission intensity. It is clearly seen that for further analysis of the results one should take into account the change in the rate of transfer of the solution components in plasma, as well as the mechanisms of excitation and deactivation of the emitting states of analyzed atoms. Changing the efficiency of the liquid cathode components transfer in the plasma can be explained by the dependence of the cathode voltage drop (U_c) on the composition of the electrolyte cathode. For example, the value of cathode voltage drop is (615±20) V in discharge with liquid cathode without sodium chloride but at a concentration of NaCl 0.05 mol l^{-1} U_c is (490±10) V and weakly increases with increasing concentration of additional electrolyte.

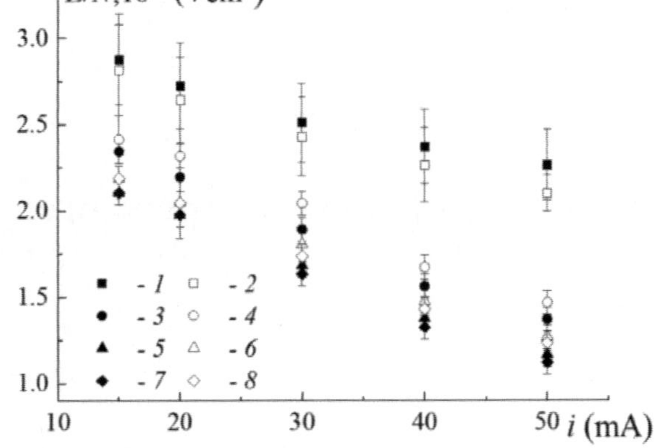

Figure 3. The dependences of the reduced electric field strength in the plasma of the discharge current at different concentrations of electrolyte solutions: *1, 3, 5, 7* – cathode – solution of $CuCl_2$, *2, 4, 6, 8* – cathode – solution of $NiCl_2$. Concentration of NaCl – 0 (*1, 2*), 0.05 mol l^{-1} (*3, 4*), 0.1 mol l^{-1} (*5, 6*), 0.5 mol l^{-1} (*7, 8*).

Table 1. The rate coefficients of the electron-impact excitation of Cu ($4p^2P^o_{3/2}$) и Ni ($4s\,^2D$)

i (mA)	*k* (cm^3 s^{-1})					
	Cu			Ni		
	Concentration of NaCl in solution (mol l^{-1})					
	0	0.05	0.5	0	0.05	0.5
30	2.68×10^{-10}	1.72×10^{-10}	1.49×10^{-10}	2.64×10^{-10}	1.69×10^{-10}	1.46×10^{-10}
40	2.16×10^{-10}	1.42×10^{-10}	1.21×10^{-10}	2.12×10^{-10}	1.41×10^{-10}	1.19×10^{-10}
50	2.07×10^{-10}	1.24×10^{-10}	1.09×10^{-10}	1.97×10^{-10}	1.22×10^{-10}	1.06×10^{-10}

Acknowledgments
The study was supported by Russian Foundation for Basic Research according to the research project 16–33–60061 mol_a_dk.

References
[1] Mezei P and Cserfalvi T 2007 *Appl. Spectr. Rev.* **42** 604
[2] Yang C, Wang L, Zhu Z, Jin L, Zheng H, Belshaw N S and Hu S 2016 *Talanta* **155** 314
[3] Shekhar R, Karunasagar D, Ranjit M and Arunachalam 2009 *J. Anal. Chem.* **81** 8157
[4] Bencs L, Laczai N, Mezei P and Cserfalvi T 2015 *Spectrochim. Acta, Part B.* **107** 139
[5] Webb M R, Andrade F J and Hieftje G M 2007 *J. Anal. At. Spectrom.* **22** 766
[6] Titov V A and Sirotkin N A 2014 *Proc. VII International Symposium on Theoretical and Applied Plasma Chemistry* (Ivanovo: ISUCT) p 160
[7] Titov V A, Rybkin V V, Smirnov S A, Kulentsan A L and Choi H S 2006 *Plasma Chem. Plasma Process* **26** 543
[8] BOLSIG+ 2005 CPAT: http://www.cpat.ups-tlse.fr
[9] Plasma Data Exchange Project: http://fr.lxcat.net
[10] Smirnov Yu M 2013 *Opt. Spectrosc.* **115** 787

Properties of the particles deposited from a low-temperature plasma arc discharge

N A Smolanov
Ogarev Mordovia State University

E-mail: smolanovna@yandex.ru

Abstract. The possible mechanisms for the formation of fractal particles of the plasma arc have been studies. The existence of dust structures of two types strongly and weakly magnetic was found. Paramagnetism of particles is the result of the synthesis of titanium carbonitride with embedded (encapsulated) magnetic substrate elements. The magnetization of particles indicates the presence in the test nanostructure particles. inhomogeneous distribution of electric charges in motion conducting bodies (drops) in a magnetized plasma is the reason for the formation of dust particles.

1. Introduction

Thin films are the main product in electric deposition. Dust particles and solid deposits in low-temperature plasma units are considered "hazardous" materials. They were found in HID devices back in 1874 and were seen simply as the mud [1]. Studies of structures and properties of these materials allow the understanding of their mechanisms of formation. Similar dust particles are formed in the tokamak as a result of the erosion of graphite and metal walls [2-3]. Processes in the cathode spot are the basis of their education. Therefore, the structure and properties of these materials provide an understanding of the mechanisms of their occurrence and the processes occurring in the plasma. Analysis of the results of recent work has revealed the structure of the cathode spot [4] and the properties of dusty plasma [5, 6]. The same nature of the processes occurring in the cathode spot and dusty plasma in an external magnetic field was set.

Products cathode spots are the result of explosive emission and secondary ion bombardment [7-9]. It was found that at a certain value of the external magnetic field, the cathode spot of a vacuum arc moves in the opposite direction of the force of Ampere. The phenomenon of changing the direction of rotation of the cathode spot is known for a long time, but not completely understood. A number of studies found similar behavior of dusty plasma in a magnetic field. The presence of dust granules of the magnetic moment due to the elemental composition and surface loop current generation of rotating charged particles [6, 10]. The magnetic properties of dusty plasmas are determined by various factors, depending on the magnitude of the magnetic induction. In a weak magnetic field is magnetized only the electronic component of the plasma, in the middle field (0.1-0.3 T) magnetized ions, and only in strong fields (2-3 Tesla) themselves are magnetized dust granules. As the external magnetic field can be dusty plasma rotation stop, followed by a change of its direction (retrograde). However, the reason of occurrence of threshold character rotation, as in the case of reverse movement of the cathode spot is still unclear [4-6]. In this paper, the possible mechanisms of dusty plasma evolution from the cathode spot before its recombination on the substrate and the walls of the vacuum chamber were considered. Particular attention is paid plasma mass - separation of particles in the plasma system of magnetic fields cathode assembly [11]. In a plasma accelerator as in all ion-plasma systems can develop different types of plasma instability. Possible trajectories of dust particles of different charges and masses examined. It was found that when driving a drop in the plasma flow inside the convective mass transfer can occur, drop under pressure flows of ions and electrons, and in some cases there is a rotation of the drops and their destruction ("Coulomb" explosion). Metal droplets leaving the cathode and evaporates rapidly converted to plasma formation in the immediate vicinity of the cathode surface. Some of them, however, is deposited near the cathode. Drops, crystallized and deposited near the cathode plane, have the elemental composition heterogeneity (Figure 1). We have shown that in an arc discharge in the presence of magnetic and electric fields, the structure of condensed matter near the cathode has a helical character due to the rotational movement of dust (drip) of the particles as a whole [11].

Dust particles can be divided into 2 groups. The first group (slow) is held gas-dynamic plug and settles near the cathode and substrates in the chamber. The second group is the fast ions and droplets formed in the cathode spot with the emission from the excited state create nanostructures with rapid (adiabatic) cooling the walls of the vacuum. The presence of 2 types of particles in the plasma emission arcs supported in a number of effects [11-14].

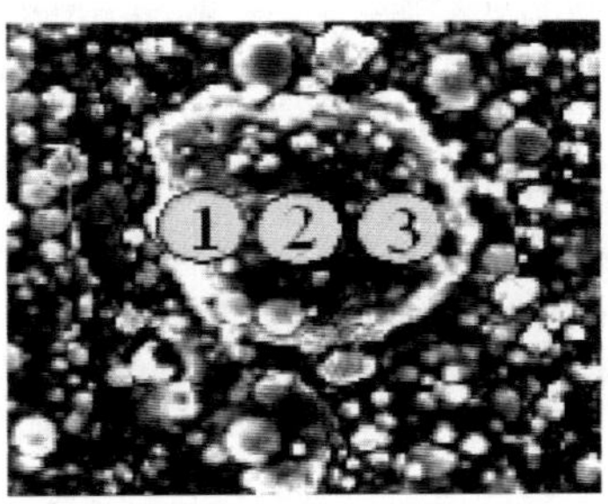

Элемент	Содержание, ат.% 1	2	3
C	9.56	16.76	18.15
N	34.87	34.46	43.81
O	1.61	3.71	3.00
Ti	53.50	44.44	33.57

Fig.1. Electron microscopy of drop and table content elements in it (droplet size 6 mcm)A

2. Work purpose

In this paper, which is a continuation of studies [11-14], an attempt to explain the formation of dust structures condensed from the plasma arc discharge of particles and their magnetic properties were made. Of particular interest to the processes in arc discharge plasma arose after the discovery of our fractal structures on the walls of the vacuum chamber [13]. This makes it relevant to study mechanisms of films and microparticles in a plasma arc. There is reason to believe that inter-process droplet erosion in the fusion plasma and the formation of droplets from the cathode spot of the arc discharge, there are some similarities. The structure of the cathode plasma jet in a vacuum arc in the presence of magnetic and electric fields is the subject of many studies, but the structure and properties of plasma deposited particle flux often are often unexplored.

Therefore the aim of this work was a detailed study of fine powder, deposited on the walls of the vacuum chamber by spraying a titanium cathode (VT1-0 alloy) arc method. At the same time we are based on generalization of ideas and the results obtained in the study of current sheets in plasma [15, 16], the plasma rotation [4-6, 8]. Taken into account the results of the study of the spectral composition of the plasma arc, the nature of the cathode spot drip stains [17, 18] (glowing drip stains).

3. Results. EPR - spectroscopy

Particulate deposited on the walls of the vacuum chamber of HHB-6 in the process of ion-plasma deposition of multilayer coating films on the substrate (steel 12X18H10T). selection of fractions with typical particle sizes [16] is produced by magnetic separation of the starting powders in the form of soot. EPR spectra were recorded on rf PS100X (frequency 9.3 GHz). The particle sizes obtained by applying potential on a submarine to magnetic separation, are within 5-100mkm. When the particle deposition without applying a potential (anode - of the vacuum chamber wall) of the basic number is in the range 70-130 microns.

Table 1. The elemental composition of the powders (substrate potential of -250 V)

Fraction	The elemental composition of the powders at. %					
	Ti	N	O	C	Fe	Cr
non-magnetic	44.23	9.69	35.29	10.30	0.12	0.03
magnetic	51.60	7.40	29.97	10.21	0.26	0.10

It was found that the change in the substrate potential, which produced the precipitation leads to the appearance of the EPR signal with a g-factor is probably due to the presence of nanoparticles. in the form of fullerenes with inclusions of atoms of magnetic elements. After dissolving with toluene, and then filtering the signal in the ESR spectrum disappears

Analysis of the EPR spectra showed (Fig. 2, 3) that the powders are characterized by magnetic properties with different g-factors. A feature of the EPR spectrum of the test powder is its asymmetric line shape. Note also that with a decrease in the size fractions in the absorption spectrum there are additional lines. Magnetic separation and leads to the appearance of new lines in the absorption spectrum of the magnetic particles and weakly magnetic fractions is not observed in the initial state.

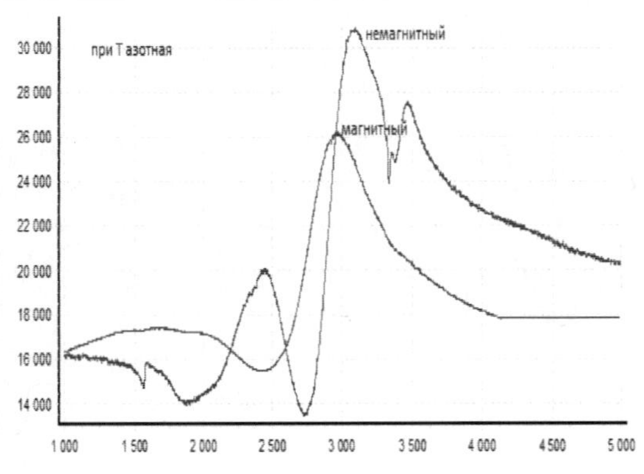

Fig. 2. EPR - spectra of the powders at liquid nitrogen temperature

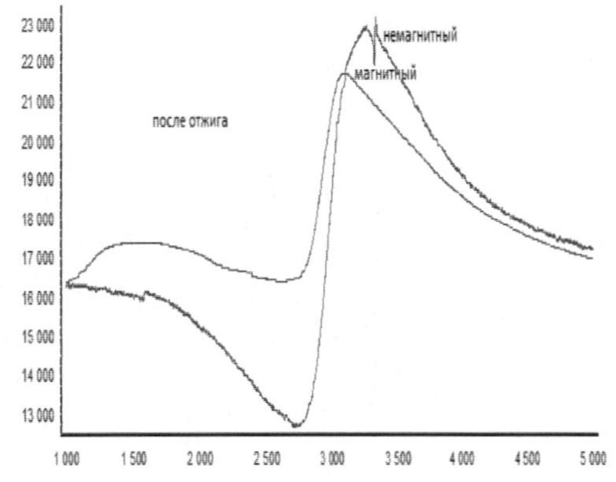

Fig. 3. EPR - spectra of the powders after annealing

At T = 78 K in a non-magnetic powder with a size of 30-60 microns appears hyperfine structure and additional absorption line (N = 1590 kOe, g = 4,18). The number of absorption lines of the magnetic fraction obtained in the absence of potential on the substrate - one (g = 2,47), while in the non-magnetic - 5 (g = 4.21; 3.08; 2.28; 1.99; 1 94) (Figure 2). Annealing in air at 300C during 1 hour leads to a change in the spectra. In the magnetic fraction is a broad line (g = 2,33), while the non-magnetic fraction - two (g = 2,22; 2,00) (Figure 3). When you change modes produce particles observed a shift of one of the absorption lines in the region of the weak, and the other lines - in the region of strong fields. This changes the g factor.

4. The magnetization of the particles.

Magnetic properties (hysteresis) of particles of different factions and states (starting, after magnetic separation) were tested for vibration magnetometer MicroSense (EZ11 model). As an example, Figure 4 shows the non-normalized magnetization curves of particles with a diameter less than 20 microns in the initial state (curve 1) and after magnetic separation (highly magnetic fraction - line 2 and weakly - curve 3). A similar behavior of electric arc spray products was observed in [13]. The appearance of the magnetization curve of the hysteresis loop 2 is the result of the formation of dust deposited in small ferromagnetic particles. It should be noted that in an external applied magnetic field with a strength of up to 8 kOe, magnetization curve reaches its saturation. Thus, it was found that the magnetic properties of the magnetic powders after their separation are determined by the presence of ferromagnetic impurities. Their content depends on deposition conditions and the size fractions. The elemental composition of the particles is shown in Table 1.

IOP Conf. Series: Journal of Physics: Conf. Series **789** (2017) 012055 doi:10.1088/1742-6596/789/1/012055

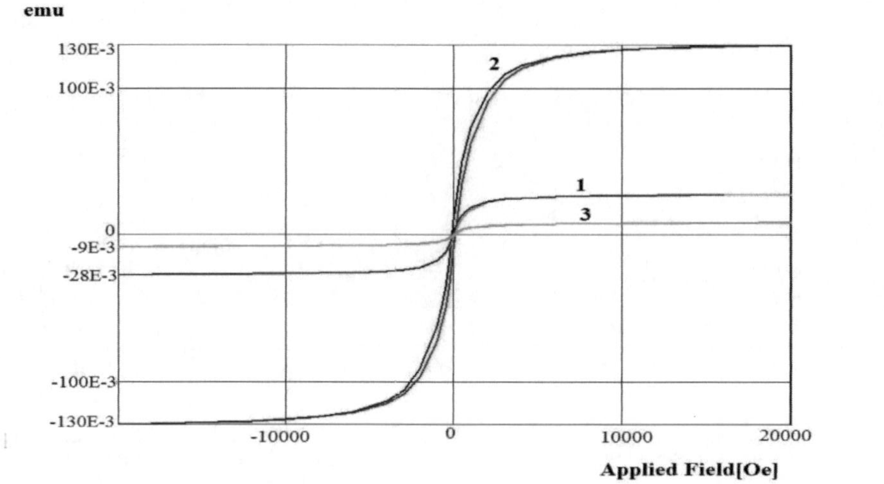

***Figure 4**. Hysteretic magnetization curves of particulate structures of the plasma ar.*

5. Discussion.

Thus, the question about the reasons for the formation of the fractal structure of the dust particles from the arc in the magnetic field is reduced to the study of the electric field caused by the uneven distribution of electric charge in a moving conducting bodies (drops) in a heterogeneous low-temperature plasma. Unsteady collective electromagnetic field leads to the appearance in the plasma containing charged particles, oscillatory processes. The presence of the electrons and ions of their own magnetic moments may lead to unsteady oscillatory complex collective processes in the plasma. Emerging with stationary magnetic structure caused by electrically charged particles, electrons. Therefore, the movement of the conductive liquid droplets in the plasma flow, their instability and decay, largely determined by the density of the charge distribution on the surface of the droplets (droplet rotation).

Several studies have shown that the maximum curvature of the surface of the conductor and the surface charge density may have different locations on the surface of the conductive body (drops) [19]. The area where the distribution of the surface charge density of the conductor is greatest substantially different from the region of maximum curvature of the body surface. well developed (fractal) surface hardening of droplets of varying dispersion, size, shapes (from cubic to spherical), and the structure of the nanoparticles (homogeneous and composite particles with the hard magnetic material of the core and the soft magnetic shell) is appeared [20]. Such variations lead to a change in the hysteresis loop (Figure 4). It is possible to non-central location of the titanium atom in the lattice of the resulting compounds. The resulting asymmetry of the EPR signal indirectly confirms this. Also keep in mind that in a dusty plasma rejecting the gyromagnetic ratio of 2 is split ion-cyclotron mode (at $g < 2$) and split-polarized electromagnetic branch ($g > 2$) [21].

References

[1] Morozov A I 2008 Vvedeniye v plazmodinamiku. M.: FIZMATLIT – 616 (in Russian)
[2] http://ufn.ru/ru/articles/2010/10/c/

[3] Zalavutdinov R. H., Gorodetsky A. E., Zakharov A. P., GUYS, it is gray. Thermonuclear synthesis, 2011, issue 1, page 39-47.
[4] Bobrov Yu K , Bystrov V P , Rukhadze A A 2006 *Journal of Technical Physics*, **76** 35(in Russian)
[5] Diachkov L G. 2011 FNTP-2011. **2** 152(in Russian)
[6] Karasev V Yu, Dzliyeva E S 2011 FNTP-2011. **1** 147(in Russian)
[7] Luchkin A G, Kashapov N F and Luchkin G S 2013 *J. Phys.: Conf. Series* **479** 012019
[8] http://ufn.ru/ru/articles/1978/8/c/
[9] Barengolts, S. A., Litvinov E. A. 1998 *Journal of Technical Physics*, **68** 60(in Russian)
[10] Saifutdinov A I, Fairushin I I and Kashapov N F 2016 *JETP Lett.* **104** 180–185
[11] Smolanov N.A., Pan'kin N.A. // Journal of Surface Investigation. X-ray, Synchrotron and Neutron Techniques. 2014, V.8, №.5, p. 1089–1092.
[12] Smolanov N A , Pankin N A , Batin V V , Pavkin E P 2014 *Prikladnaya fizika* **1** 30 (in Russian)
[13] Smolanov N A , Neverov V A 2015 *Pisma o materialakh* **5** 179 (in Russian)
[14] Smolanov N..Physics and chemistry of materials processing. 2015. No. 3. S. 18-23.
[15] http://ufn.ru/ru/articles/2010/9/h/
[16] http://ufn.ru/ru/articles/2015/2/a/
[17] Batrakov A.V., Juttner B., Popov S. A., Proskurovsky D. I., Vogel N. //Letters to JETP. – 2002. - vol. 75. - vol. 2. - P. 84 – 91.
[18] Mesyats G. A., Barengolts S. A. //JETP Letters. – 2002. - vol. 75. - vol. 6. - P. 306 – 308.
[19] N A Smolanov. //Journal of Physics: Conference Series 669 (2016)
[20] S.-H. Noh, W. Na, J.-t. Jang, J.-H. Lee, E. J. Lee, S. H. Moon, Y. Lim, J.-S. Shin, J. Cheon, Nano Lett. 2012, 12, 3716−3721
[21] Poles P. A., Rusakov N. E., Samukhina Yu. V. News of RAS, сер.физ., 2016, 80. Page 604.

Numerical investigation the dynamics of vaporization at the flow of liquid methane in channel with variable section

B A Snigerev[2], A L Tukmakov[1] and V G Tonkonog[1]

[1]Kazan National Research Technical University of A. N. Tupolev-KAI, Russia
[2]Institute of Mechanics and Engineering, Kazan science center, Russian Academy of sciences, Russia

E-mail snigerev@imm.knc.ru

Abstract. Numerically transient dynamics of vaporization of the liquid bubbling up methane flow in the Laval nozzle are analyzed. The equation conservation of mass, momentum and energy with averaging of Favre are used. For numerical modeling of cavitation the additional equation for transport the volume fraction of vapour with special source terms for evaporation and condensations is used. Non-stationary process of evaporation is analyzed in various checkpoints in diffuser portion of the nozzle are investigated. By the study of the unsteady dynamics of changes the characteristic hydrodynamic quantities of two-phase gas-liquid flow, the main jet stream pattern of destruction of the process of liquid methane at the expiration of the nozzle throat are identified.

1. Introduction

Two-phase flows are the primary working fluid in power plants, chemical engineering, in different elements of the new technology. The processes of formation and evolution of bubbles play an important role in physics, chemistry, technology, medicine and many other areas of life. The great practical and the fundamental interest in the problem of boiling liquids, in which the decisive role played by the emergence and evolution of steam bubbles are important and still far from the final solution. The raging streams are used in many technological processes. Among them two-phase flow in boilers and steam generators, flow saturated and subcooled liquid in channels of the cooling system, a special role is played by the two-phase expires in accidents at nuclear reactors, and many others [1-2]. The increase in the rate of intensification technological processes, further increasing the technical level of production leads to the fact that you are using the ultimate and critical thermal load in the power equipment. Wherein the liquid the coolant may be in a nonequilibrium or metastable state, when is rapid reduction of the pressure in the liquid at the end of the boiling liquids from vessel high pressure through short channels or of the Laval nozzle to the atmosphere. The initial parameters of the liquid, close to the thermodynamic critical point, overheating can achieve the limit values and the flow can be substantial homogeneous vaporization. The relevance of the study the critical conditions of the flow and explosive boiling in superheated liquid flows is connected with the problem of safety of energy equipment, in particular, with the objectives description of gap elements cooling systems, in relation to the needs of nuclear power engineering, cryogenic engineering, environment [1-2].

2. Governing equations

To simulate the motion of a mixture of liquid and vapor phases Favre-averaged Navier-Stokes equation for a homogeneous medium with a variable density are used, supplemented with the equation for the enthalpy transfer in the form [2-4]:

$$\frac{\partial \rho_m}{\partial t}+\frac{\partial(\rho_m u_j)}{\partial x_j}=0, \tag{1}$$

$$\frac{\partial(\rho_m u_i)}{\partial t}+\frac{\partial(\rho_m u_i u_j)}{\partial x_j}=-\frac{\partial \hat{p}}{\partial x_i}+\frac{\partial}{\partial x_j}\left[(\mu_m+\mu_t)\left(\frac{\partial u_i}{\partial x_j}+\frac{\partial u_j}{\partial x_i}-\frac{2}{3}\frac{\partial u_k}{\partial x_k}\delta_{ij}\right)\right], \tag{2}$$

$$\frac{\partial}{\partial t}\left[\rho_m(h+f_V L)\right]+\frac{\partial}{\partial x_j}\left[\rho_m u_j(h+f_V L)\right]=\frac{\partial}{\partial x_j}\left[\left(\frac{\mu_m}{\mathrm{Pr}_L}+\frac{\mu_t}{\mathrm{Pr}_t}\right)\frac{\partial h}{\partial x_j}\right], \tag{3}$$

$$\frac{\partial \alpha_L}{\partial t}+\frac{\partial(\alpha_L u_j)}{\partial x_j}=\frac{1}{\rho_L}(\dot{m}^+ + \dot{m}^-) \tag{4}$$

There x_i - Cartesian coordinates, u_i - the components of the velocity vector, t -time, $\hat{p}=p+(2/3)\rho_m k$, p - pressure, k - kinetic energy of turbulent pulsations, $\rho_m=\alpha_L\rho_L+(1-\alpha_L)\rho_V$ - the density of the mixture, ρ_L - the density of the liquid, ρ_V - vapor density, α_L - the volume fraction of the liquid phase in the mixture, δ_{ij} - the Kronecker symbol, $\mu_m=\alpha_L\mu_L+(1-\alpha_L)\mu_V$ - dynamic coefficient viscosity of the mixture, μ_L, μ_V - the coefficients of dynamic viscosity of the liquid and vapor respectively, $h=c_p T$-specific enthalpy, T - temperature, c_p - thermal capacity coefficient at constant pressure, f_V - is the mass fraction of vapor, defined by the relation $f_V=\rho_V(1-\alpha_L)/\rho_m$, L - latent heat of vaporization, Pr_L, Pr_t - molecular and turbulent ($\mathrm{Pr}_t=0.9$) Prandl number, $\dot{m}^-$, $\dot{m}^+$ - model source components, taking into account the process of vapor condensation and liquid evaporation. Assuming the presence of non-condensable gases in the liquid with a mass content f_g, in the paper [4] are represented equations for $\dot{m}^-$, $\dot{m}^+$ in the form of

$$\dot{m}^+=C_e\frac{\max(1.0,\sqrt{k})(1-f_V-f_g)}{\gamma}\rho_L\rho_V\sqrt{\left(\frac{2}{3}\frac{p_V(T)-p}{\rho_L}\right)}, \quad p<p_V(T), \tag{5}$$

$$\dot{m}^-=C_c\frac{\max(1.0,\sqrt{k})f_V}{\gamma}\rho_L\rho_V\sqrt{\left(\frac{2}{3}\frac{p_V-p}{\rho_L}\right)}, \quad p\geq p_V(T) \tag{6}$$

In relations (5) - (6) $p_V(T)$ - saturation pressure of methane at the temperature T, γ - coefficient of surface tension of liquid methane. Empirical constants C_e, C_c that determine the rate of evaporation and speed of condensation and have values $C_e/\gamma=1.665\cdot 10^3$, $C_c/\gamma=1.225\cdot 10^3$. The coefficient of turbulent viscosity is calculated according to the formula of Kolmogorov-Prandl $\mu_t=C_\mu\rho_m k^2/\varepsilon$, where to determine the kinetic energy of turbulence k and rate of dissipation ε two-parametrical model of turbulence $k-\varepsilon$ is applied [5].

3. Results

The study of flushing flows of liquid methane at various modes for a plane Laval nozzle are investigated, the design scheme of which is shown on figure 1. Minimum cross-section of the nozzle is located at $x_1 =$ 0.1 m ($l_6 =$ 100 mm in figure 1), the height is $l_4 =$ 4 mm. The widening of the nozzle ($0.1 \le x_1 \le 0.25$ m) has a length of $l_3 =$ 150mm, width of the nozzle at the cut in a section $x_1 =$ 0.25 m have the size $l_2 =$ 46 mm. The cone angle of the diffuser nozzle is $\gamma = 6^0$. To study the dynamics of the jet breaking the unsteady changing of physical quantities, characterizing the gas-vapor stream, was investigated, namely, the value of the longitudinal rate u_1 , the value of the local void fraction α_V , density mixture ρ_m , the static P_{st} and the total pressure P_{tot} in the three control points with coordinates $x_p^i (i = 1,3) =$ 0.11m, 0.15m, 0.23m. The control points are located on the axis of the nozzle, as shown in figure 1. For simulation of flow in a Laval nozzle a grid with 124000 finite volumes with a significant concentration in the area of throat constriction was used. In the calculation of turbulent flows special attention should be paid to mesh quality near the walls, in regions of large gradients of flow parameters.

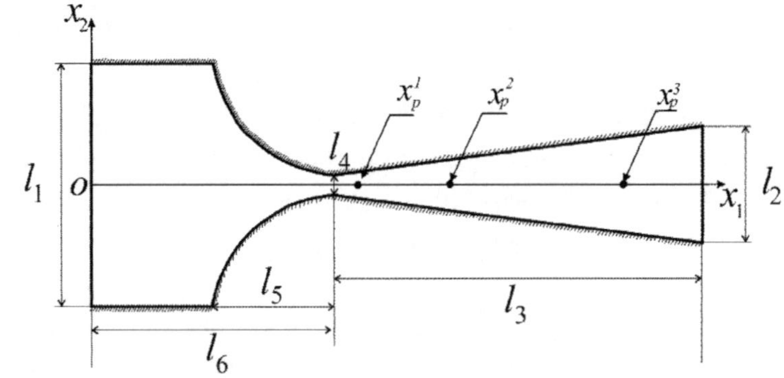

Figure 1. Computational domain of the Laval nozzle with control points $x_p^i (i = 1,3)$.

For low pressure flow the inlet pressure is $P_0 = 0.57$ MPa, the outlet of $P_b = 0.30$ MPa and the inlet temperature of liquid methane $T_0 =$ 124 K. On figure 2 the distribution over time at the control points $x_p^i (i = 1,3)$ of variables α_V, u_1 are shown. The curves marked with numbers corresponding to the respective number of control points. From the distribution of void fraction α_V , shown in figure 2, this demonstrates that the dynamics of the jet breaking happens over time as the waves of pressure in diffuser part of the nozzle are moved. Upon reaching the control point 1, in the neighborhood of a given point the phase transitions processes are intensified, leading to changes the liquid medium to the gas-droplet state in a very short period of time $\Delta t \le 0.001$ s. Volumetric gas content in the given point rises sharply to the value of α_V = 0.15 and further reaches steady-state value. As the distance from the throat of nozzle the destruction of the jet increases and volumetric gas content increases in the points 2, 3 to α_V = 0.55, 0.75 respectively. After the passage the pressure wave the value of α_V in all the control points reaches a certain steady-state value and does not change. The figure 2 shows changes the axial velocity distribution at three points as the distance from the throat of the nozzle. Dynamics of change the speed over time in the diffuser part of the nozzle into three sections can be divided. The first part is a linear plot of growth rate, then at the passage of the pressure wave and the strong intensification of phase transitions of the second portion with a sharp speed increase is formed.

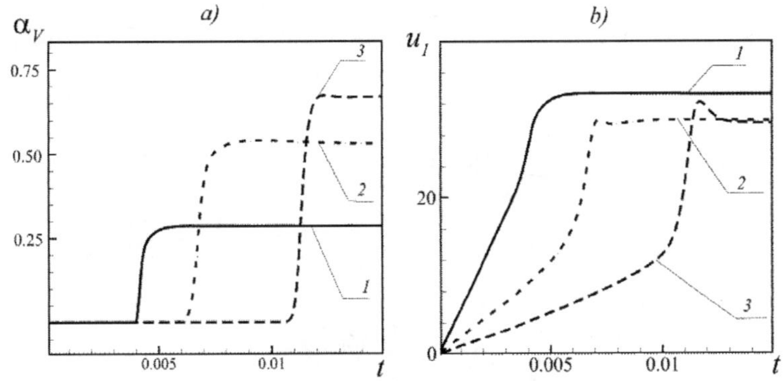

Figure 2. The distributions over time of different variables in control points $x_p^i (i = 1,3)$:
$a)\ \alpha_V$; $b)\ u_l$

The third section characterizes the speed for steady state. Then upon reaching a certain time when a steady state is established in the flow, the speed also changes not much. The some differences in the dynamics of speed in different control points, that describe the processes of destruction of the jet we will note. Closer to the throat the gradients increasing of speed in the initial stage of destruction are greatest, but as the distance from the nozzle throat they are reduced. The figure 3 shows contours for the case of low-pressure flow at P_0 = 0.57 MPa, $\varepsilon_a = 0.76$, $\Delta t_H = 4.0^0$K for three points time t_i = 0.005, 0.025, 0.2s. From the figure 3a shows that the process of destruction the jet is initiated, when an intensive formation of bubbles near the walls of the channel in the diffuser part of the nozzle is started. The intensification of the process of growth of bubbles near the throat accompanied by the formation of the pressure zone equal to the vapor pressure of the saturated methane. The area of low pressure (figure 3*b*) is displaced downward stream, accompanied by the destruction of the jet liquid methane and increase in the content of bubbles in the flow core. The complete destruction of the jet occurs at time t_3 = 0.2 s., as illustrated in figure 3*c*. This figure shows that in the minimum section of the nozzle incompressible fluid is moving, and the process of vaporization develops directly after passing the nozzle throat. The characteristic "stretching" of the contours of the volumetric gas phase in the central part of the nozzle shows that vapor phase are intensively formed in the boundary layer along the channel walls and then extends to the axis of the nozzle. At some distance from the minimum section the liquid kernel disappears and bubble transitional structure will be transformed to the vapor-bubble flow. In this paper to simulate the vaporization process a model with an additional transport equation of volume of phase with the source terms for modeling the speed of condensation and evaporation is used [5,6]. The model assumes that the rate of change of phase state is proportional to the characteristic speed, as which the relative velocity between the liquid and bubbles is selected. The relative velocity V_{ch} of the same order as the velocity of turbulent fluctuations, for which the estimate $V_{ch} \square \sqrt{k}$ is fair. Comparing the calculated data on the distribution of the void fraction of vapor, is shown in figure 2 it can be concluded that the structure of wet steam flow in the diffuser part of the nozzle for low-pressure have the some characteristic features. The vapor phase initially is greater formed in the boundary layer near the walls of the channel and then will extend with the flow to the nozzle axis. The characteristic feature seen in the fact that the increase the velocity of bubbling flow in the Laval nozzle is accompanied by a decrease in the volumetric void fraction in the central part of the expanding portion of nozzle, but and in the near-wall regions it increases. For high-pressure flow rate of vaporization in the region near the wall increases due to the increase of turbulent kinetic energy near the walls and consequently the growth of the characteristic speed. Small nucleation particles that appeared near the walls, growing in size and are carried by flow to the central portion of the nozzle.

Due to the increased flow rate, the time for the growth of bubbles decreases, and they more quickly propagate downstream, which leads that the volumetric void fraction near the axis of the nozzle is decreased.

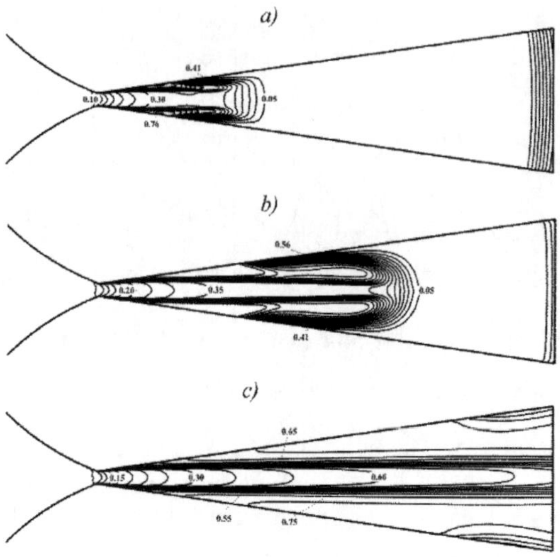

Figure 3. The contour of distributions of α_V for different times *a*) t = 0.005 s; *b*) 0.025 s; *c*) 0.2 s

The transient dynamics of the flow of liquid methane bubbling up in the Laval nozzle is numerically analyzed, specific processes of evaporation in the various control points in the diffuser section of the nozzle are studied. By the study of non-stationary dynamics of characteristic hydrodynamic values of two-phase gas-liquid flow changing over time, the basic laws of the flow of process failure methane liquid stream flowing out from the nozzle throat is revealed.

Work is performed with financial support of the Ministry of Education and Science of the Russian Federation, the Agreement on a subsidy No. 14.577.21.0151 of 28.11.2014. The identifying number of the project RFMEFI57714X0151.

References

[1] Zusin B A, Baranov G A, Barilovich B A and Parfenova I H 1976 *Flushinag adiabatic flows* Atomizdat

[2] Nigmatulin R I 1978 *Fundamentals mechsnics of geretogenous media* Science

[3] Snigerev B A, Tukmakov A L and Tonkonog V G 2016 *Proceedings of XV International Simposium of Heat and Mass Transfer* Minsk 390-394

[4] Singhal A K, Athavale M M and Jiang U 2002 Mathematical basis and validation of the full cavitation model *J. Fluids Eng ASME* **124** 617-624

[5] Kataoka I and Serizawa A 1989 Basic equations of turbulence in gas-liquid two-phase flow. Parts I. Theory *Int. J. Multiphase flows* **15**(5) 843-855

[6] Saifutdinov A I, Fadeev S A, Saifutdinova A A and Kashapov N F 2015 *JETP Lett* **102** 637–642

Application of low pressure capacitively coupled rf hydrogen plasma for low temperature reduction of iron clusters in structure of fe-pillared materials

V L Starshinova[1], V E Gorelysheva[1], A A Shinkarev (Jr.) [1], S G Gnevashev[1] G N Kulevtsov[1] and A A Shinkarev[2]

[1]Department of Plasma Technology and Nanotechnology of High Molecular Weight Materials, Kazan National Research Technological University, 68 Karl Marx street, Kazan, 420015, Russian Federation
[2]Department of soil science, Kazan Federal University, 18 Kremlyovskaya street, Kazan, 420008, Russian Federation

E-mail: starshinovavl@gmail.com

Abstract. The unique properties of pillared materials determine their use in catalysis, purification and separation. The paper studies the reduction of composite catalysts, Fe-pillared materials. The authors compare their reduction in low temperature capacitively coupled RF hydrogen discharge of low pressure to their conventional direct hydrogen reduction in a tubular muffle furnace. X-ray diffraction analysis was used to characterize the iron-bearing phases. The results show that the reduction of iron hydro/oxide clusters associated with an aluminosilicate matrix to metallic iron is very challenging due to the degree of the pore space availability for hydrogen.

1. Introduction

Chemical behavior in extreme and unconventional conditions has been widely studied in applied research papers. Innovative approaches, however, promise to increase the reaction rate, product yield, selectivity and to reach milder reaction conditions in chemical synthesis. In recent decades, the use of plasma chemical reactions has been a subject of considerable attention and research in this field.

The chemistry of hydrogen reduction of metals from oxides is well known. In accordance with the principle of Baikov's transformation sequence [1] the process of iron reduction from oxides proceeds stepwise by successive transition from higher oxides to lower ones. Also, according to Baikov, hydrogen is the best option of the three gases (H_2, CO_2, CH_4), which can be used for reduction processes.

In experimental practice, hydrogen reduction of metal oxides is usually carried out using high-temperature plasma. Non-equilibrium plasma possesses an advantage of high energy, which is sufficient to generate atomic hydrogen, excited hydrogen molecules and other active particles, stimulating reduction process at low gas temperatures and solid oxides. The main problem of non-thermal plasma metallurgy is associated with the secondary surface reaction kinetics and gas-phase products (particularly water vapor) desorption efficiency to provide a sufficient number of active plasma particles penetrating into the deeper layers of solid oxides. Another problem is the discharge power and, therefore, productivity of non-thermal plasma metallurgy.

Atomic hydrogen generated in non-thermal plasma reaches the solid metal oxide surface by stimulating the 'cold' reduction process. The process can be influenced by vibration-excited H_2 molecules by such factors as its surface decomposition and H atom diffusion in a crystalline structure.

The surface temperature in non-equilibrium discharges can be higher than the gas temperature, which is the effect of non-equilibrium surface heating. This effect can stimulate the surface oxide reduction process while keeping low gas temperature, therefore, the H_2 decomposition efficiency in the plasma is high [2]. In the presence of hydrogen atoms and ions, generated in the plasma discharge, the reduction of metals from oxide can be observed at room temperature [3].

The use of easily available and inexpensive natural layered minerals with a high specific surface area and catalytic activity for petrochemical industry processes is promising. The fundamental limitation for the use of natural layered minerals as catalysts and carriers is its low thermal stability (dehydration leads to the interlayer space collapse).

This problem can be solved by modifying the natural layered minerals using different chemical and physical methods. The intercalation of large size polyvalent metals, oligo-/polycations (with a structure similar to Keggin cation), in the interlayer space of layered silicates allows for obtaining composite materials (pillared materials) with a defined porosity, high specific surface and a defined catalytic activity depending on the selected cation.

Theoretically, plasma chemical technologies can ensure reduction of the iron compounds (of oxide or hydroxide forms) in the interlayer space to +2 or 0 rate of oxidation. If proven experimentally, the use of these technologies may be promising for the production of pillared structures based on layered silicates and iron compounds in different oxidation rates.

2. Methods

The object under investigation is a sample of dispersed powder pillared material (Fe-PM) obtained using a definite method [4].

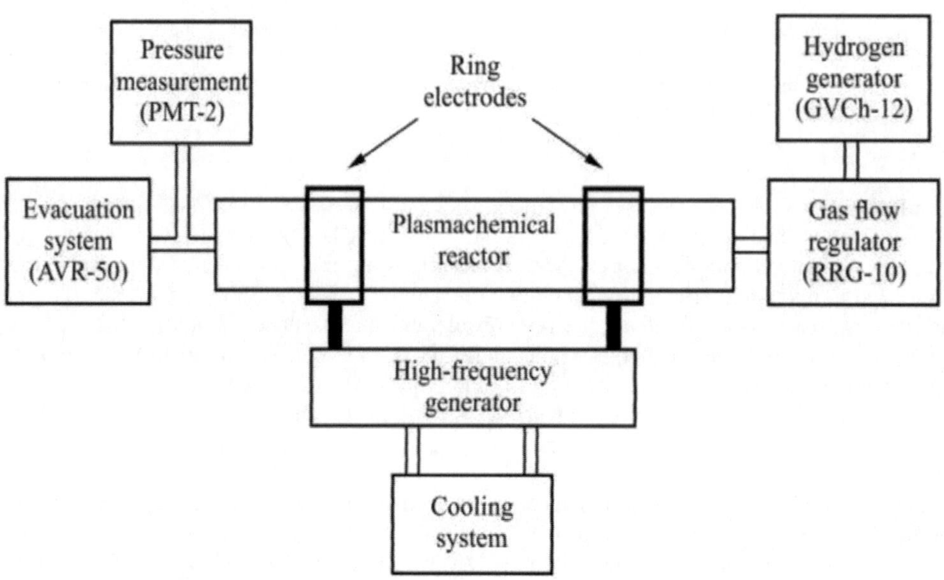

Figure 1. Circuit diagram of plasma chemical hydrogen setup.

Classical direct hydrogen reduction was carried out in a tubular muffle furnace. 0.2 g of the sample was placed in a silica boat and kept at 450°C for 3 hours in a hydrogen stream at atmospheric pressure.

Plasma chemical reduction of the Fe-PM powder was performed in a cylindrical hydrogen plasma chemical quartz glass reactor under conditions chosen by the authors previously [5]. The circuit diagram is shown in Figure 1.

The diffraction spectra of powder and oriented samples were recorded on the D2 Phaser diffractometer produced by Bruker Axs GmbH with the standard Bragg-Brentano (θ-θ) geometry on reflection and the 141.4 mm goniometer radius. CuKα radiation (30 kV, 10 mA), fixed slits, and a linear position-sensitive LYNXEYE detector and sample rotation in its own plane were used. The obtained diffraction spectra were processed in DIFFRAC.SUITE package software. The phases were identified in the diffraction data ICDD PDF-2 Release 2013 with the DIFFRAC.EVA module software. Quantitative calculation of the individual phases was carried out by Rietveld method using the DIFFRAC.TOPAS module software.

Samples: Fe-PM-AD – air dried at room temperature sample, Fe-PM-T350 – heat-treated at 350 °C, Fe-PM-450H2 conventional hydrogen reduction at 450 °C, Fe-PM-P – plasma chemical reduction.

3. Results and Discussion

Figure 2 shows the diffraction patterns of air-dried and heat-treated oriented Fe-PM samples. After thermal treatment at 350°C, the structure keeps stability, showing the interlayer distance value of 6.6 nm.

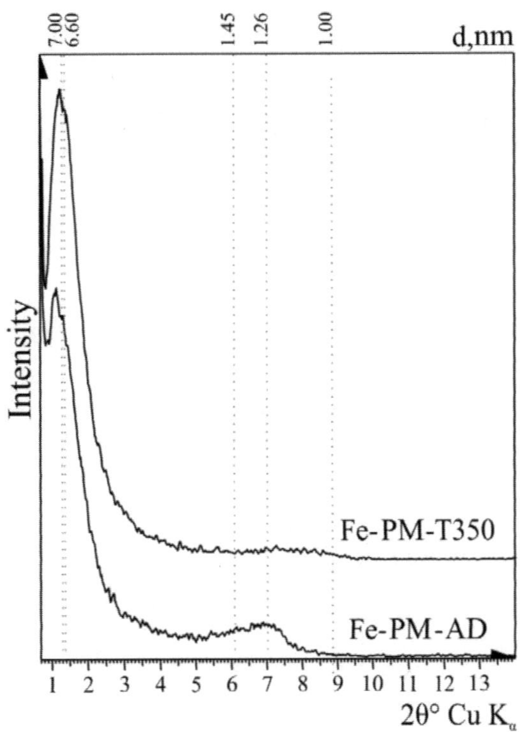

Figure 2. X-ray diffraction patterns of oriented Fe-pillared materials samples before and after thermal treatment.

The increase of the distance between silicate layers compared to Na^+-form (d=1.25 nm) and the stability of the structure during thermal treatment have to ensure the availability of porous space and iron oxide phase clusters associated with the silicate matrix for the penetration of hydrogen molecules and iron reduction.

IOP Conf. Series: Journal of Physics: Conf. Series **789** (2017) 012057 doi:10.1088/1742-6596/789/1/012057

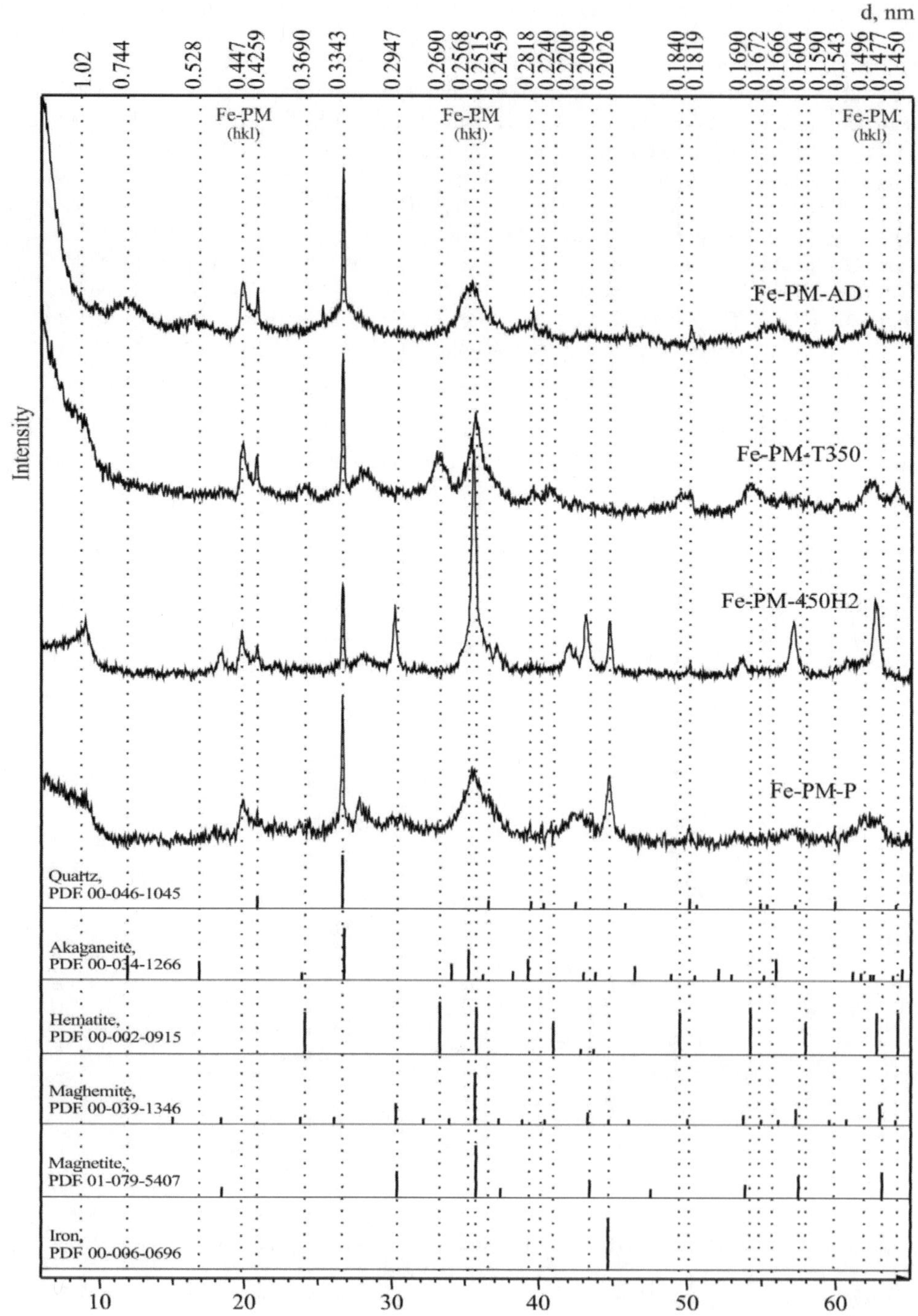

Figure 3. X-ray diffraction patterns of powder Fe-pillared materials samples under different exposure.

Figure 3 shows the diffraction patterns of the powder Fe-PM samples under different exposures. The iron in air dried sample is in akaganeite phase. Fe-PM thermal treatment in the air leads to the formation of hematite and maghemite phases. Conventional reduction by hydrogen during 3 hours leads to the formation of the magnetite, wustite and metallic iron phases.

Plasma chemical hydrogen reduction also leads to a more reduced oxide phases and metallic iron, but the diffraction peaks shape has a lower intensity and a larger FWHM. Table 1 shows the percentage ratio of iron-containing phases (calculated for 100% without the silicate phases) and their average coherent scattering region sizes (CSR).

Table 1. The ratio of iron-containing phases and their average CSR size for Fe-PM samples.

Sample	Phase content, % / CSR, nm					
	Akaganeite (β-FeOOH)	Hematite (α-Fe_2O_3)	Maghemite (γ- Fe_2O_3)	Magnetite (Fe_3O_4)	Wustite (FeO)	Iron (α-Fe)
Fe-PM-AD	100 / 5.0±0.2	-	-	-	-	-
Fe-PM-T350	-	63 / 12.0±0.6	37 / 7.0±0.7	-	-	-
Fe-PM-450H2	-	-	-	73 / 32.7±0.7	21 / 12.7±0.4	6 / 57.0±5.6
Fe-PM-P	-	-	-	67 / 5,7±0.4	21 / 9.0±1.2	12 / 26.6±2.3

The table shows that in case of conventional reduction in a hydrogen stream at atmospheric pressure and 450°C, the crystallite size of the iron metal phase is more than two times higher for metallic iron and six times higher for magnetite than that after low pressure capacitively coupled RF hydrogen plasma treatment. Therefore, plasma chemical reduction occurs in milder conditions.

4. Conclusion

According to the experimental results, the use of plasma chemistry for the reduction of clusters of iron-containing compounds in the composite catalysts is reasonable, because this treatment does not lead to intense crystallization of metal phase due to milder reaction conditions (low temperature).

In spite of the higher temperature and hydrogen concentration in conventional recovery, it gives the same amount of iron as the plasma chemical method. In consideration of the substantial differences between the methods, it is obvious, that the reduction process of iron in the structure of the Fe-PM is limited by the gas diffusion in the microporous space.

The problem of iron-containing clusters availability for hydrogen cannot be solved only by the synthesis of Fe-pillared materials with sufficiently large (6.6 nm) interlayer distances, and require technological solutions to increase the diffusion rate and provide minimal high reduction temperature.

Acknowledgments

The work has been performed under the support of the Ministry of education and science of the Russia Federation, project No. 2196 of 01.02.2014.

References

[1] Baikov A 1961 Izbrannye trudy (Moscva: Metallurgizdat) p 327
[2] Fridman A 2008 Plasma chemistry (Cambridge: Cambridge University Press) p 978
[3] Atamanov V, Elizarov L et al 2003 *VANT* **4** pp 213-216
[4] Mandalia T, Crespin M, Messad D and Bergaya F 1998 *Chem. Commun.* pp 2111-12
[5] Shinkarev A (Jr.), Starshinova V, Gnevashev S and Abdullin I 2015 *Vestnik tehnologicheskogo un-ta* **13** pp 122-126

The material balance of process of plasma-chemical conversion of polymer wastes into synthesis gas

A Kh Tazmeev and R N Tazmeeva

Kazan Federal University Naberezhnye Chelny Institute, 68/19 Mira Street, Naberezhnye Chelny, 423810, Russian Federation

E-mail: tazmeev@gmail.com

Abstract. The process of conversion of polymer wastes in the flow of water-steam plasma which are created by the liquid electrodes plasma generators was experimentally studied. The material balance was calculated. The regularities of the participating of hydrogen and oxygen which contained in the water-steam plasma, in formation of chemical compounds in the final products were revealed.

1. Introduction

It is known that the water-steam plasma is the most attractive option of energy carrier of plasma-chemical processes, including plasma waste treatment processes [1, 2]. It provides a number of advantages. Water-steam plasma does not contain the ballast components (such as nitrogen in the composition of the air plasma). Therefore, the thermal efficiency of energy carrier becomes substantially higher. Formation mechanisms of harmful oxides such as oxides of nitrogen and sulfur are suppressed in the water-steam plasma. This ensures the most favourable environmental conditions. The water-steam plasma is enriching synthesis gas by hydrogen due the carbon oxidation of raw materials by water-steam: $C + H_2O \rightarrow H_2 + CO$. This increases the amount of the final product. There are other positive effects that accompany the practical application of water-steam plasma which are may be noted [3, 4].

Currently arc plasma is used mainly in the processes of recycling [5]. Energy carrier plasma is generated by blowing out the arc discharge using various gases, including water-steam. However, in practice, there are additional technical difficulties at using water-steam in arc plasma torches. In real industrial settings in the first place required an efficient steam generator for obtaining superheated steam. The protection of the refractory electrodes from direct exposure to water-steam needed. And also to take steps to prevent moisture condensation on the surfaces of the conductive elements.

Applying of liquid electrodes plasma generators based on gas discharge with liquid-electrolyte cathode eliminates the above-mentioned negative factors. The plasma stream is forming from electrolytic fumes. Evaporation of electrolyte takes place under the influence of heat coming from the gas discharge.

2. Experiment

In this work, the liquid-cathode plasma generator was used, allowing to create a plasma flow of electrolyte vapor at temperature up to 1800°C and the mass flow rate of up to 3,0 g/s [6, 7]. Previously used pieces of plastic film (PE material) and bottles made of polyethyleneterephthalate (PET material) were as raw material. Processing of raw materials was in a sequence: 1) thermal decomposition in the environment of water-steam at a temperature of 770-800 K; 2) conversion to synthesis gas in plasma-chemical reactor in the temperature range 1670-1720 K. Some results of experimental studies published in [8-13].

The adopted simplifications and source data. In the calculations the balance of the main chemical elements that make up the reagents was considered. These are: carbon C, hydrogen H and oxygen O. Material balance implies equality of the masses of individual chemical elements in the primary components (polymer, raw materials and water) and in the end product in all physical phases: solid, liquid and gaseous.

The limited options which available direct measurement, do not allow the calculation material balance in full, taking into account changes of the chemical composition of a reacting mixture during the whole duration of the process. In this regard, for the calculations selected the steadiest regime, which in time is located at a sufficient distance from the initial and final stages of the process. However, even in this case there is an uncertainty in the numerical values of several process parameters. In particular, remain unknown the exact values of the mass arrival rate of raw material components (chemical elements C, H and O) in the plasma chemical reactor, where are formed the final products of plasma chemical reactions. Therefore, the calculation of the material balance is only possible under certain simplifying assumptions. As such, in this work, we adopted the following simplifications.

1. Thermal decomposition of the raw material occurs uniformly. The source reagents are received in the plasma-chemical reactor with mass velocity $\dot{m} = k \cdot m/\Delta t$. Here k – is the coefficient taking into account the formation of solid residue; m – is the mass of raw material; Δt – is the period of time during which there is an intensive thermal decomposition of raw materials of volatile components.

2. The mass rate of the chemical elements C, H and O in a plasma-chemical reactor is constant and their ratio to each other is equal to the atomic mass ratio in a conditional formula of raw materials. For PE-raw materials (conditional formula C_2H_4): $\dot{m}_C : \dot{m}_H = 24 : 4$.

It is necessary to consider when compiling a conditional formula of the raw material polyethyleneterephthalate about forming of a solid residue of carbon. In the plasma-chemical reactor is receives only the volatile components of the thermal decomposition of polyethyleneterephthalate. Therefore, in the composition of the reagents that acts in plasma-chemical reactor, C atoms less than the original feedstock. This leads to changes in the stoichiometric coefficients in the chemical formula. In the experiments, the mass of the solid residue amounted to an average of 10% from the initial weight of raw materials. With this in mind, conditional formula is written as $C_{8.4}H_8O_4$. Thus, for PET-raw materials: $\dot{m}_C : \dot{m}_H : \dot{m}_O = 100.8 : 8 : 64$.

3. Mass rate of $\dot{m}_C$, $\dot{m}_H$ and $\dot{m}_O$ make up a part from proportional of the atomic masses of the corresponding chemical elements. For polyethylene: $\dot{m}_C = (24/28) \cdot \dot{m}$ и $\dot{m}_H = (4/28) \cdot \dot{m}$. For polyethyleneterephthalate: $\dot{m}_C = (100.8/172.8) \cdot \dot{m}$, and $\dot{m}_H = (8/172.8) \cdot \dot{m}$ and $\dot{m}_O = (64/172.8) \cdot \dot{m}$.

4. The compositions of the gas phase of end-products and a liquid condensation are unchanged for the entire duration of the process. These compounds such as those obtained in the analyses of gas and liquid samples.

5. Mass and volumetric rates of formation of individual substances in the final product are constant, mass rates of appearance of chemical elements C, H and O in the gas phase ($\dot{m}_{gC}$, $\dot{m}_{gH}$ и $\dot{m}_{gO}$) are constant respectively.

Table 1. Phase composition of the final products.

Raw material	Solid phase	Liquid phase	Gas
PE-raw	0 %	0,6 %	99.4 %
PET-raw	10 %	2.8 %	87.2 %

The following are the calculations on the example of two experiments, characteristic for PE-raw and PET-raw materials. In tables 1 - 3 shows the results of chemical analyses of the final products, as well as the calculated values of mass rates $\dot{m}_{gC}$, $\dot{m}_{gH}$ и $\dot{m}_{gO}$.

IOP Conf. Series: Journal of Physics: Conf. Series **789** (2017) 012058 doi:10.1088/1742-6596/789/1/012058

From table 1 follows that the end products in plasma-chemical process are mainly generated in the gas phase. Significant amounts hydrogen, carbon oxides and methane are formed (tables 3 and 4). In small amounts are present acetylene, ethylene and benzene. Thus in the total mass of gaseous end products the content of heavy chemical elements O and C more, than lightweight H.

Table 2. Gaseous end products of the conversion process of PE -raw.

Component	Volumetric content, vol.%	$\dot{m}_{g}$, g/min	$\dot{m}_{gC}$, g/min	$\dot{m}_{gH}$, g/min	$\dot{m}_{gO}$, g/min
H_2	66.3	0.3966	—	0.3966	—
CO	9.6	0.8040	0.3446	—	0.4594
CO_2	12.0	1.5793	0.4307	—	1.1486
CH_4	7.8	0.3733	0.2800	0.0933	—
C_2H_2	2.3	0.1789	0.1651	0.0138	—
C_2H_4	0.30	0.0251	0.0215	0.0036	—
C_6H_6	0.021	0.0049	0.0045	0.0004	—
Summarily	98.321[a]	3.3621	1.2464	0.5077	1.6080

[a] Samples for analysis contain impurities.

Table 3. Gaseous end products of the conversion process of PET-raw.

Component	Volumetric content, vol.%	$\dot{m}_{g}$, g/min	$\dot{m}_{gC}$, g/min	$\dot{m}_{gH}$, g/min	$\dot{m}_{gO}$, g/min
H_2	51.1	0.2601	—	0.2601	—
CO	14.7	0.5990	0.2570	—	0.3420
CO_2	25.4	2.8430	0.7750	—	2.0680
CH_4	4.8	0.1948	0.1460	0.0488	—
C_2H_2	0.340	0.0225	0.0208	0.0017	—
C_2H_4	0.019	0.0013	0.0011	0.0002	—
C_6H_6	0.210	0.0417	0.0385	0.0032	—
Summarily	96.569[a]	3.9624	1.2384	0.3140	2.4100

[a] Samples for analysis contain impurities.

For comparative analysis more informative is the number of atoms, i.e. the molar content, rather than mass. These data are obtained by the recalculation are shown in figures 1 and 2 in the form of diagrams.

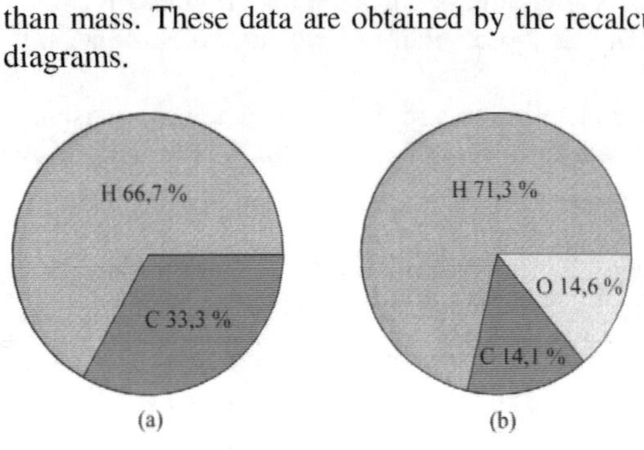

Figure 1. The mole composition of raw materials at input in plasma-chemical reactor. (a) – PE-raw; (b) – PET-raw.

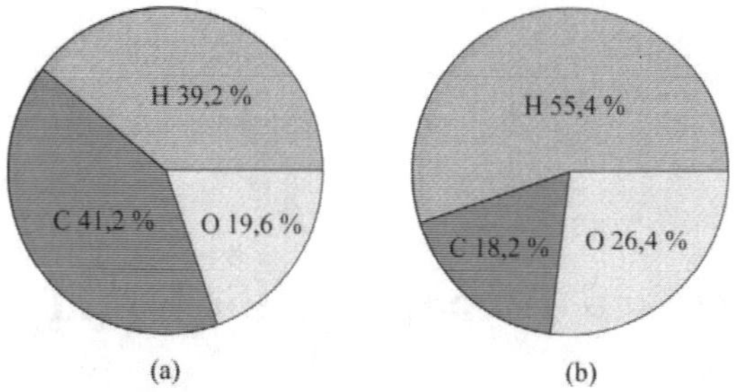

Figure 2. The mole ratio of chemical elements in the gaseous end products. (a) – PE-raw; (b) – PET-raw.

As seen from the presented diagrams that PE-raw does not contain atoms O, and in the final products they appear. At the same time there is a significant addition of atoms H. Similar patterns are present in the case of PET-raw. In both cases, the atoms H and O are transferred from the steam.

3. Conclusions

Significant amount of hydrogen from water-steam is added to the hydrogen coming to the reaction zone of the plasma-chemical reactor from composition of raw. Therefore, hydrogen volume content is the largest among of all the components in the part of produced synthesis gas.

References

[1] Altovsky G S, Bernadiner M N and Ivanov V V 2011 *Ecology ahd Industry of Russia* No. 2 8
[2] Gudim Y A and Golubev A A 2009 *Ecology ahd Industry of Russia* No. 2 4
[3] Khafizov A A, Valiev R I, Shakirov Yu I and Valiev R A 2014 *J. Phys.: Conf. Ser.* **567** 012026
[4] Valiev R A, Shakirov Yu I and Iliukhin A N 2014 *J. Phys.: Conf. Ser.* **567** 012040
[5] Artemov A V, Pereslavsev A V, Krutyakov Y A, Voschinin C A, Kudrinsky A A, Bulba V A and Ostryi I I 2011 *Ecology ahd Industry of Russia* No. 10 18
[6] Tazmeev Kh K and Tazmeev A Kh 2005 *Heat Transfer Research* **36** 623
[7] Miftachov M N, Tazmeev A Kh, Tazmeev Kh K and Fridland S V 2006 *Journal of Engineering Physics and Thermophysics* **79** (3) 532-538
[8] Fridland S V, Tazmeev A Kh and Miftachov M N 2006 *Vestnik Kazanskogo technologicheskogo universiteta* No. 6 10
[9] Tazmeev A Kh, Fridland S V and Miftachov M N 2006 *Vestnik Kazanskogo technologicheskogo universiteta* No. 6 43
[10] Kashapov L N, Kashapov N F and Kashapov R N 2013 *J. Phys.: Conf. Ser.* **479** 012011
[11] Denisov D G, Kashapov N F and Kashapov R N 2015 *IOP Conference Series: Materials Science and Engineering* **86** 012005
[12] Kashapov L N, Kashapov N F and Kashapov R N 2014 *J. Phys.: Conf. Ser.* **567** 012025
[13] Tazmeeva R N, Ziganshin R R and Tazmeev Kh K 2008 *Vestnik mashinostroeniya* No. 10 87

Study of the binding zone of electrical discharge to the liquid cathode by high-speed visualization

G Kh Tazmeev[1], B A Timerkaev[1] and Kh K Tazmeev[2]

[1]Kazan National Research Technical University named after A.N. Tupolev, 10 K. Marx str., Kazan, 420111, Russia
[2]Branch of Kazan Federal University in Naberezhnye Chelny, 10A Sujumbike av., Naberezhnye Chelny, 423812, Russia

E-mail: tazmeevg@mail.ru

Abstract. This work is devoted to the experimental study of the properties of the binding zone of gas discharge to the liquid electrolyte cathode. Detailed description of the processes on the border "plasma-liquid" was analysis by of high-speed shooting. Were installed the regularities of phenomena of transfer substance and electric charges from the liquid electrolyte in the plasma column.

1. Introduction
In recent years, electrical discharges tied to the liquid attract the attention of researchers due their great potential for practical applications [1-5]. In the presence of fluid in the near areas of electrode occur are quite complex processes. This work is devoted to a detailed study of these processes using high-speed video.

2. Experiment
In experimental practice, different electrode systems are used to creating a gas discharge with liquid electrolyte cathode. In the open air at atmospheric pressure, typically, the electrolyte is pour in a container with a large geometric volume and circulates through it, or is expire from the tube.

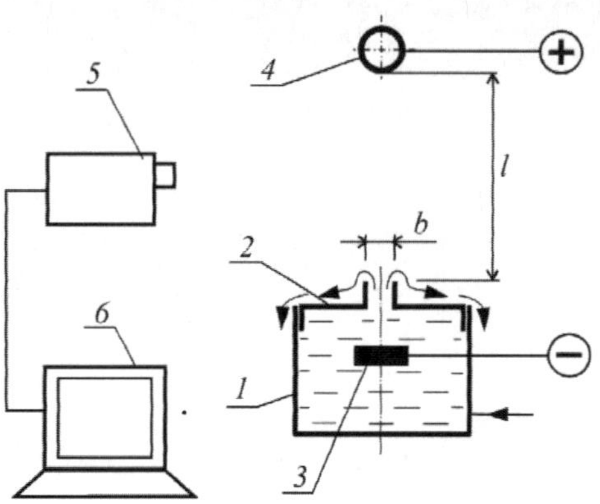

Figure 1. Experimental setup (electric discharge unit).

In such embodiments, a discharge formed in the form of a body of cylindrical shape. Near its axis, figuratively speaking, the discharge is "thickens", and spatial structure is "smeared". In this work, the electrolyte is flowing out from a slit gap. The discharge formed as a vertical glowing wall and relatively small thickness. Thereby, made possible study of its spatial structure more detailed.

The experimental set-up used to generate and study of electrical discharge, which schematically depicted in figure 1.

Liquid electrolyte fed into the cylindrical container *1*. It closed by a cover *2*, which had a slit gap of width b = 2.5 mm. Length of slit gap was 35 mm. On the edges of the gap was done shoulder of height of 5 mm. Sodium chloride solution in distilled water as electrolyte was used. It was prepared with a concentration of 0.55% by mass. Specific electric conductivity was in the range of 10.00 ± 0.20 mSm/cm. The mass flow rate m of electrolyte through the gap was 5-6 g/s. Inside of the container was mounted graphite plate *3*, which was connected to the negative pole of the power source. The discharge was burning between the electrolyte and the water-cooled metallic anode 4. The anode was made in the form of a long tube. It was oriented along the slit and located above it at a distance of l = 4 cm. The power supply was a three-phase full-wave rectifier. Voltage ripple was smoothed out by C-L-C-filter.

High-speed videography was filmed by camera Photron FASTCAM SA4. Its technical characteristics allow to study in detail processes in the system with the liquid medium [6]. Accumulated in the camcorder *5* information was being transferred to the computer *6*.

3. The results of the experiments and their analysis.

In figure 2 shows the video footage obtained under the regime of 10000 frames per second. As seen, closer to cathode the glowing column is split into multiple channels. These channels rest against on the surface of the electrolyte, forming a small luminous cathode spots.

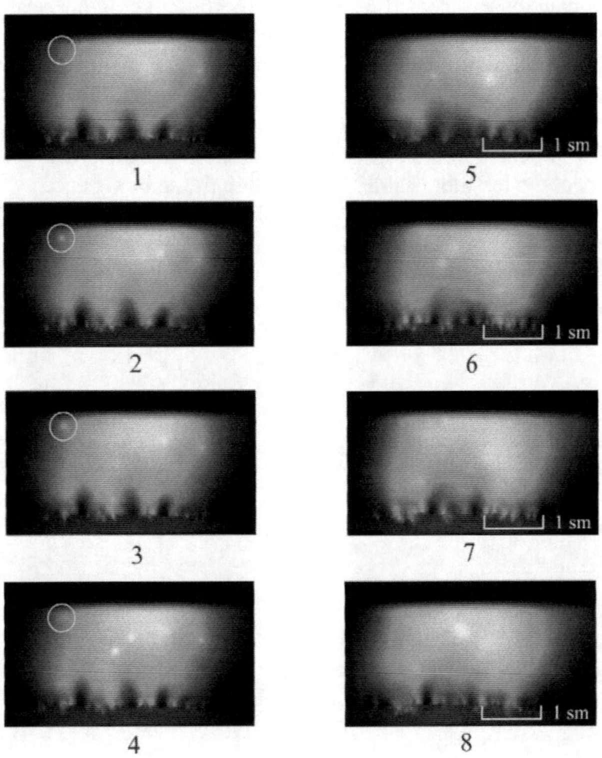

Figure 2. Video frames. The cathode at the bottom, the anode at the top and it covered with an opaque plate. Current 1.3 A; the terminal voltage 1100 V. 1, 2, 3, 4 - interval between the frames 10^{-4} s; 5, 6, 7, 8 – 10^{-3} s. One of many outbreaks isolated with white circle.

High-speed shooting is allowed for trace the movement of cathode spots and reveal some of their properties.

1. Separately taken cathode spots existed not long. Some of them appeared and quickly disappeared, others lasted a little longer. Minimum lifetime τ was ~ 1 ms. The maximum value τ was difficult to determine. It was more than 10 ms. During this time, the cathode spots grouped and merge with others, i.e. becomes difficult to discriminate in the video frame.

2. During the time τ brightness of the cathode spot was varied. In the moment of maximum brightness of the spot expanded to the largest size. The maximum value of the spot diameter was in the range of 0.15-0.25 mm.

3. The number of cathode spots with the maximum brightness was constantly changing. This happened due to the fact that the spots appeared and disappeared, and as well "merged". However, in each frame they could be counted with some error. In particular, at a current of 1.3 A the number was 35-50 pieces.

4. There is a continuous movement of the cathode spots on the surface of the electrolyte. Moreover, the cathode spots moved much faster than electrolyte flow, which could be seen visually and it confirmed by calculations. The average speed of the cathode spots was ~ 1m/s.

The results of observations were used to estimate the current density on the liquid electrolyte cathode by the formula $j_k = I/(N \cdot S)$, where N – is the number of cathode spots, S – sectional area of individual cathode spot. It was found that j is in the range of A/sm^2. This is several orders of magnitude smaller than the current density on the metallic cathode during combustion of electric arc [7]. However, in the case of the gas discharge with liquid electrolyte cathode is a relatively large value in comparison with the known data [8-10]. Traditionally, the current density on the liquid cathode is measured as the ratio of the amperage to the area of the luminous region on the surface of the liquid. For small currents not exceeding ten milliamps, this method is acceptable because the nonuniformity of the binding zone is not observed, at least, cannot be detected by visual inspection. The numerical value of j_k is obtained ~ 1 A/sm^2 [9]. At currents over 1 A nonuniformity binding zone of discharge to the liquid cathode becomes apparent. It is noteworthy that in these conditions the calculation according to the traditional methods gives the same result. [10]

Another property of electric charge, which is clearly evident in studies of high-speed shooting is the appearance of outbreaks. In the videos they allocated in the form of bright circles with different diameters (figure 2). They are increased and become less noticeable. The duration of their existence was a few tenths of a millisecond.

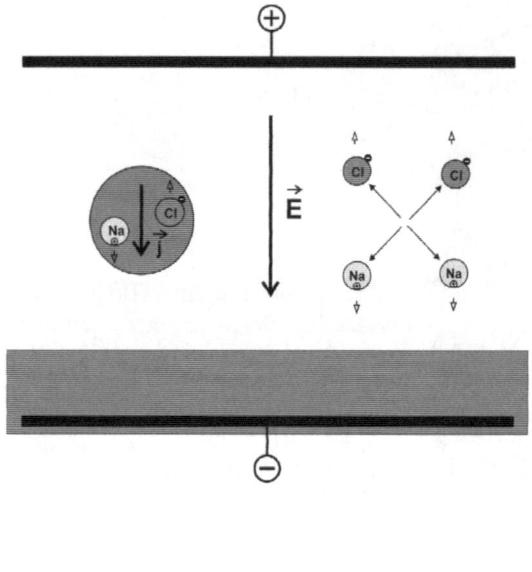

Figure 3. A drop of the electrolyte in the electric field of the plasma column before and after the explosion.

We can assume that microbursts electrolyte droplet happening in the plasma column. At the site of the explosion concentration of atoms is much larger than in its environment. Accordingly, radiation emitted from atoms observed in the form of bursts.

Droplets of electrolyte contain ions. After the explosion, they scatter. Further, under the influence of an electric field, they drift to the electrodes, creating a diffuse electric current. This can explain the formation of significant volume of the plasma column above the liquid electrolyte.

The electric field creates directional movement of ions inside the drop. The positive sodium ions move towards the cathode, and the negative chlorine ions - to the anode (figure 3). Thus is arises ionic current. Probably the Joule heat generated during the flow of ion current resulting to an explosion of drop. It is possible that an explosion occurs in a different way. However, what would not have been an mechanism of explosion, the ion current plays a role in this process.

4. Conclusions

Binding zone of electrical discharge to the liquid electrolyte cathode is composed of a lot small cathode spots, with dimensions constituting a few tenths of a millimeter. The current density in the cathode spots is large and reaches a value up to 100 A/sm^2.

The liquid cathode is spray into the cathode spots. Small droplets getting in discharge area is heat, vaporize and explode. Ions carried out from the liquid cathode by electrolyte droplets fly away. Diffuse ion current is formed the plasma column.

References

[1] Titov V A, Stokozenko V G, Titova Y V and Moryganov A P 2015 *High Energy Chemistry* **49** (6) 459

[2] Kashapov L N, Kashapov N F and Kashapov R N 2016 *J. Phys.: Conf. Ser.* **669** 012029

[3] Shakirov Yu I, Valiev R I, Khafizjv A A, Valiev R A and Khakimov R G 2016 *J. Phys.: Conf. Ser.* **669** 012064

[4] Zaripov R G, Kashapov N F, Tkachenko L A and Shaydullin L R 2016 *J. Phys.: Conf. Ser.* **669** 012053

[5] Barinov Y A, Kuzikova I L, Zinov'eva S V, Shkol'nik S M, and Medvedeva N G 2015 *Technical Physics Letters* **41** (17) 68

[6] Popov I A and Shchelchkov 2014 *Journal of Engineering Physics and Thermophysics* **87** (6) 1420

[7] Zhukov M F, Koroteev A S and Yuryukov B A 1975 *Prikladnaya dinamika termicheckoy plasmu* (Novosibrirsk: Nauka) p 298

[8] Cserfalvi T and Mezei P 1996 *Fresenius J. Anal. Chem.* **355** 813

[9] Petrov A E, Kalbenin D A, Titov V A, Kulentsan A L and Smirnov S A 2011 *Gorenie i plazmochimiya* **9** (3) 160

[10] Miftachov M N, Tazmeev A Kh, Tazmeev Kh K and Fridland S V 2006 *Journal of Engineering Physics and Thermophysics* **79** (3) 532

Study of gas discharge with a liquid cathode at maximum thermal load to the cathode

G Kh Tazmeev[1], B A Timerkaev[1], Kh K Tazmeev[2], I M Arslanov[2], B K Tazmeev [2] and F S Sarvarov[2]

[1]Kazan National Research Technical University named after A.N. Tupolev, 10 K. Marx str., Kazan, 420111, Russian Federation
[2]Kazan Federal University Naberezhnye Chelny Institute, 68/19 Mira Street, Naberezhnye Chelny, 423810, Russian Federation

E-mail: tazmeevh@mail.com

Abstract. Thermal phenomena were experimentally studied in the atmospheric pressure gas discharge between the electrolyte liquid cathode and the metal anode under conditions in which the electrolyte temperature is close to the boiling temperature. It is shown that electrolyte mass discharge can only be reduced to a certain limit, while maintaining stable mode of burning discharge.

1. Introduction

Gas discharges with liquid electrodes is of concern due to the fact that they have large possibilities of practical application. It is also important that they can be obtained by relatively simple technical means in a wide range of current and power. When burning of gas discharge between the electrodes, liquid plasma is in contact with the liquid. There are complex processes of heat and mass transfer between the electrolyte and the plasma. The electrolyte is heated, evaporated and subjected to sputtering. Under these conditions, the gas discharge characteristics largely depend on the cooling conditions of the electrolyte. In works [1, 2] experimentally have been investigated the energy characteristics of a gas discharge with electrolyte cathode flow at relatively high currents and capacities, respectively (7 - 22) A and (8 - 33) kW. It showed a reduction of heat losses by reducing the mass flow rate of the electrolyte. In this study, the study continued in a small electrolyte entering the nip with the discharge.

2. Experiment

Fig. 1 schematically shows a cathode assembly of gas discharge device. Key elements of its structure: pin *1*, a housing *2* and an outlet *3*. The pin *1* is copper. Its diameter is 10 mm, the rounded radius is 5 mm. The housing *2* and the outlet *3* are made of a dielectric material. The outlet *3* is removable. In experiments three different branches' diameters were used: 12.7; 20.0 and 28.0 mm. The wall thickness was 2 mm.

The electrolyte *4* was supplied inside the outlet *3*. It is poured through the edge of the outlet *3* into the groove *5* and there from merged into the cooling system. Immersion depth *h* of a copper rod *1* into the electrolyte *4* was varied by moving it vertically. To measure the voltage drop within the electrolyte ΔU_k electric probe *6* was used. It consisted of a tungsten wire with a diameter of 0.15 mm, inserted into a tube of quartz glass. In the experiments the working end of the probe was placed on the edge of the outlet *3* as shown in figure 1. Power of Joule heat in the electrolyte was calculated by the formula:

$$Q_{\mathrm{j}} = I \cdot \Delta U_{\mathrm{k}}, \quad (1)$$

where I – discharge current.

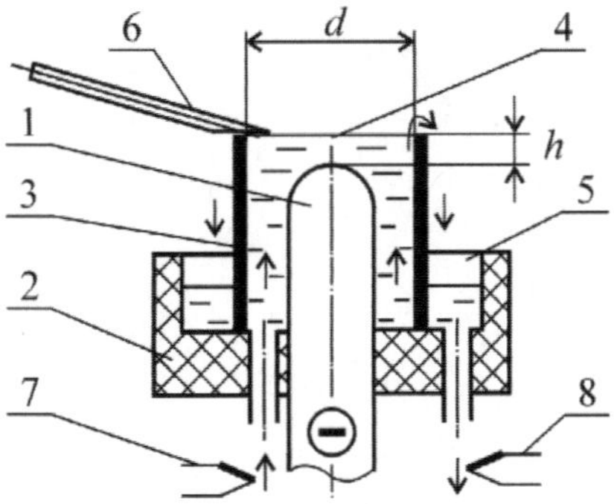

Figure 1. Cathode assembly.

Heat loss at the cathode is determined by a calorimetric method.

$$Q_{\mathrm{k}} = m_{exit} \cdot c \cdot \Delta t . \quad (2)$$

There m_{exit} - electrolyte flow through the cathode assembly; c - heat capacity of the electrolyte; Δt - the difference between the indications of chromel-alumel thermocouples installed at the entrance to the cathode assembly *7* and at its exit *8*.

Electrolyte flow was defined as the proportion of electrolyte amount leaked from the cathode assembly to the time interval during which the electrolyte is flowed. Electrolyte flow measurements were carried out before the gas discharge ignition (m_{ent}) and during its burning time (m_{exit}). Flow difference was taken as the mass carry over of electrolyte G, occurring under the influence of the gas discharge.

$$G = m_{ent} - m_{exit} . \quad (3)$$

As an electrolyte weakly concentrated solution of sodium chloride in distilled water was used. Specific electrical conductivity σ of the electrolyte was measured by conductometer ANION 4150. Solution was prepared with σ = 10.00 mCm/sm. It is known that electrolytic processes and the impact of the plasma discharge cause changes in the electrical properties of the electrolyte. Thus, after each small series of experiments adjusting of the value σ was carried out. Deviations after each initial value does not exceed ± 0.20 mCm/sm.

Current was measured by needle multirange indicator of type M 2015 of accuracy class 0.2. A power source based on inverter converter was used. The choice was dictated by the fact that the sources of this power type have protection against short-circuit and provide sustained combustion gas discharge at any point in the current-voltage characteristics. In this constant current regulation errors does not exceed 1%.

3. The experimental results and their analysis.

The graphs, described in figure 2 shows how the heat losses of the cathode Q_{k} are changing and the power Joule heat Q_{j} within the electrolyte at low constant electrolyte flow $m = const$. With the current increase in the intensity of heat and mass transfer processes between the electrolyte cathode and gas plasma enhanced. Accordingly, heat loss at the cathode Q_{k} is increasing. Along with this comes an increase in the power of Joule heat Q_j. As can be seen, Q_k and Q_j graphics virtually converge with increasing current. However, in practice, we cannot achieve intersection of Q_k and Q_j graphics as when approaching such regimes there is a breakdown of the electrolyte layer over the current lead.

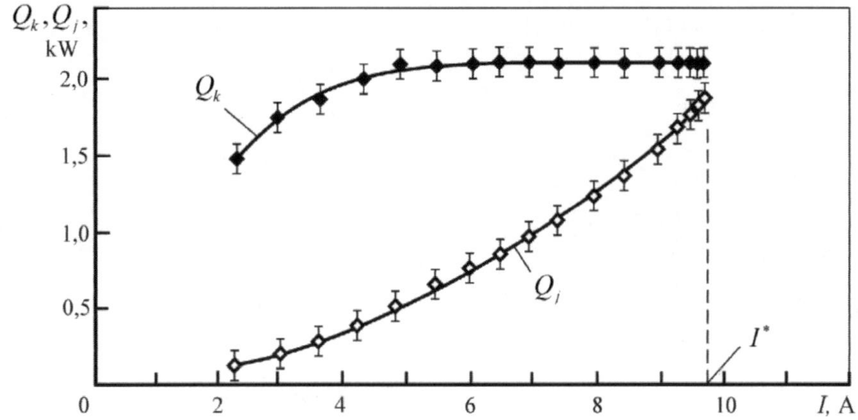

Figure 2. Dependence on current: heat loss on cathode Q_k и and capacity of power Joule heat in electrolyte Q_j. volume electrolyte velocity m = 10 g/s.

At higher currents, the electrolyte hardly heats up and reaches a boiling state. The bubbles of steam break the electrolyte layer. Apparently due to such phenomena there comes a breakdown.

Breakdown stops occurring with increasing thickness of electrolyte layer. In this variant electrolyte boiling leads to complete rupture of the current channel, and the discharge is extinguished by achievement of current to some certain limit. As the experiments showed the smaller the mass of the electrolyte flow rate is, the lower the numerical value of the current limit. Occurrence of instability of the discharge burning at a low mass flow rate of the electrolyte was observed on the oscillograms of currents and voltages, as well as on instant photographs of the gas discharge.

In the limit, the heat loss at the cathode and the power of Joule heat in the electrolyte can be considered numerically equal to:

$$m^*_{exit} \cdot c \cdot \Delta t \approx I^* \cdot \Delta U_k . \quad (4)$$

Here, I^* and m^*_{exit} – limits of current and flow through the electrolyte through cathode assembly.

Using equation (4) it is possible to determine the minimum flow of the electrolyte, which is necessary for sustaining constant combustion gas discharge.

$$m^*_{exit} \approx \frac{(I^*)^2 \cdot R_k}{c \cdot \Delta t} . \quad (5)$$

Here R_k – the ohmic resistance of the cathode assembly. The numerical value can be taken as water equal to 4.19 кJ/(кg·К), and Δt as the difference 100 – $t_{ent.}$, where t_{ent} - electrolyte temperature at the inlet of cathode assembly.

From formula (5) it follows that the electrolyte flow required for work near critical thermal conditions, along with it, the heat loss at the cathode it can be minimized by taking measures to reduce R_k.

On figure 3 are graphs constructed from the results of experiments. The parabolic shape of the curves indicates the legitimacy of the assumptions on which the formula (5) is obtained. The same figure shows the results of mass carry over of electrolyte calculation formula (3) for critical conditions. As can be seen the mass carry over of electrolyte is increasing substantially proportional to the current. This principle can be regarded as a characteristic feature of the gas discharge with liquid electrolyte cathode, as it manifests itself in other experimental conditions [3-4]. The proportionality factor varies within narrow limits depending on the properties of the electrolyte and its cooling conditions. In this work you can take it as $k = 0.09 \cdot 10^{-3}$ kg/(s·A).

Thus, given the mass flow carry over of the electrolyte estimated value into the cathode at the input node may be defined by the formula

$$m^*_{exit} \approx \frac{(I^*)^2 \cdot R_k}{c \cdot \Delta t} + k \cdot I^* . \tag{6}$$

Formula (6) can be used to determine the minimum flow of the electrolyte through the cathode assembly at the design stage of discharge devices.

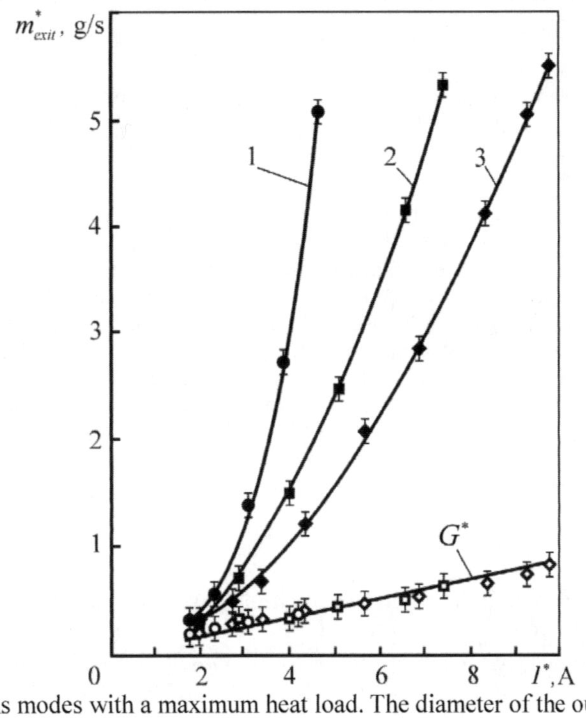

Figure 3. Diagrams modes with a maximum heat load. The diameter of the outlet of the cathode assembly *d*: *1* – 12.7 mm; *2* – 20.0; *3* – 28.0.

4. Conclusions

The mass flow of the electrolyte can be reduced only to a certain limit. In this case should be taken actions to prevent boiling of electrolyte battery into the zone of discharge binding.

References

[1] Tazmeev Kh.K., Arslanov I.M., Tazmeev A.Kh. 2014 *J. Phys.: Conf. Ser.* **567** 012001
[2] Tazmeev Kh.K., Arslanov I.M., Tazmeev A.Kh. 2016 *J. Phys.: Conf. Ser.* **669** 012058
[3] Tazmeev Kh K and Tazmeev B Kh 2003 *Inzhenerno-Fizicheskii Zhurnal* **76** (4) 107
[4] Tazmeev Kh K and Tazmeev A Kh 2005 *Heat Transfer Research* **36** (8) 623
[5] Miftachov M N, Tazmeev A Kh, Tazmeev Kh K and Fridland S V 2006 *Journal of Engineering Physics and Thermophysics* **79** (3) 532
[6] Tazmeeva R N and Tazmeev B K 2014 *Applied Physics* No. 1 35

Physical effects in electrical discharge of atmospheric pressure caused by external flow of copper cathode by water

Kh K Tazmeev[1], B A Timerkaev[2] and G Kh Tazmeev[2]

[1]Kazan Federal University Naberezhnye Chelny Institute, 68/19 Mira Street, Naberezhnye Chelny, 423810, Russian Federation
[2]Kazan National Research Technical University named after A.N. Tupolev, 10 K. Marx str., Kazan, 420111, Russian Federation

E-mail: tazmeevh@mail.ru

Abstract. This article experimentally studies gas discharge between copper electrodes in the current range of 5-20 A. The electrode spacing was varied in the range of 45-70 mm. The cathode was a rod with a diameter of 10 mm and was located in the water stream, poured from a dielectric tube. Three variants that differed in the working end of the cathode arrangement were studied: 1) over the water; 2) on the surface of the water; 3) underwater. It was established that in the second variant in the plasma column near the cathode an extended area is formed with of intense radiation of hydrogens. In the third option, a significant erosion of the cathode was detected. It was revealed that in the process of erosion fine particles of copper metal are formed.

1. Introduction

In recent years there has been a significant increase of interest to the gas discharge excited in the presence of a water medium. We study different types of gas discharges: high-voltage, low-voltage, pulse wise and others. Some research is directed to the use of gas discharges for the synthesis of nanoparticles underwater and in aqueous solutions. In review articles dozens of ways to implement this process are listed [1-3].

The aim of this work is the experimental study of gas discharge of constant current in the air flow under condition of the external flow of metal cathode by water. More often, a solid state cathode is completely immersed in the water flow. This type of cathode arrangement is found in many studies [4-14]. In cases where a metal cathode extends from the water flow or is close to this state, the gas discharge isn't studied enough. Research in this work is aimed to study these variants.

2. Experiment

Gas discharge was ignited in the air between copper electrodes at currents of 5-15 A. The power supply was rectifier with smoothing filter. The output voltage was of 1760 V. The current was limited by ballast resistor. Discharge was oriented in the vertical direction. The cathode was located at the bottom and the anode - at the top. The anode was a water-cooled disk with a diameter of 50 mm. It was placed horizontally. The cathode was made as a rod with a rounded end and was mounted within a dielectric tube (figure 1). Water was fed into the gap between the metallic cathode and the internal wall of the tube. Distilled water was used to prevent scale formation in the cooling system. Its specific electric conductivity was in the range of 100 to 600 μSm/sm. These slightly higher values were obtained because of the content of sodium chloride in the units and aggregates of the cooling system.

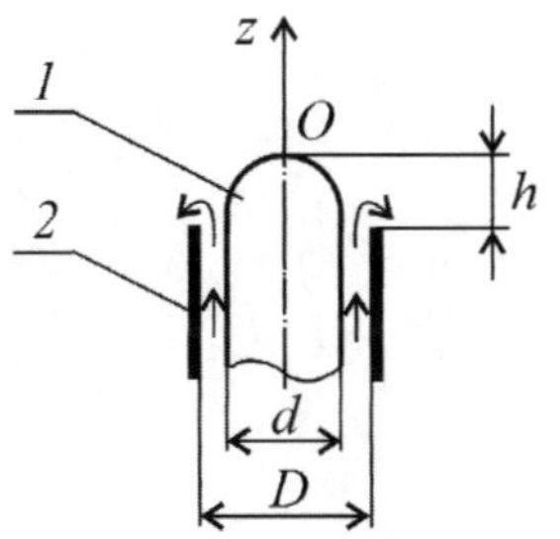

Figure 1. Cathode assembly. *1* – copper bar; *2* – dielectric tube. *d* – 10 mm; *D* – 12.7. The arrows indicate the direction of water flow.

Voltage *U* and current *I* were recorded by digital storage oscilloscope AKIP-15/1, with guidance band of 25 MHz. Photographing and spectroscopy of a gas discharge were carried out with speed camera Videoscan-415 and high-speed fiber-optic spectrometer AvaSpec-3648. Camcorder let receive frames with an exposure of up to 1 μs. Spectrometer recorded radiation in the wavelength of range 484-708 nm with a resolution of 0.15 nm (diffraction grating 1200 lines / mm, the input optical gap is 10 microns). The radiation of the individual sections of the discharge area was projected on the entrance of the spectrometer through an assembly, combined of collecting lens and the optical gap.

Experiments were performed at various projections *h* of copper rod cathode. Three options have been identified. A characteristic feature, distinguishing these options from each other, was the end location of the upper edge of the cathode relative to the water level in the insulating tube (figure 2).

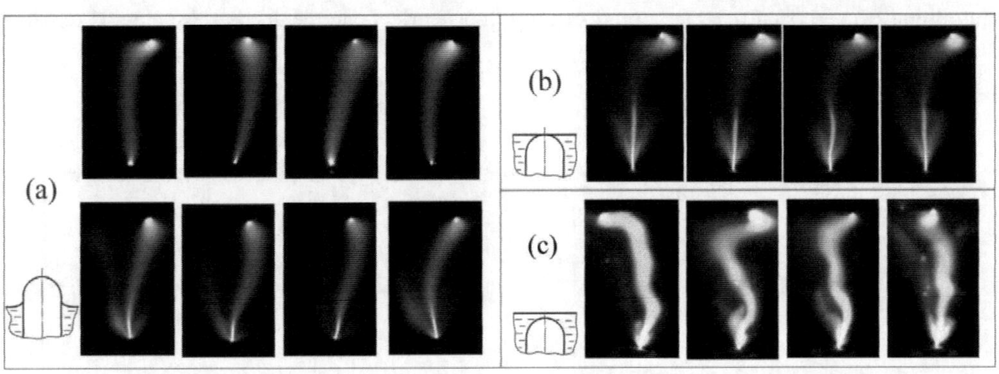

Figure 2. Instant photos of gas discharge at different variants of cathode water flow. Exposure 200 μs. The end of the metal cathode: (a) – above the water *h* = 5 mm; (b) – on the surface of water; (c) – underwater on the depth of 0.5 mm. Water consumption: (a) top row – 10 r/c; (a) lower row, (b), (c) – 1,5. Average current: (a) top row – 10.0 A; (a) lower row, (b) – 14.7: (c) – 8.7. Anode-to-cathode distance: (a) – 45 mm; (b), (c) – 50 mm.

In option (a) at low flow of water there was a narrowing of the discharge channel near the cathode (figure 2(a), lower row). On instant photos this discharge portion manifested in a scarlet color, which is relevant to the Balmer line H_α in spectrum of the hydrogen emission.

Having switched to option (b) picture near the cathode slightly changed. Bright narrow channel near the cathode lengthened and stretched vertically (figure 2(b)). The yellow background formed around it.

A characteristic feature in option c was that the discharge channel was colored green lengthwise. This color indicates the intensive introduction of copper atoms in the discharge region. Apparently, copper atomic is formed in the cathode electrolysis processes.

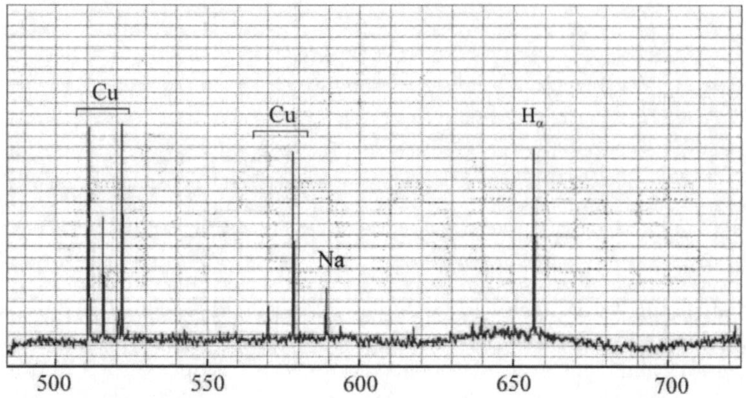

Figure 3. Radiation spectrum of a gas discharge (option *b*). Distance from the cathode – 30 mm. Anode-to-cathode distance – 50 mm. Optical gap width 0.50 mm. Water consumption – 1.5 g/s. Current– 10.0 A.

The presence of hydrogen, sodium and copper in a gas discharge plasma column confirmed spectral studies (figure 3). Distributions of emitting atoms along the discharge channel, received from research results, to comply fully with the images on the instant photography.

In options (a) and (b) discharge burned steadily. Its electrical parameters were stable and remained virtually unchanged over time (figure 4).

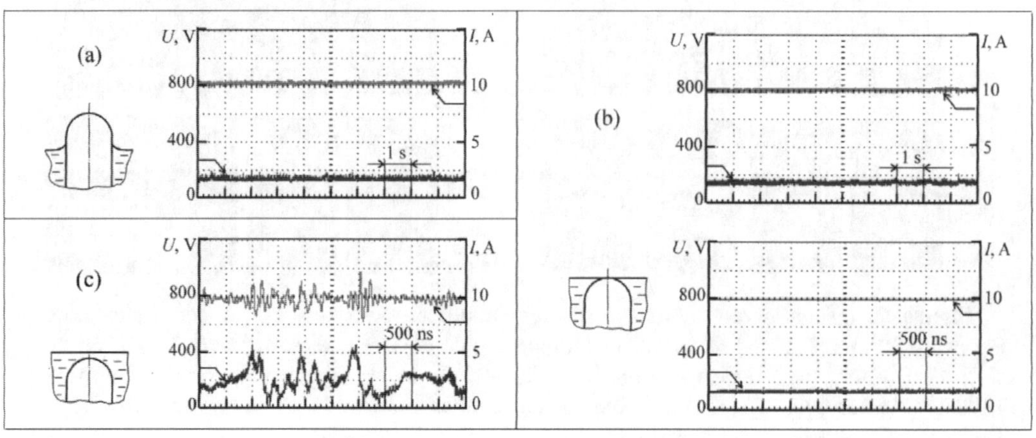

Figure 4. Oscillograms of gas discharge at different options of cathode water flow. Water consumption – 1.5 g/s. Electric conductivity – 370 μSm/sm. Anode-to-cathode distance – 50 mm.

In the option (c), i.e. when copper cathode was underwater, there was a significant erosion of its end. Some results of these processes are shown in table. In the experiments specific electric conductivity of water was within 200 - 225 μSm/sm. The water flow was 1.5 g/s.

Specific erosion γ of copper cathode was enough large quantity. It is noteworthy that the numerical value of its order of magnitude is the same as a in a vacuum pulse discharge at multiple inclusions of current [15].

Table 1. Specific erosion of copper cathode.

<*I*>, A	γ, μg/Kl	<*I*>, A	γ, μg/Kl
9	22.3	15	21.6
12	20.6	18	19.1

In the products of erosion under the microscope in reflected light are clearly visible particles of metallic copper, which also indicates of the probability of micro-explosions' origin. Technical capabilities of the microscope allowed observing particles with sizes of tens of micrometers. It is possible that smaller size particles exist.

3. Conclusions

Water flow around a copper cathode, causes significant changes of the gas discharge properties. These changes are most clearly manifested in two options of finding a copper cathode in the aqueous stream. These options differ in the working end of the cathode: on the surface of water and underwater. In the option, when the cathode end face is on the surface of water in the plasma column near the cathode is formed zone where hydrogen atoms radiation prevails. It can be assumed that under the influence of the gas discharge formed a stream of steam-plasma directed from the surface of the water vertically up.

References

[1] Saito G and Akiyama T 2015 *J. Nanomaterials* **2015** 123696
[2] Chen Q, Li J and Li Y 2015 *J. Phys. D.: Appl Phys.* **48** (42) 42400
[3] Tepanov A A, Krutyakov Yu A and Lisichkin G V 2014 *Russian J. General Chem.* **84** 986
[4] Valiev R A, Shakirov Yu I and Iliukhin A N 2014 *J. Phys.: Conf. Ser.* **567** 012040
[5] Tazmeev G Kh, Timerkaev B A, Tazmeev Kh K and Arslanov I M 2016 *Applied Physics* No. 1 72
[6] Smirnov S A, Shutov D A, Bobkova E S and Rybkin V V 2016 *J. Plasma Chem. Plasma Process.* **36** 415.
[7] Shakirov Yu I, Valiev R I, Khafizjv A A, Valiev R A and Khakimov R G 2016 *J. Phys.: Conf. Ser.* **669** 012064
[8] Tazmeeva R N and Tazmeev B K 2014 *Applied Physics* No. 1 35
[9] Kashapov L N, Kashapov N F, Kashapov R N and Denisov D G 2016 *J. Phys.: Conf. Ser.* **669** 012029.
[10] Tazmeev Kh K and Tazmeev A Kh 2014 *J. Phys.: Conf. Ser.* **567** 012035
[11] Titov V A, Rybkin V V, Smirnov S A, Kulentsan A L and Choi H.-S. 2006 *J. Plasma Chem. Plasma Process.* **26** (6) 543.
[12] Valiev R I, Gaisin A F, Gaisin F M, Gumerov A Z, Nasibullin R T, Sadriev R S, Sarimov L R and Khafizjv A A 2014 *Izvestiy Vuzov. Fizika* **57** (3/3) 66
[13] Zaripov R G, Kashapov N F, Tkachenko L A and Shaydullin L R 2016 *J. Phys.: Conf. Ser.* **669** 012053
[14] Denisov D G, Kashapov N F and Kashapov R N 2015 *IOP Conference Series: Materials Science and Engineering* **86** 012005
[15] Mesyats G A 1993 *Jektony. Chast 1* (Ekaterinburg: UIF Nauka) p 298

IOP Conf. Series: Journal of Physics: Conf. Series **789** (2017) 012062 doi:10.1088/1742-6596/789/1/012062

The influence of fluid - flexible particle interaction on fluid flow optical non-homogeneity in channel bifurcation

F Kh Tazyukov, E R Kutuzova and F A Garifullin

Kazan national research technological university, 68 Karl Marx street, Kazan, 68, 420015, Republic of Tatarstan, Russian Federation

E-mail: elvira.kutuzova@list.ru

Abstract. In the present paper the peculiar properties of convergent fluid flow in T-junction channel is considered. There is no interaction between flexible particles in the flow. Such kind of situation is described by rheological FENE-P and Oldroyd-B models. The first one predicts viscosity anomaly, dependence of longitudinal viscosity on longitudinal strain rate and elastic properties; the last one – existence of longitudinal viscosity depending on longitudinal strain rate $\dot{\varepsilon}$ and having a physical sense only for $\dot{\varepsilon} \ll 1/(2\lambda)$ and elastic properties. The model's governing parameters are the Weissenberg number (*We*), the Reynolds number (*Re*), the ability of flexible particle to change its orientation and stretching degree (L^2) in the main flow. The bifurcation area is of great importance due to possibility of high stresses and velocities existence not only in central area, but also on the walls and near the corners. The symmetry-loss effect at creeping flows regime (*Re*<<1) is investigated. It has been showed that at certain set of *We* and L^2 values the symmetrical shape of fluid flow turns to asymmetrical shape.

1. Introduction

Many scientists all over the world investigate on microfluidic flows since the previous century [1], including non-Newtonian flows through T-junction channels. These fluids such as suspensions, emulsions, liquid bioassays, polymer melts and solutions play a special part in nowadays. Studying and understanding its behavior is of great importance due to its common usage: conducting chemical and medical analyses, microfluidic devices production, industrial production [2-4]. Flexible particles and fluid flow interaction is responsible for the effects emerging in channel bifurcation. The aim of the present paper is to investigate the fluid flow structure in a planar T-junction symmetrical channel with cavity.

2. Governing equations

Fluid flow may be described by the momentum equations and mass conversation:

$$\rho\left(\frac{\partial \vec{v}}{\partial t} + \vec{v} \cdot \vec{\nabla}\vec{v}\right) = -\vec{\nabla}p + \vec{\nabla} \cdot \tilde{\tau}, \quad (1)$$

$$\vec{\nabla} \cdot \vec{v} = 0, \quad (2)$$

For non-Newtonian fluids the extra-stress is the sum of polymer and solvent stress contributions:

$$\tilde{\tau} = \tilde{\tau}^{p} + \tilde{\tau}^{s}, \quad (3)$$

where $\tilde{\tau}^S = 2\eta^S \tilde{D}$.

Non-Newtonian stress contribution for the FENE-P model could be written as follows:

$$\tilde{\tau}^p = \frac{\eta^p}{\lambda}\left[\frac{\tilde{A}}{1-\dfrac{tr(\tilde{A})}{3L^2}} - \frac{\tilde{I}}{1-\dfrac{1}{L^2}}\right], \quad (4)$$

$$\frac{\tilde{A}}{1-(tr\tilde{A})/(3L^2)} + We\overset{\nabla}{\tilde{A}} = \frac{\tilde{I}}{1-1/L^2}, \quad \overset{\nabla}{\tilde{A}} = \frac{\partial \tilde{A}}{\partial t} + \vec{v}\cdot\vec{\nabla}\tilde{A} - \vec{\nabla}\vec{v}\cdot\tilde{A} - \tilde{A}\cdot(\vec{\nabla}\vec{v})^T, \quad (5)$$

Non-Newtonian stress contribution for the Oldroyd-B model could be written as follows:

$$\tilde{\tau}^p + \lambda\overset{\nabla}{\tilde{\tau}^p} = 2\eta^s\tilde{D}, \quad (6)$$

$$\overset{\nabla}{\tilde{\tau}^p} = \frac{\partial\tilde{\tau}}{\partial t} + \vec{v}\cdot\vec{\nabla}\tilde{\tau} - \nabla\vec{v}\cdot\tilde{\tau} - \tilde{\tau}\cdot(\vec{\nabla}\vec{v})^T. \quad (7)$$

where $\vec{v}$ - velocity vector; ρ - fluid density; λ - stress relaxation time; $\eta^0 = \eta^s + \eta^p$ - total viscosity of fluid; η^p - dynamic viscosity of polymer contribution at zero shear rate; η^s - dynamic viscosity of polymer contribution at zero shear rate; $\tilde{\tau}^p$ - non-Newtonian polymer stress contribution; $\tilde{\tau}^S$ - Newtonian solvent stress contribution; $\tilde{A} = \frac{3\langle\vec{Q}\vec{Q}\rangle}{Q_{eq}^2}$ - configuration tensor, where $\vec{Q}\vec{Q}$ - dyadic multiplication of configuration vectors; $\langle\vec{Q}\vec{Q}\rangle = \int_{-\infty}^{\infty}\int_{-\infty}^{\infty}\int_{-\infty}^{\infty}\vec{Q}\vec{Q}\cdot P_N(\vec{Q})\cdot d\vec{Q}d\vec{Q}d\vec{Q}$ - ensemble averaging, where $P_N(\vec{Q})$- probability that a randomly selected macromolecules has a predetermined size in the interval from $\vec{Q}$ to $\vec{Q}+d\vec{Q}$; $\tilde{D} = \frac{1}{2}\left(\vec{\nabla}\vec{v} + \left(\vec{\nabla}\vec{v}\right)^T\right)$ - - strain rate tensor, where $(\cdot)^T$ - transpose procedure.

Dimensionless form of governing equations contains following dimensionless parameters:

$$We = \frac{\lambda U}{l}, \quad Re = \frac{\rho U l}{\eta^0}, \quad \beta = \frac{\eta^s}{\eta^0}, \quad L^2 = 3\left(\frac{Q_0}{Q_{eq}}\right)^2$$

3. Boundary conditions

Schematic representation of the channel with boundaries is sketched on figure 1. The length of the channels was set to 10 channel's width for the formation of the velocity profile in the inlet flow and to establish the outlet flow.

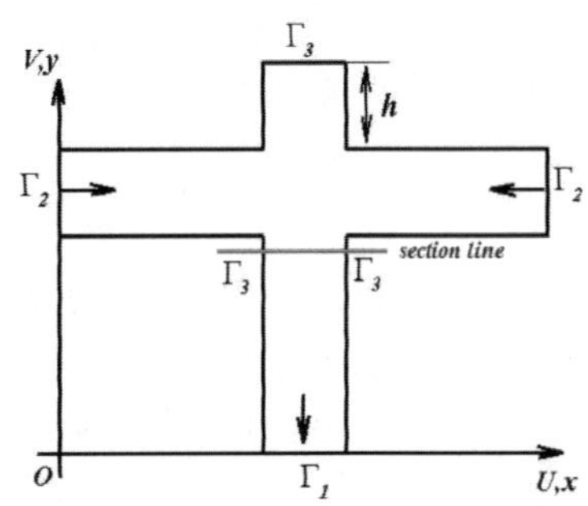

Figure 1 – Scheme of the branched channel geometry

4. Results

To get an idea about fluid - flexible particle interaction is possible by means of *Re, We* numbers and retardation coefficient β characterizing the particles concentration for Oldroyd-B model. For the fluid models such as FENE-P taking into account the stretching of flexible particles, one of the most important parameter is L^2. Combination of *We* and L^2 (at fixed value retardation coefficient β=0.1 and *Re*=0.01) is certainly characterizing the fluid behavior performed by the FENE-P model [5, 6].

The asymmetrical form of fluid flow is emerging in the channel bifurcation for the Oldroyd-B and FENE-P models at L^2=100 (fig. 2). From streamline patterns picture follows that the symmetry loss effect is observed for the Oldroyd-B model at *We=0.6* and for the FENE-P model at *We=0.6; L^2=100.* As it might be seeing that for the set of values *We=0.6; L^2=10* flexible particle's stretching is not enough for breaking symmetry behavior.

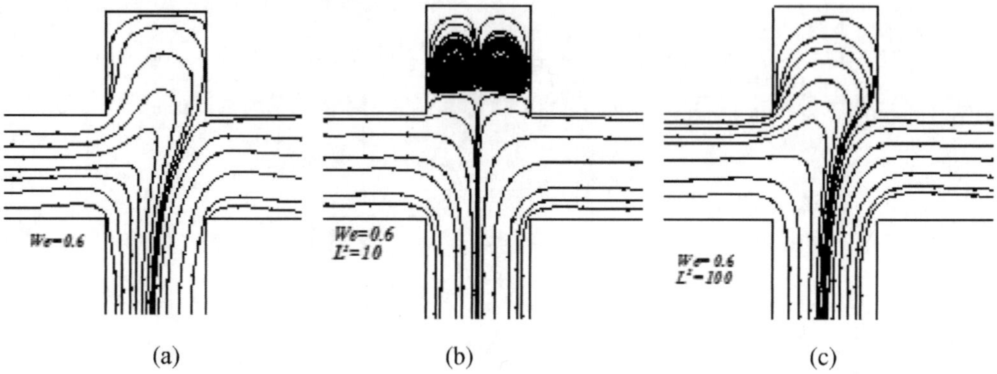

Figure 2 - Viscoelastic fluid flow through a channel with cavity:

(a) –Oldroyd-B, (b, c) – FENE-P

The distribution of shear stresses along the symmetry line of outlet channel's part one can see on fig.3. For the FENE-P model at values of *We=0.6; L^2=10* this distribution is symmetrical that corresponds to fig.2b. For the symmetry loss cases the patterns of τ_{xy} are similar for both models. However, the area of maximum values is more shifted from the symmetry line compared to FENE-P model's results and the values are greater approximately for 3 times.

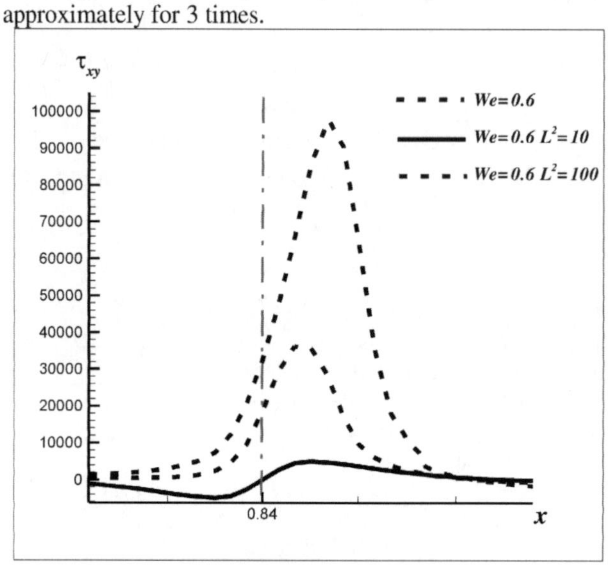

Figure 3 – Shear stresses distribution

From fig. 4 follows that for the first normal difference *N1* the maximum change of values is observed for the fluid presented by Oldroyd-B model. This behavior caused by the ability of flexible particles to infinitely extension. Due to the linear relation between its elongation and interaction force between the ends of flexible particles. Large elongation leads to increase of interaction force and significant change in tension state in the fluid flow.

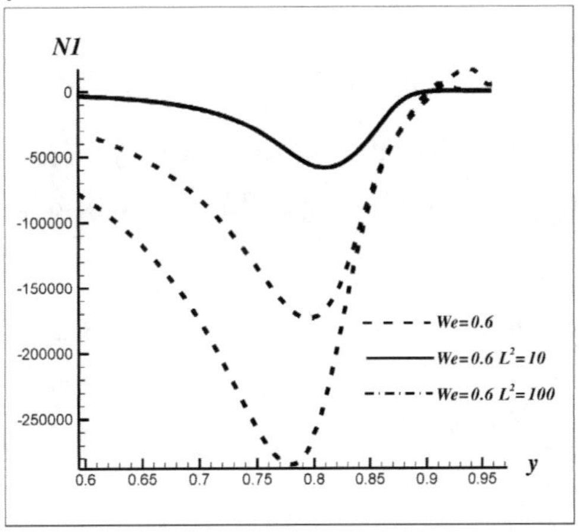

Figure 4 – Distribution *N1* along the channel's symmetry line

5. Conclusions

In the present paper the numerical results for convergent viscoelastic fluid flow in branched channel have been presented. The symmetrical shape of flow turns to asymmetrical at *We*=0.06 for both models. This effect of the loss-symmetry patterns is associated with the interaction of flexible particle and the main flow. The differences of presented viscoelastic fluids lies in consideration of flexible particles behavior. For the Oldroyd-B model the particle theoretically could infinitely elongate and for the FENE-P model the elongation characterizing by the L^2 parameter. It means that for the same values of *We* number the particles stretching will be different. The changing of the flow direction leads to changes in conformation of flexible particles associated with their stretching and orientation in the flow. This non-equilibrium configuration, in turn, leads to change in normal stresses that also affect on the flow patterns.

References

[1] Mukhopadhyay S. Computation of Viscous Flow Field in a Tapered Artery with an Overlapping Constriction. / S. Mukhopadhyay, C.Midya, G. C. Layek // International Journal of Fluid Mechanics Research. 2009. T. 36. №. 4. pp. 343-356

[2] Soulages J. et al Investigating the stability of viscoelastic stagnation flows in T-shaped Microchannels //Journal of Non-Newtonian Fluid Mechanics. 2009. T. 163. №. 1. c. 9-24.

[3] Tazyukov F.Kh. et al. Non-Newtonian flow of blood through a symmetric stenosed artery // Russian Journal of Biomechanics. 2012. v. 16(1) p. 46-57

[4] Stone H. A., Kim S. Microfluidics: basic issues, applications, and challenges //AIChE Journal. 2001. T. 47. №. 6. c. 1250-1254.

[5] Tazyukov F.Kh., Kutuzova E.R., Layek G.C. Symmetry-loss effect of viscoelatic fluid flow through a T-juncion channel // Transaction of Academenergo (Russian) – 2016. – № 2. – pp. 20-28.

[6] Tazyukov F.Kh., Kutuzova E.R., Snigerev B.A. Divergent non-Newtonian fluid flow // Ecological bulletin of research centers of the Black sea economic cooperation (Russian) – 2016. – № 3. – pp. 59-64.

IOP Conf. Series: Journal of Physics: Conf. Series **789** (2017) 012063 doi:10.1088/1742-6596/789/1/012063

Discharge creeping along the surface in the process of cleaning and strengthening of the materials surface

Timerkaev B. A.[1], **Andreeva A. A.**[1,2], **Sofronitskiy A. O.**[1,2]

General Physics Department, Kazan National Research Technical University – KAI, 420111, K. Marx St, 10, Kazan, Russia[1]
Kazan Federal University, Institute of Engineering, 420008, Kremlevskaya St. 18, Kazan, Russia[2]

btimerkaev@gmail.com

Abstract. This paper studied the creeping surface discharge and its interaction with conductive and dielectric surfaces.

1. Introduction

Magnetron glow discharge is widely applied in the processes of cathode sputtering of the materials and coating of conductive materials. The mechanism of ejection of individual atoms from the cathode is associated to the interaction of fast positive ions of the plasma with the atoms of matter, in which the atoms gain additional energy and leave the crystal. However, this method is absolutely not acceptable for dielectric materials so as in the direction to them the ions have no acceleration, and there is almost no interaction. Random diffusion interaction of ions with dielectric walls does not give the necessary intensity for the process of deposition of dielectric materials. The theory of glow discharge with account of nonlocal nature of the ionization processes by electron impact in the cathode regions was developed in [1-4]. A magnetron discharge with special organization of magnetic field is considered in this work. The magnetic field is organized near the dielectric surface so that the Lorentz force pushes the moving charges to the dielectric surface and provides close interaction of plasma glow discharge with the dielectric surface. The glow discharge creeping on the surface with the purpose of decomposition of heavy hydrocarbons and growing the carbon nanoparticles was studied in [5,6].

2. Experiment

The experimental setup (Fig.1) was designed to study the effect of spraying of a dielectric material in a glow discharge. The experimental setup has a dielectric target between the cathode and the anode. A permanent magnetic field is arranged along the dielectric surface. The magnetic field is directed parallel to the dielectric surface and perpendicular to the direction of the electric field. The surface for the coating is on the opposite side. To exclude the ingress of the cathode material to the substrate the experimental setup contains a screen to prevent direct access of atoms, sputtered from the cathode, to the substrate.

After the ignition of the discharge in this setup, the discharge is pressed against the surface of the dielectric. Ions of the plasma accelerated by an electric field, change their direction due to magnetic field action and collide with the atoms on the dielectric surface. This cause the evaporation of dielectric atoms and the coating of the substrate.

Glow discharge was organized between copper electrodes. Electrodes had the form of rectangular parallelepipeds with geometric dimensions of 50x20x3 mm. Electrodes were put against each other at a distance of 50 mm. The target was placed at a distance of 15 mm below the axis on which were the barycentre of the electrodes. The substrate was disposed above this axis at a distance of 10 mm. A magnetic system was arranged perpendicular to the direction of the electric field below the substrate with the possibility to vary the distance from the axis on which were the barycentre of the electrodes depending on the magnetic permeability of the target. It was used two neodymium magnet (type N38) of cylindrical shape (D=45mm, h=15 mm) with axial magnetization, attached to the steel magnetic core having a geometric size 100x45x5 mm and magnetic permeability 1000 Gn/m. The coercive force of the magnets was equal 955 кA/m. The values of magnetic field induction at various distances near the magnetic system was calculated by the model which was created in the program Elcut Student 6.1. The result is shown in Fig.2.

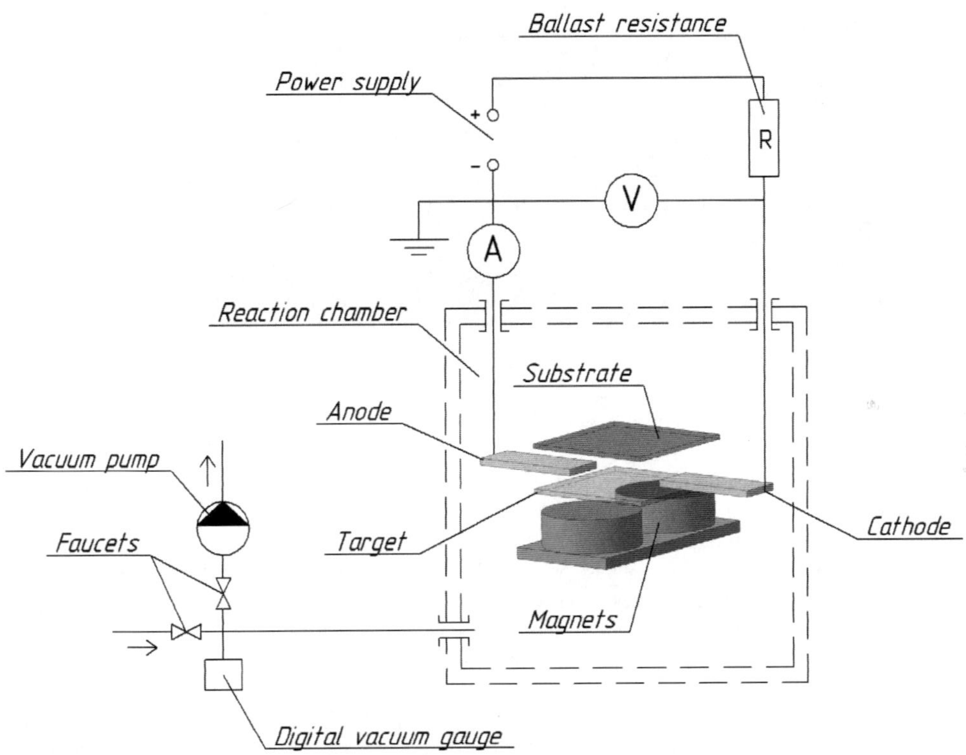

Figure 1. Schematic diagram of the experimental setup.

Many different dielectric and conductive materials were tried in the experiments. In one case the polymethyl methacrylate was used as a dielectric target. It has been successfully sprayed on a glass substrate. The target was sprayed in the reaction chamber at a pressure of 1 Torr. The values of voltage and current were in the ranges typical for glow discharge.

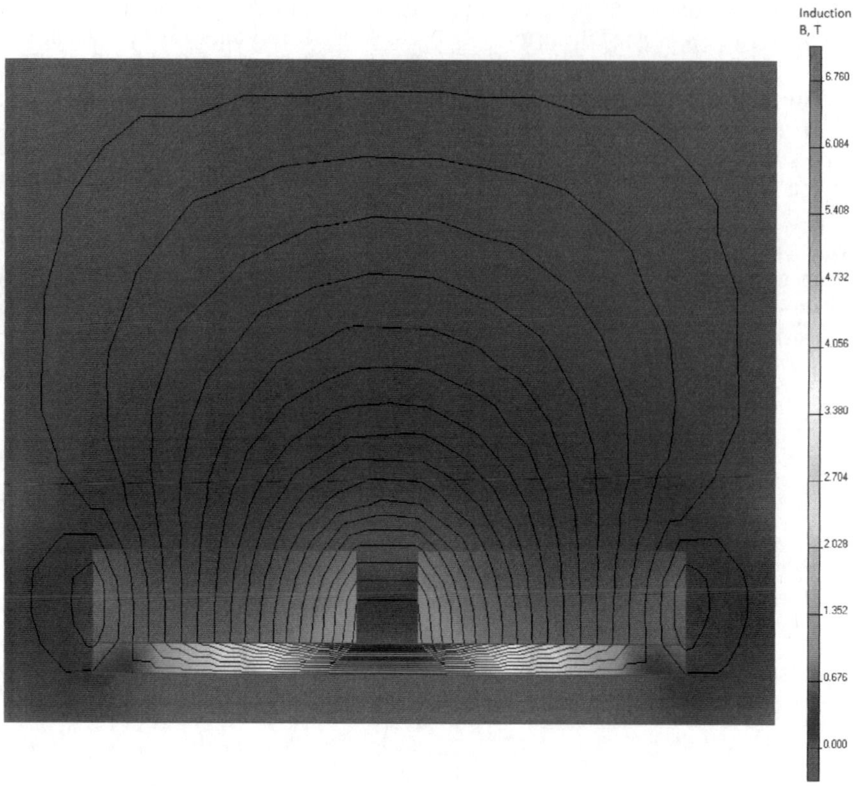

Figure 2. The distribution of the magnetic field, calculated with the use of program Elcut 6.1 Student.

At low values of current the sputtering of the target material did not happen, but there was observed a process of target surface cleaning from various contaminants. The spraying rate was increasing with the increase in the current value. Soon after the start of sputtering the coating of white color could be observed on the glass substrate. To obtain a stable uniform coating of the substrate is required to observe the corresponding mode current and pressure. Low current values are corresponding the efficient cleaning, and high currents are corresponding the effective deposition. With a larger increase in current the allocation of the soot particles is observed.

3. Conclusion

The uniqueness of the observed phenomena lies in the fact that in this implementation, the sputtering of the target material occurs as a bombardment of heavy ions and electrons. The energy of the ions is not as great as in the case of the bombardment of the cathode. It just gives the possibility of obtaining gentle spray modes, when from the surface of the sputtered material can be ejected not only individual atoms but entire molecules which are preserving the properties of the substance.

References

[1] Saifutdinov A I Timerkaev B A Drift model of glow discharge with account of nonlocal value of the electric field in the ionization source Engineering-physical journal vol 85 No 5 2012 pp 1104-1109

[2] Timerkaev B A Petrova O A Saifutdinov A I Self-organization of a layered structure of the normal glow discharge Journal of engineering physics vol 89 2015 No 2 pp 477-482

[3] Saifutdinov A I Timerkaev B A Zalyaliev B R Control of the glow discharge parameters at low pressures by means of transverse supersonic gas flow Thermophysics of high

temperatures 2016 vol 54 No 5 pp 669-675
[4] Saifutdinov A I Timerkaev B A Nonlocal model of a longitudinal structure of the glow discharge and the role of diffusion of charges along the current Vestnik KGTU im A N Tupolev No 2 2012 pp 188-193
[5] Timerkaev B A Sofronitzkiy A O Andreeva A A Creeping on the surface of the glow discharge and the prospects for its application Vestnik of Kazan state technical University. A N Tupolev No 3 2015 pp 5-9
[6] Timerkaev B A Sofronitskiy A O Andreeva A A Carbon nanotubes formation in the decomposition of heavy hydrocarbons creeping along the surface of the glow discharge VII Conference on Low Temperature Plasma in the Processes of Functional Coating Preparation IOP Publishing Journal of Physics: Conference Series 669 2016 012062

Effect of radiation intensity pulsations of plasma on its temperature

R Toukhvatoulline, F Bashirov and I Timerbaev

Abstract. Effect of radiation intensity pulsations of non-stationary plasma on its radial temperature is studied. Consequence of these pulsations on accuracy of the measurement of plasma optical characteristics is discussed. The case of the plasma is stationary in space whereas temperature makes pulsations obeyed both harmonious and random laws in each its points is also investigated.

1. Introduction

It is well-known fact that both electroarc and high-frequency plasmas are not stationary [1-2]. The electric current and voltage, as well as speed and luminance of a stream of plasma are subjects of oscillation. There are many reasons causing instability of arc arrangement in space, such as interacting of arc with intrinsic magnetic field, circumambient convection, strong electric fields, turbulent flow fluctuations and others. It has been proven that both oscillations of luminance intensity of a stream on the plasmatron exit and voltage of arc display a random character, and an analytical relation between them has been derived [3]. Investigating the dependence of pulsations in a plasma stream – of plasma stream turbulence, fine-scale shunting, lateral oscillations of a positive column (and some other reasons) resulted in the fact that temperature oscillations submit to normal distribution [1]. In [4], pulsations of speed, electric current and voltage, and an optical signal in a stream of direct current arc plasma were studied. It was proven that fluctuation of these parameters attains 10-20 % of its stationary values. In [5-8], the research of space oscillations of an electric arc in plasmatron with fixed arc length resulted in the fact that the space displacement of arc can alterate considerably the temperature of plasma and its radial distribution. High-frequency argon plasma can be also non-stationary. Anisotropic plasma's hydrodynamic-type instability produced by interacting of plasma kernel with a flow of blown in cold gas is the principal cause of instability. In [9], it was studied the dependence of temperature into the plasma axis of ICP-AES (Inductively Coupled Plasma Atomic Emission Spectrometry) of the inductive frequency and it was shown that frequency increasing leads to temperature decrease on the discharge axis. Besides, it was shown that the space pulsations in ICP-AES plasma are not significant by comparison to that of arc plasma, whereas time oscillations can be essential.

2. Plasma model.

Let's axisymmetric plasma, which is stationary in space, plus the temperature exhibits simultaneously two kinds of oscillations, harmonic and random, in each its points about average temperature T_c.

Harmonious temperature oscillations can be expressed by:

$$T = T_c + a\cdot\cos(\omega t), \tag{1}$$

where ω – the cyclic frequency and a – the amplitude value of temperature oscillations. The distribution function of the temperature difference T - T_c related to equation (1) can be presented as:

$$\varphi_2(T - T_c) = \frac{1}{\pi\sqrt{a^2 - (T - T_c)^2}}. \tag{2}$$

The distribution function of random temperature oscillations is taken as Gauss distribution function, which can be written by:

$$\varphi_1(T - T_c) = \frac{1}{\sigma\sqrt{2\pi}} e^{-\frac{(T-T_c)^2}{2\sigma^2}}, \tag{3}$$

where σ is the root-mean-square deviation of temperature.

By accounting for expressions (2) and (3) and replacing $\xi = (T - T_c)$, the generalized distribution function takes the form:

$$\varphi(\xi) = \int_{-a}^{a} \varphi_1(\varsigma - \xi)\varphi_2(\varsigma)d\varsigma = \frac{1}{\sigma\sqrt{2\pi^3}} \int_{-a}^{a} \frac{\exp(-\frac{(\varsigma-\xi)^2}{2\sigma^2})}{\sqrt{a^2 - \varsigma^2}} d\varsigma. \tag{4}$$

During the integration prescribed by the function (4) the limits of the quantities a and □□has been taken from 1000 K up to 4000 K in this study. In figure 3, we are presenting the graphs of the function $\varphi(\xi)$ drawn for σ =2000 K and a = 1000 K (1), 2000 K (2), 3000 K (3) и 4000 K (4).

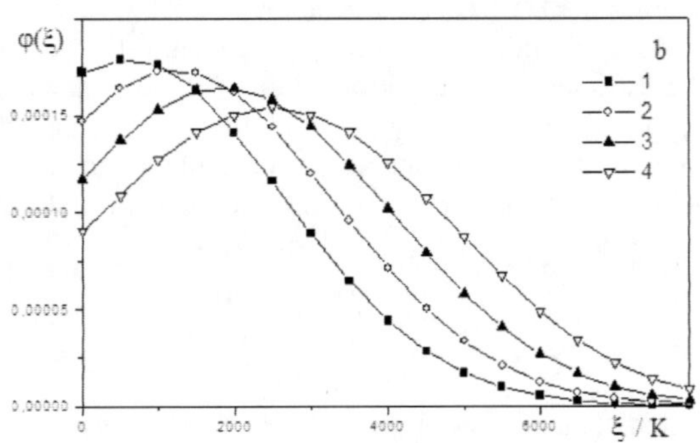

Figure 1.

It is seen from the graphs presented in figure 1 that the maximal values of distribution functions are displaced to right side from average temperature by rising a for a given σ.

3. Radial radiation intensity and temperature

The oscillations of temperature followed the distribution function (4) act to radiation intensity and the radial temperature. Radiation intensity relates to temperature by relationship [12]:

$$\varepsilon_\nu(T) = \frac{1}{4\pi} A_{ij} n_0(T) \frac{g_i}{Z_a(T)} e^{-\frac{E_i}{kT}} h\nu, \tag{5}$$

where $Z_a(T) = \sum_i g_i e^{-\frac{E_i}{kT}}$ – the statistic sum, A_{ij} – the probability of atom transition from i-th energy state with energy E_i to j-th energy state with energy E_j, g_i – the weight of i-th state.

Taking into account the distribution function (4) average radiation intensity can be calculated by the formula:

$$\bar{\varepsilon}(T_c) = \int \varepsilon(T)\varphi(T - T_c)\mathrm{d}T \tag{6}$$

We calculated radiation intensity for a chosen spectral line Ar I 415.8 nm. The dependence of radiation intensity of the temperature for this line was given in [6]. Integration (6) was performed for the temperature region T_c = 4000 K – 16000 K, the oscillation amplitudes were taken equal to a = 500 K – 4000 K and root-mean-square deviations – □□= 500 K – 4000 K.

The results of integrating (6) testify that the real values of $\varepsilon(T)$ differ from those averaged $\bar{\varepsilon}(T)$. This difference depends on the amplitude of temperature oscillation a, root-mean-square pulsations σ, and the temperature region. When a is constant, this difference increases with growing σ. However, in the temperatures region of 12000 K – 13000 K, this difference is insignificant. At constant σ, with growing a this difference also increases, and for chosen line Ar I 415.8 nm difference between $\bar{\varepsilon}(T)$ and $\varepsilon(T)$ is maximal in the vicinity of temperature 8000 K and it is minimal in the vicinity of 12500 K – 13000 K.

Spectroscopic determination of the temperature of axisymmetric plasma was performing by experimental founding the lateral intensity distribution of the arc, then, the radial radiation intensity was deriving by means of Abel's integral transformation [13] and, finally, the temperature distribution upon arc radius was determined by using temperature dependence of the radiation intensity (5).

To determine the radial temperature, we are choosing lateral intensity distribution on arc in the form:

$$I(x) = I(0)\exp(-\alpha x^2), \tag{7}$$

where $I(0)$ – the value of radiation intensity on the arc axis and □ – a factor. Substituting (7) in Abel's integral equation gives us the radial distribution $\varepsilon_\nu(r')$, where $r' = r/R$ – relative radius of plasma flow.

The radial radiation intensity and the radial temperatures have been calculated by taking constant values of σ and variable amplitude a, which were equal successfully 1000 K, 2000 K, 3000 K and 4000 K, then – for constant a and variable □ equal successfully 1000 K, 2000 K, 3000 K and 4000 K.

Calculation of temperature distribution on arc radius has been performed for three temperatures of arc axis equal to T_0 = 8000 K, 10000 K and 12000 K and for the range of σ and a equal to 1000 K – 4000 K by taking into consideration the values of radial intensity and dependences of radiation intensity of temperature.

The results of calculating the radial temperatures T for the axis temperature 10000 K are drawn in figure 2. The curves of dependences $T(r')$ for σ = 1000 K and a = 1000 K (2), 2000 K (3) and 3000 K (4) are presented in figure 2-a, and for a = 1000 K and σ = 1000 K (2), 2000 K (3) and 3000 K (4) – in figure 2-b. The curves labeled by the number 1 correspond to the radial temperatures at the absence of oscillations.

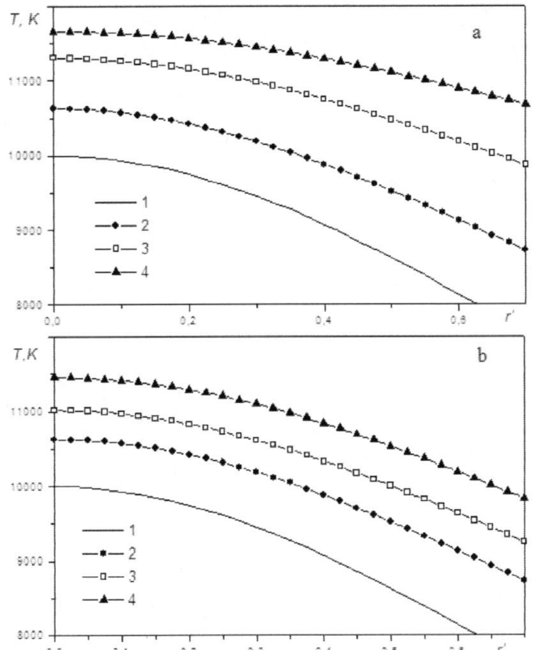

Figure 2

We can see in figure 2 that pulsations result in temperature growing on entire profile. Effect of pulsations depends on σ and a. By growing σ difference between the true and observed profiles increases and can attain 1500 K at arc axis and 3000 K at periphery. Growing of oscillation amplitude a also results in increase of difference between the true and observed profiles.

References

[1] Moshkin B B 1967 *TVT* **5** 75-9
[2] Ghorui S and Das A K 2004 *Phys. Rev. E* **69** 2571–9
[3] Anshakov A S, Dautov G Yu and Mustafin G M 1967 **PMTF** 35-41
[4] Planche M P, Coudert J F and Fauchais P 1998 *Plasma Chem. Plasma Process.* **18** 263–83
[5] Thoukhvatoulline R and Feldmann G 2000 *J. Phys. D* **33** 2420-4
[6] Thoukhvatoulline R, Dautov G and Feldmann G 2004 *J. Phys. D* **37** 1058-64

[7] Zakirov I M, Zalyalieva F F, Timerkaeva D B and Tukhvatullin R S 2011 *High Temperature* **49**, No 3, 338–42.
[8] Toukhvatoulline R S, Dautov G Yu, Zalyalieva F F and Ashrapov T F 2015 *TVT* **53,** No 5, 136-42
[9] Zakirov I M, Zalyalieva F F, Toukhvatoulline R S and Ashrapov T F 2013 *High Temperature* **51**, No 6, 742–6.
[10] Thoukhvatoulline R, Feldmann G, Bonadiman H and Blumke M R 2002 *Winter Conference on Plasma Spectrochemistry, Scottsdale, Arizona*, **1** 239-40.
[11] Mostaghimi J and Boulos M I 1990 *J. Appl. Phys.* **68** 2643-8
[12] *Piasma Research Methods*, ed. by Lohte-Holtgreven (Russ. transl.), Mir, 1970, 446 p
[13] Larkina L T *Application of plasmatron in spectroscopy. Frunze. ILIM*, 1970. 212 p

Evaluating space and time oscillations of plasma radiant intensity by studying beam radial temperature

R S Tuhvatullin[1], I R Timirbaev[1] and F I Bashirov[2]

[1]Department of General Physics, Kazan State Technical University, K. Marx, 70, Kazan, Russia
[2]Department of General Physics, Kazan Federal University, Kremlevskaja St, 18, Kazan, Russia

Abstract. Effects of space and time pulsations of non-stationary plasma on radiant intensity and the radial temperature of plasma are studied. The case, when plasma simultaneously non-stationary both in time and in space is examined. The influence of these perturbations on accuracy of definition of plasma optical features is shown.

1. Introduction

It is well-known fact that both electroarc and high-frequency plasmas are not stationary [1-2]. The electric current and voltage, as well as speed and luminance of a stream of plasma are subjects of oscillation. There are many reasons causing instability of arc arrangement in space, such as interacting of arc with intrinsic magnetic field, circumambient convection, strong electric fields, turbulent flow fluctuations and others. By studying the space oscillations of electric arc in a plasma generator with fixed arc length [3-4], it has been established that the space displacement of an arc can influence temperature of plasma and its radial distribution considerably. High-frequency argon plasma also can be non-stationary. By examining the temperature on a discharge axis in plasma ICP-AES (Inductively Coupled Plasma - Atomic Emission Spectrometry) as a function of the inductive frequency, it was shown that frequency growth leads to temperature reduction on the discharge axis [5]. Besides, it was shown that the space pulsations in ICP-AES plasma are insignificant by comparison to that in arc plasma, whereas time dependent oscillations can be essential. Accounting for the fact that both the space and time pulsations are always presented in plasma, this work has a deal to studying simultaneously space-time pulsations in plasma.

2. Studying technique

Spectroscopic determination of the temperature of axially symmetric plasma is carrying out by experimental establishment of the lateral intensity distribution of the arc. Then, the radial radiation intensity is deriving by means of Abel's integral transformation [6]. Finally, one fixes a temperature distribution on arc radius by using temperature dependence of the radiation intensity [6]. At the same time, if the plasma displaces in the space, the intensity distribution on the arc altitude is observing with a crude error. To reduce a lapse, we are using the Fredholm equation of the first kind [4]:

$$I_v(x) = \int_{-R}^{R} q_v(x)\varphi(x-x')\mathrm{d}x'. \quad (1)$$

Here, $I_\nu(x)$ is the observed distribution function and $q_\nu(x)$ is the true distribution function that is that corresponded to the absence of oscillations, $\varphi(x)$ – a distribution function of oscillations of a radiant. This equation belongs to the class of mathematical physics inverse problems. When $\varphi(x)$ is presented by following Gaussian law:

$$\varphi(x) = \frac{1}{\sigma_x\sqrt{2\pi}}\exp(-\frac{x^2}{2\sigma_x^2}), \tag{2}$$

the exact solution of equation (1) takes the form [6]:

$$q_\nu(x) = \frac{1}{2\pi}\int_{-\infty}^{\infty}\int_{-\infty}^{\infty} I_\nu(\theta)\,\mathrm{e}^{\frac{\sigma_x{}^2\lambda^2}{2}-\mathrm{i}\lambda'(x-\theta)}\,\mathrm{d}\lambda'\mathrm{d}\theta. \tag{3}$$

Here, σ_x – arc root-mean-square deviations. Observed distribution of $I_\nu(x)$ is often expressed by

$$I_\nu(x) = I_\nu(0)\mathrm{e}^{-\sigma x^2}. \tag{4}$$

Therefore, the solution (3) will equal to

$$q_\nu(x) = \frac{I_\nu(0)}{\sqrt{1-2\alpha\sigma_x^2}}\exp(-\frac{\alpha x^2}{1-2\alpha\sigma_x^2}). \tag{5}$$

In the case of the temporary pulsations of plasma presented by the following normal distribution function:

$$f(T-T_c) = \frac{1}{\sigma_T\sqrt{2\pi}}\exp[-\frac{(T-T_c)^2}{2\sigma^2}], \tag{6}$$

where σ_T – temperature root-mean-square deviations, the average emittance follows to expression:

$$\bar{\varepsilon}(T_c) = \int\varepsilon(T)f(T-T_c)\mathrm{d}T. \tag{7}$$

3. Plasma model

Let each point of axially symmetric plasma exhibits simultaneously two oscillation modes: a) the space oscillations of arc described by the function (2) and b) temporary temperature variations yielded to expression (6). During computing, we were using Fredholm equation (3) with intensity distribution on arc altitude taken in the form (4). By numerical calculations of $\varepsilon_\nu(r)$, intensity distribution on arc radius, we applied Abel's integral transformation [7]:

$$\varepsilon_\nu(r_k) = -\frac{N}{40R}\sum_{j=0}^{N-1} b_{jk}I_\nu(x_j),\ x_j=j/N;\ r_k=k/N;\ j,k = 0...N\text{-}1. \tag{8}$$

Coefficients b_{jk} are tabulated in [7]. Calculations were performed for $q(0)=1$, $\alpha = 10$; $\sigma_x = 0.1$, 0.2, and 0.3; $\sigma_T = 1000$, 2000, and 3000, temperature of arc axis $T_{CO} = 8000$, 12000, and 15000 K in argon plasma. Argon spectral line Ar I 415.8 nm has been chosen during computing. Dependence of the relative emittance on temperature for the yielded line is presented in fig. 1 by the curve 1. The curves 2, 3, and 4 of figure 1 are displayed in the presence of pulsations for $\sigma_T = 1000$ (2), 2000 (3) and 3000 K (4). Accounting for known radial emittances, temperature distributions on arc radius were determined. The results of calculation of the radial temperatures for axis temperature T_{CO} = 12000 K at simultaneously presence the space oscillations well as temporary ones of plasma, at σ_x = 0.2, σ_T = 1000, 2000 and 3000 K are pictured in figure 2-*a*, and those at σ_T = 2000 K, σ_x = 0.1, 0.2, and 0.3 – in figure 2-*b*. The curve 1 of figure 2-*a* corresponds to the true distribution of the temperature. It is seen from presented graphs that pulsations of temperature and the space oscillations of arc can distort noticeably a temperature profile. Temperature pulsations lead to rise the radial temperature on whole profile and to appreciable rise of temperature on periphery. Transverse vibrations of arc disturb a temperature profile so that the temperature is downgraded in paraxial areas and it rises on peripheries (see figure 2-*b*) in order to profiles become more fill.

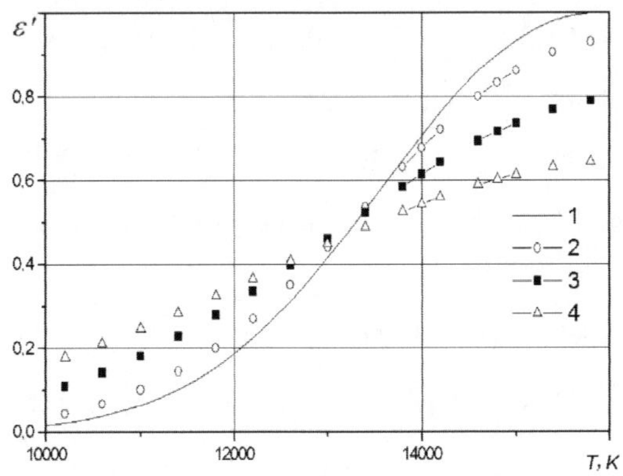

Figure 1

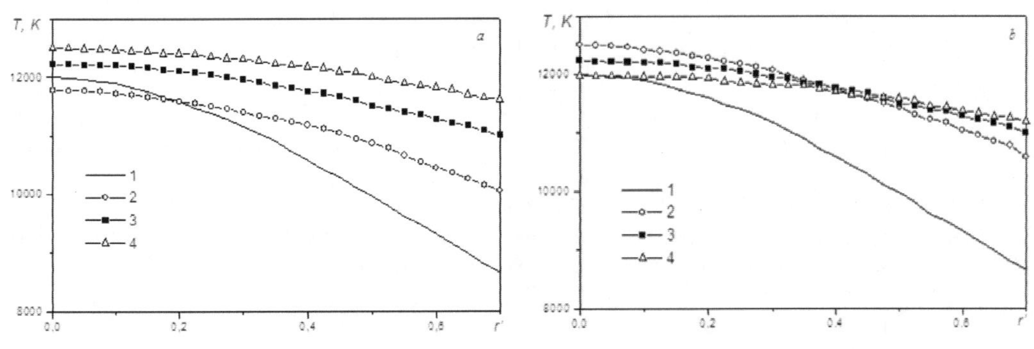

Figure 2

4. References

[1] Moshkin B B 1967 *TVT* **5** 75
[2] Ghorui S and Das A K 2004 *Phys. Rev. E* **69** 2571
[3] Thoukhvatoulline R and Feldmann G 2000 *J. Phys. D* **33** 2420
[4] Thoukhvatoulline R, Dautov G and Feldmann G 2004 *J. Phys. D* **37** 1058
[5] Mostaghimi J and Boulos M I 1990 *J. Appl. Phys.* **68** 2643
[6] *Plasma Research Methods*, ed. by Lohte-Holtgreven (Russ. transl.), Mir, 1970, 446
[7] Larkina L T *Application of plasmatron in spectroscopy. Frunze. ILIM*, 1970. 212

The mathematical model of dynamics of dust plasma considering polydisperse structure of the condensed phase

D A Tukmakov[1], **N A Tukmakova**[2]

[1] Kazan (Volga Region) Federal University, 18 Kremlyovskaya str., Kazan, 420008, Russia
[2] Kazan national research university named after A.N. Tupolev, 10 K. Marks str., Kazan, 420111, Russia

E-mail: tukmakovda@imm.knc.ru

Abstract. Work is devoted to studying of non-stationary processes in the complex plasma representing a suspension of solid particles in the uncharged gas. This model constitutes the closed system of the multiphase mediums dynamics equations in two-dimensional statement which involves the bearing medium and the condensed phase movement equations.

1. Introduction

During the research the mathematical model of quasi-neutral dusty plasma condensed phase of which has multifraction structure is presented by the particles having various size and consisting of substances with various physical properties. The offered model of dusty plasma is developed on the basis of the polydisperse multi-speed and multitemperature gas-suspension dynamics theory taking into account speed and temperature delay of the condensed fractions particles. At the same time it is supposed that disperse inclusions represent the spherical particles differing in the size and materials which particles consist of can have various density and heat conductivity.

2. Mathematical model

The movement of the bearing medium is described by system of the Navier-Stokes' equations with the accounting of interphase power interaction, heat exchange and electric field [1-4]:

$$\frac{\partial \rho_1}{\partial t}+\frac{\partial(\rho_1 u_1)}{\partial x}+\frac{\partial(\rho_1 v_1)}{\partial y}=0,$$

$$\frac{\partial(\rho_1\, u_1)}{\partial t}+\frac{\partial}{\partial x}\left(\rho_1\, u_1^2+p-\tau_{xx}\right)+\frac{\partial}{\partial y}\left(\rho_1 u_1 v_1-\tau_{xy}\right)=-\sum_{i=2,n}F_{xi}+\sum_{i=2,n}\alpha_i\frac{\partial p}{\partial x},$$

$$\frac{\partial(e_1)}{\partial t}+\frac{\partial}{\partial x}\left(\left[e_1+p-\tau_{xx}\right]u_1-\tau_{xy}\,v_1+\lambda\frac{\partial T_1}{\partial x}\right)+\frac{\partial}{\partial y}\left(\left[e_1+p-\tau_{yy}\right]v_1-\tau_{xy}\,u_1+\lambda\frac{\partial T_1}{\partial y}\right)=$$

$$=\sum_{i=2,n}Q_i-\sum_{i=2,n}\left(\left|F_{xi}\right|(u-u_i)-\left|F_{yi}\right|(v-v_i)\right)+\sum_{i=2,n}\alpha_i\left(\frac{\partial(pu)}{\partial x}+\frac{\partial(pv)}{\partial y}\right),$$

$$\tau_{xx}=\mu\left(2\frac{\partial u_1}{\partial x}-\frac{2}{3}D\right),\ \tau_{yy}=\mu\left(2\frac{\partial v_1}{\partial y}-\frac{2}{3}D\right),\ \tau_{xy}=\mu\left(\frac{\partial u_1}{\partial y}+\frac{\partial v_1}{\partial x}\right),\ D=\frac{\partial u_1}{\partial x}+\frac{\partial v_1}{\partial y}.$$

The system of the equations including the continuity equations for average density, the conservation equations of impulse components and the conservation equation of internal energy of the condensed phase fractions is applied to the description of the solid phase fractions movement. Dynamics of the each fraction of the disperse phase is described by the conversation equation of fractions' average density, the conversation equations of impulse components and the conservation equation of energy [1-5]:

$$\frac{\partial \rho_i}{\partial t}+\frac{\partial(\rho_i u_i)}{\partial x}+\frac{\partial(\rho_i v_i)}{\partial y}=0,$$

$$\frac{\partial(\rho_i u_i)}{\partial t}+\frac{\partial}{\partial x}\left(\rho_i u_i^2\right)+\frac{\partial}{\partial y}\left(\rho_i u_i v_i\right)=F_{xi}-F_{exi}-\alpha_i\frac{\partial p}{\partial x},$$

$$\frac{\partial(\rho_i v_i)}{\partial t}+\frac{\partial}{\partial x}\left(\rho_i u_i v_i\right)+\frac{\partial}{\partial y}\left(\rho_i v_i^2\right)=F_{yi}-F_{eyi}-\alpha_i\frac{\partial p}{\partial y},$$

$$\frac{\partial(e_i)}{\partial t}+\frac{\partial}{\partial x}\left(e_i u_i\right)+\frac{\partial}{\partial y}\left(e_i v_i\right)=-Q_i,$$

$$\rho_i=\alpha_i\rho_{i0},\ e_i=\rho_i C_{vi}T_i,$$

$$F_{xi}=\frac{3}{4}\frac{\alpha}{(2r)}C_d\rho_1\sqrt{(u_1-u_i)^2+(v_1-v_i)^2}\,(u_1-u_i)+\alpha_i\rho_1\left(\frac{\partial u_1}{\partial t}+u_1\frac{\partial u_1}{\partial x}+\right.$$

$$\left.+v_1\frac{\partial u_1}{\partial y}\right)+0.5\alpha_i\rho_1\left(\frac{\partial u_1}{\partial t}+u_1\frac{\partial u_1}{\partial x}+v_1\frac{\partial u_1}{\partial y}-\frac{\partial u_i}{\partial t}-u_i\frac{\partial u_i}{\partial x}-v_i\frac{\partial u_i}{\partial y}\right),$$

$$F_{yi}=\frac{3}{4}\frac{\alpha_i}{(2r)}C_d\rho_1\sqrt{(u_1-u_i)^2+(v_1-v_i)^2}\,(v_1-v_i)+\alpha_i\rho_1\left(\frac{\partial v_1}{\partial t}+u_1\frac{\partial v_1}{\partial x}+\right.$$

$$\left.+v_1\frac{\partial v_1}{\partial y}\right)+0.5\alpha_i\rho_1\left(\frac{\partial v_1}{\partial t}+u_1\frac{\partial v_1}{\partial x}+v_1\frac{\partial v_1}{\partial y}-\frac{\partial v_i}{\partial t}-u_i\frac{\partial v_i}{\partial x}-v_i\frac{\partial v_i}{\partial y}\right),$$

$$C_{di}=\frac{24}{Re_{i1}}+\frac{4}{Re_{i1}^{0.5}}+0.4,\ M_{i1}=\left|\bar{V}_1-\bar{V}_i\right|/c\ ,\ ,$$

$$Re_{i1}=\rho_1\left|\bar{V}_1-\bar{V}_i\right|2r/\mu,\ Pr=\gamma\,C_p\mu/\lambda$$

$$Q_i=6\alpha_i Nu\lambda\left(T_1-T_i\right)/(2r)^2\ i=2..n.$$

The Coulomb force components per volume unit of a gas-suspension are determined through its specific charge, volume density of a solid phase and electric field strength:

$$F_{iex}=-q_i\rho_i\,\partial\varphi/\partial x,\ \ F_{iey}=-q_i\rho_i\,\partial\varphi/\partial y,$$

where q_i – specific charge of i fraction of solid phase mass unit, φ - the potential of electrostatic field [6,7]. The electric field potential in interelectrode space is determined from Poisson's equation with boundary conditions of the first and second kind: potentials are given on electrodes and Neumann's conditions are set on the free borders of the rated area. The gas-suspension's charge density is contained in the right side of Poisson's equation carried to absolute dielectric permeability of the bearing medium:

$$div\mathbf{E}=\frac{\rho_e}{\varepsilon\varepsilon_0},\ \mathbf{E}=-\bar{\nabla}\varphi,\ \Delta^2\varphi=-\frac{\rho_e}{\varepsilon\varepsilon_0},\ \rho_e=\rho_1\cdot q,\ \varepsilon_0=\frac{10^{-9}}{36\pi}\ \mathrm{F}/m.$$

The internal energy of the solid phase suspended in gas is determined as $e_2 = \rho_2 C_{p2} T_2$. The system of the equations was supplemented with the corresponding initial and boundary conditions. Adhesion conditions are set on rated area borders for bearing and disperse phase speed and boundary conditions of Neumann for other gas-dynamic functions of both phases. The described system of the bearing and disperse phase movement equations together with the equation for turbulence is decided by an obvious Mac-Cormac's finite-difference method [8,9] with splitting in the spatial directions and the scheme of nonlinear correction of the numerical solution. The computer system allows to receive more monotonous decisions [10].

3. Results of calculations.

The scheme of rated area is shown on figure 1. The gas-suspension consists of five fractions of solid particles with radiuses of R = 1, 5, 10, 15 and 20 microns. Density of particles' material isρ_0 = 1000 kg/m^3. All fractions particles are motionless in initial timepoint. The average density of each fraction is distributed along an axis of the channel under the linear law $\rho(x) = \rho_0\alpha_0(0.2+x)$, α_0=0,0004. The equilibrium average density which corresponds to uniform spatial distribution of each fraction composes ρ_I = 0,28 kg/m^3 (i = 1-5). It is supposed that charge of gas-suspension disperse fraction #i is proportional to its average density and specific electric charge of unit of mass: $q_i = \rho_i q_0$. Temperatures of phases are equal, density of the bearing medium(air)is ρ = 1,29 kg/m^3, specific electric charge of unit of mass is q_0 = 0.0004 C/kg in initial timepoint.

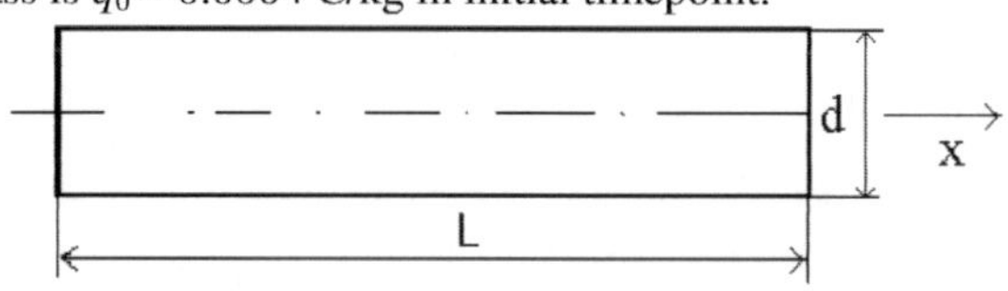

Figure 1. The scheme of rated area. L – length of the channel, d – width of the channel.

Figure 2, a shows the gas-suspension volumetriccontent along area axis in initial timepoint. The disperse system of charged particles with linearly increasing average density creates electric field with linearly increasing potential. Distribution of potential is shown on figure 2, b in timepoint t = 0,001 s. Electric field strength is negative, particles begin to be displaced in the negative direction of 0x axis.As a result redistribution of average density begins on area borders close to x = 0, x = L. It leads to decrease of potential on the right border at x = L and his growth on the left border of area at x = 0.

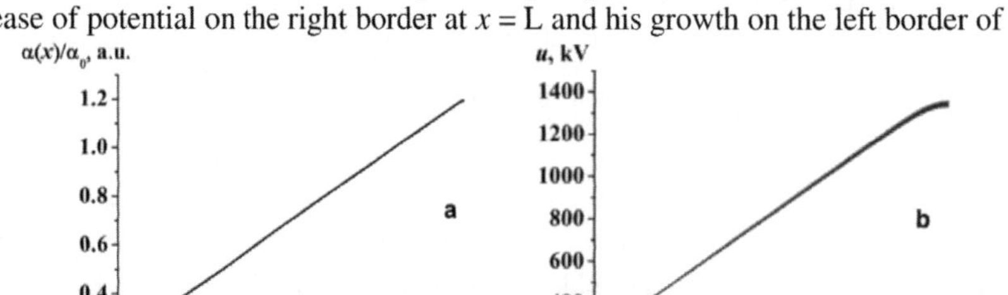

Figure 2. The volumetriccontent of the gas-suspension on area axis in initial timepoint (a) and distribution of electric field potential on axis in timepoint t = 0.001 s (b).

4. Conclusions

Numerical modeling showed that transition of the charged gas-suspension to equilibrium state includes two consistent stages proceeding with different speeds. At the first fast-proceeding stage particlesare raced to the largest speedsunder the influence of the Coulomb force. At the second slow stage the gradual formation spatially uniform distribution of average density of fractions and potential of electric field at the decreasing strength of electric field and speed of a disperse phase occurs. Time of establishment of equilibrium distribution depends on dispersion of the system.

Acknowledgments
Work is performed according to the financial support of Academy of Sciences of RT (agreement No. 03-127-f G 2016).

References

[1] Nigmatulin R I 1991 *Dynamics of Multiphase Media* (New York: Hemisphere)
[1] Loytsyansky L G 2003 *Mechanics of liquid and gas*(Drofa)
[2] Nigmatulin R I 1991 *Dynamics of Multiphase Media* (New York: Hemisphere)
[3] Kutushev A G 2003 *Mathematical modeling of wave processes in aero disperse and powdery environments*(St. Petersburg:Subsoil)
[4] Sternin L E1980*Two-phase mono- and polydisperse currents of gas with particles*(Mechanical engineering)
[5] Tukmakov A L 2011Numerical modeling of fluctuations of a monodisperse gas-suspension in the nonlinear wave field *Applied mechanics and technical physics* vol 52 pp 36-43
[6] Kadomtsev B B 1988*The collective phenomena in Plasma* (Moscow:Nauka) p 303
[7] Salyanov F A 1997*Fundamentals of physics of low-temperature plasma, plasma apparatus and technologies*(Moskow:Nauka)p 240
[8] Fletcher K 1991*Computing methods in dynamics Liquids*vol2 (Moscow: World) p 551
[9] Steger J L 1978*Implicit Finite-Difference Simulation of Flow about Arbitrary Two-Dimensional Geometries* AIAA J. vol 16 p 679–686
[10] Muzafarov I F, Utyuzhenikov S V 1993Application of compact differential schemes to a research of non-stationary currents of the compressed gas*Mathematical modeling*vol5 pp 74-83

IOP Conf. Series: Journal of Physics: Conf. Series **789** (2017) 012067 doi:10.1088/1742-6596/789/1/012067

Generalized current-voltage characteristics of electric discharge liquid cathode

R I Valiev[1], Yu I Shakirov, A A Khafizov, R A Valiev and I M Nuriev

Kazan Federal University Naberezhnye Chelny Institute, 68/19 Mira Street, Naberezhnye Chelny, 423810, Russian Federation

E-mail: rivaliev87@mail.ru

Abstract. The experimental and calculated current-voltage characteristics of the electric discharge between the metal anode and liquid cathode was got. As the liquid electrode process water, copper sulfate solution and various concentrations of sodium chloride were used, a solid cylindrical electrode rods were made of copper, iron and steel of different diameters. The influence of pressure, distance between electrodes, the anode material, electrolyte composition of the cathode on the current-voltage characteristics of the discharge was researched. The current-voltage characteristics are falling, increasing the distance between electrodes raises these curves along the voltage axis. The methods of simulation based on the similarity theory and the dimension formula is obtained for calculating the generalized current-voltage characteristics, taking into account, inter alia, the effect of pressure and electrode spacing.

1. Introduction

Concentrated energy flows is widely used for creating new materials, and to modify the existing [1, 2]. The effectiveness of the implementation of the production depends on the completeness of the research taking place in this process. In particular, for the development process of the plasma treatment of materials and its automation is important to study the characteristics of the current-voltage (CVC) of electrical discharges [3, 4]. It must be borne in mind that the nature of interdependence voltage and discharge current is also affected by the conditions and the implementation of the process, both qualitatively and quantitatively. Therefore, obtaining generalized characteristics, including those conditions, is of particular interest for the understanding of the processes occurring in this case, and for the development process.

The processes taking place under the influence on a substance of the plasma of electric discharges using fluids are the least studied. At the same time, this method is of particular interest because the use of different liquid natures gives opportunities to regulate the properties of modified and synthesized materials [5 – 9]. This makes the calculation of the generalized characteristics of this category to be actual.

2. Experiment

Experimental current-voltage characteristics obtained on the original line of own production [7]. The electrical discharge occurs between the metal anode and liquid cathode. The process water, copper sulfate solution and varying salt concentrations, served as a solid cylindrical electrode rods made of

[1] To whom any correspondence should be addressed.

copper, iron and steel of different diameter (1 mm to 10 mm) were used as the liquid electrode. Discharge was realized in chamber at different pressures (10^3 Pa to 10^5 Pa). The maximum voltage and the discharge current was 1500 V and 10 A, respectively. Inter-electrode distance was among providing an electrical discharge (maximum 100 mm).

At the experimental line the large number of current-voltage characteristics for different values of variable parameters, such as pressure, electrode spacing, anode material, electrolyte composition of the cathode are received. The current-voltage characteristics are falling, increasing the distance between electrodes raises these curves along the voltage axis. As an example, Figure 1 shows one set of current-voltage characteristics of the discharge between the cathode of the solution (20%) and copper sulfate steel anode under a pressure of 10^3 Pa for different interelectrode distances.

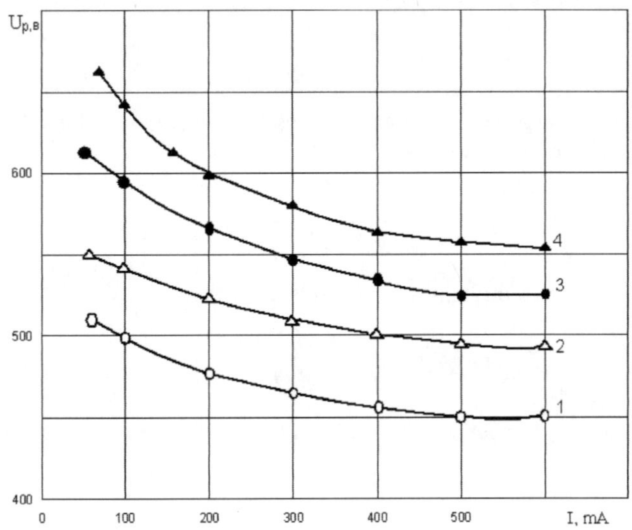

Figure 1. Current-voltage characteristics of the discharge between the solution of copper sulphate and steel for various interelectrode distances (1 – 10 mm; 2 – 20 mm; 3 – 30 mm; 4 – 40 mm).

3. Results and their discussion

Generalized characteristics obtain modeling methods based on the theory of similarity and dimension [10]. For free electric arcs, the main similarity criteria were identified [11]. There are empirical formulas which allow calculating the characteristics of the discharge for the implementation of different arc conditions. Also criterial generalization of the current-voltage characteristics of the negative corona discharge in an argon stream was held. [12] Known generalizations CVC for electric discharges using liquid electrodes was received without taking into account the pressure [13]. Since the pressure was constant and electrode spacing value changes slightly, at this situation the influence of Knudsen on generalized characteristics of the discharge was not revealed.

Generalized discharge CVC with a liquid cathode obtain solid anodes from a particular material, and take them geometrically similar. If the velocity of the gas is a certain parameter, the Reynolds number is not a determining criterion. Then, in a first approximation, the dimensionless integral characteristics of the discharge ψ are functions of defining similarity criteria and have the form

$$\Psi = f_1\left(\frac{I^2}{l^{1,5} g^{0,5} \chi_R T_R}, \frac{l\mathrm{PQ}_{II}}{\mathrm{KT_R}}\right), \tag{1}$$

here I - discharge current, P - pressure, l - electrode spacing, χ_R - thermal conductivity of the feed gas at a temperature T_R, K - Boltzmann constant, g - acceleration due to gravity, Q_l - sectional electron-molecule collision.

Distribution of dimensionless parameters ψ1 discharge will be described by criterion equation

$$\Psi = \left(\frac{I^2}{l^{1,5} g^{0,5} \chi_R T_R}, \frac{l\mathrm{PQ}_{la}}{\mathrm{KT_R}}, \frac{\tau}{l}, \frac{y}{l} \right), \tag{2}$$

where τ and y are the distance to a given point, respectively, from the axis of the discharge liquid and the surface of the cathode. Local dimensionless discharge characteristics depend on the dimensionless coordinates

$$\bar{\tau} = \frac{\tau}{l}, \; \bar{y} = \frac{y}{l}.$$

For generalizations of discharge voltage characteristics between the liquid and solid electrodes, as well as the open arcs the similarity criterion was adopted [11, 14]

$$\frac{I^2}{l^{1,5} g^{0,5} \chi_R T_R}.$$

If the mean of free path of an electron is determined by the formula

$$\lambda_l = \frac{1}{n_a Q_{al}} = \frac{kT_R}{PQ_{la}},$$

then the inverse of the Knudsen number for electrons is a similarity criterion

$$\frac{lPQ_{la}}{KT_R}.$$

In our experiments, the Knudsen number varied by several orders of magnitude, which allowed to establish its role in the generalized characteristics.

Let's define a non-dimensional voltage

$$\bar{U} = \frac{U_p l^{0,5} g^{0,5}}{I}. \tag{3}$$

If the value of g, χ_R, T_R, $Q_{\ell a}$ for our case are assumed to be constant, from (1) and (3) you can get a simple equation

$$\frac{U_p l^{0,5}}{I} = f_2\left(\frac{I^2}{l^{1,5}}, Pl \right). \tag{4}$$

The accepted range of current I (from 0.1 A to 0.6 A), the pressure P (from 10^3 to 10^5 Pa), and the interelectrode distance l (from 2 mm to 40 mm) of the discharge current-voltage characteristics of a generalized view of dependence on parameters Pl will be described by the formula:

$$\frac{U_p l^{0,5}}{I} = a_3 \left(\frac{I}{l^{0,75}}\right)^{-1,1} \cdot (Pl)^{0,42}, \; a_3 = 430 A^{0,1} \cdot B \cdot \text{Па}^{-0,42} \cdot \text{м}^{-0,745}. \tag{5}$$

The calculation shows that the standard deviation of the calculated values U_p from experimental values does not exceed 10%. Figure 2 shows a generalized current-voltage characteristics of the discharge of liquid electrodes made of process water.

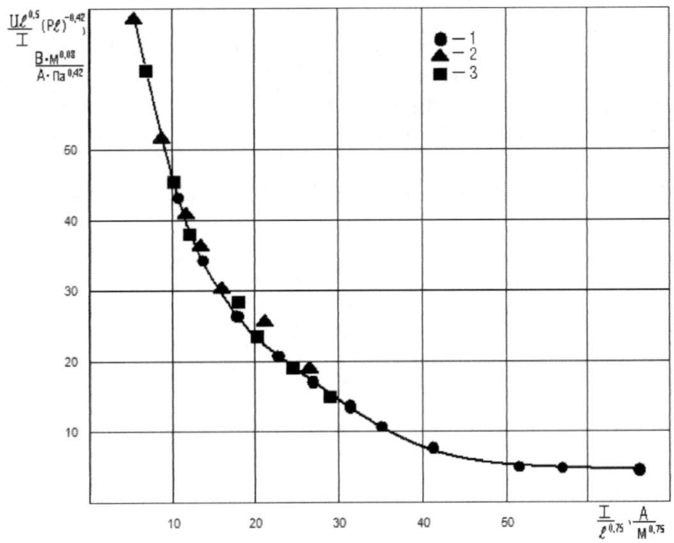

Figure 2. Generalized discharge current-voltage characteristics for different inter-electrode distances and pressures (1 – 2 mm; 10^5 Pa; 2 – 7 mm; $8{\cdot}10^3$ Pa; 3 – 15 mm; $2.5{\cdot}10^3$ Pa).

4. Conclusion

The slight value of the standard deviation calculated from the experimental values allows us to recommend a generalized formula of the current-voltage characteristics for use in the development of plasma electrothermal line with liquid electrode and the design of appropriate processes.

References

[1] Sadikov K, Fayrushin I, Shamsutdinov A, Kashapov N 2016 *J. of Phys.: Conf. Ser.* **669** 012048
[2] Valiev R A, Shakirov Yu I, Iliukhin A N 2014 Phase and structural conditions of low-temperature plasma interaction products with steel *J. of Phys.: Conf. Ser.* **567** 012040
[3] Saifutdinov A I, Fairushin I I, Kashapov N F 2016 Analysis of various scenarios of the behavior of voltage–current characteristics of direct-current microdischarges at atmospheric pressure JETP Letters **104** (3) p 180
[4] Khafizov A A, Shakirov Y I, Valiev R A, Valiev R I, Khafizova G M 2016 Study of thermal and electrical parameters of workpieces during spray coating by electrolytic plasma jet *J. of Phys.: Conf. Ser.* **669** 012030
[5] Tazmeev Kh K, Tazmeev A Kh 2014 Gas discharge with liquid electrolyte cathode in the mode of occurrence of the constricted channels *J. of Phys.: Conf. Ser.* **567** 012035
[6] Tazmeev A K, Tazmeeva R N, Sarvarov F S 2016 The features of high-current gas discharge in a narrow gap between the liquid electrolyte and solid electrode *J. of Phys.: Conf. Ser.* **669** 012056
[7] Khafizov A A, Valiev R I, Shakirov Yu I , Valiev R A 2014 Steel surface modification with plasma spraying electrothermal installation using a liquid electrode *J. of Phys.: Conf. Ser.* **567** 012026
[8] Shakirov Y I, Valiev R I, Khafizov A A, Valiev R A, Khakimov R G 2016 Erosion of electrode

metal in the electric discharge under the exposure of the electrolyte stream *J. of Phys.: Conf. Ser.* **669** 012064
[9] Guzel R. Gibadullina, Almaz H. Tazmeev, Ramilya N. Tazmeeva 2015 The creation of high temperature steam flow for plasma chemical gasification of polymer waste *International Journal of Applied Engineering Research* 10(24) p 45015-21
[10] Sedov L I 1977 *Metody podobija i razmernostej v mehanike Nauka* p 439
[11] Dautov G Ju, Zhukov M F 1965 *Prikladnaja mehanika i tehnicheskaja fizika* **2** pp 97-105
[12] Dandaron N B, Shagdarov V B, Bazarsadaev B C 2007 *Zhurnal tehnicheskoj fiziki* **77** p 3
[13] Gajsin F M, Son J E 1989 *Jelektrofizicheskie processy v razrjadah s tverdymi i zhidkimi jelektrodami* p 432
[14] Gajsin F M, Gizatullina F A, Kamalov P P 1985 *Fizika i himija obrabotki materialov* **54** p 58

Influence the loading effect on modification of PET film and fiber by Argon Plasma

D P Vasilkin[1], T G Shikova[1], V A Titov[2], S A Smirnov[1], L A Kuzmicheva[2]

[1] Ivanovo State University of Chemistry and Technology, Ivanovo, 153000, Russia
[2] G.A. Krestov Institute of Solution Chemistry of the Russian Academy of Sciences, Ivanovo, 153045, Russia

E-mail: titov25@gmail.com, d-vasilkin@mail.ru

Abstract. Poly(ethylene terepthalate) films and fabrics were modified by low-pressure argon plasma at different area of samples been treated. Contact angles for water and glycerol were measured and surface energy was calculated for film surface characterization. Height of water capillary rise was measured for fabric. The changes in chemical structure of surface layer were analyzed by ATR-FTIR method. Influence of sample area on non-homogeneity of plasma modification was shown. Some experiments were performed with polypropylene treatment in flowing plasma afterglow to confirm the reactions of oxygen active species originated from gas products of poly(ethylene terepthalate) etching in argon plasma.

1. Introduction

Plasma modification of polymers is widely used for improving their surface properties, such as wettability and dyeability, adhesion, biocompatibility etc. Treatment of polymer materials in low-pressure plasma of non-polymerized gases (O_2, N_2, NH_3, CO_2, air, Ar, He etc.) is accompanied by the formation of gaseous etching products, which alter plasma composition and properties [1, 2]. This phenomena is known as a so called "loading effect". Molecular hydrogen is the main gaseous product at the treatment of polyethylene, polypropylene and some other polyolefins in plasma of noble gases, at the same time, oxygen, carbon monoxide and water molecules were observed in mass-spectra of argon plasma at the treatment of oxygen-containing polymers: polyvinyl alcohol, polyacrilic acid, poly(ethylene terephthalate) [1]. Moreover, atomic oxygen and hydrogen lines and OH radical bands were observed in emission spectra of argon plasma reacting with polyaramide fibers and poly(ethylene terephthalate) (PET) fabric [2]. It means the noble gas plasma contains the reactive oxygen species, which origin from the polymer etching products. As a result, it can be expected not only the changing of plasma composition and properties but changing the results of polymer modification due to gas products evolution. The aim of this work is experimental study of influence of gaseous etching products on surface modification of PET film and fabric in low-pressure argon plasma.

2. Experimental

PET film with the thickness of 60 μm and PET fabric with specific surface density of 187 $g \cdot m^{-2}$ were used in our experiments. Area of samples was varied from 20 to 123 cm^2. Direct current discharge in argon was excited in cylindrical glass reactor with 3 cm inner diameter at argon pressure of was 50, 100 and 200 Pa, discharge current of 80 mA, and gas flow rate of 30 $cm \cdot s^{-1}$. Polymer samples were placed as cylinders on the reactor wall. Treatment time was 0.5, 1 and 5 min. Experimental setup and methods are described in detail elsewhere [4].

Surface modification of PET film was characterized by the contact angles (θ) for distilled water and glycerol, which were obtained using sessile drop image analysis (software ImageJ, "Drop-analysis"). The contact angles were used for calculation of polymer surface energy (γ), its polar (γ_p) and dispersive (γ_d) components by Owens-Wendt method. For fabric samples, height of water capillary rise was measured according to Russia State Standard 3816-81. The chemical changes induced by the action of plasma onto PET were studied by FTIR spectroscopy. The FTIR analysis was performed by a spectrometer (Avatar 360 FT-IR ESP, Nicolet) in an Attenuated Total Reflection mode with a ZnSe crystal.

3. Results and Discussion

Plasma treatment results in increase of hydrophilic properties of PET film and fabric (figure 1, tables 1, 2). Contact angles decreased, and surface energy increased due to increase of polar component. The polymer sample sizes slightly influence the result of PET film surface modification (table 1). Initial PET fabric has hydrophobic properties: water contact angle θ>90° (figure 1) and height of water capillary rise is only 2.66 cm after 3 min of wetting. Hydrophilicity was remarkably improved by plasma treatment: height of water capillary rise increased up to ~ 6 cm (table 2). It has to be noted that non-homogeneity of treatment results are observed for samples with large area (*S*=123 cm) as it is seen from Figure 2. The non-homogeneity is most pronounced at low treatment time and it reduces with the increase of treatment time.

To study the spatial uniformity of plasma etching, PET film samples were cut into three parts and placed in the reactor as shown in Figure 3a. The mass loss for each sample with the width of 4 cm was measured after plasma treatment and FTIR-spectra of samples were obtained. The experimental results show the sample placed in middle has minimum value of etching rate and maximum absorbance of C=O groups (ν=1710 cm^{-1}) as compared to samples located at the edges (Figure 3b, c). These data indicates the spatial non-homogeneity of plasma properties and flaxes of active species onto sample surface at high loading of reactor with material under treatment. The non- homogeneity can be caused by fast consumption of active species in heterogeneous reactions or by influence of destruction products on kinetics of active species formation in plasma. Furthermore, formation of new active species as a result of excitation and dissociation of polymer destruction products can not be excluded.

To confirm the possible reactions of reactive oxygen species, special experiments were performed with the treatment of polypropylene (PP) films in flowing afterglow of argon plasma. PP films were placed 4 cm downstream of plasma and were treated in two regimes. In the first case, PP films were treated without PET sample loading in argon plasma. In the second case, PP films were treated with PET film loading in argon plasma. In both cases ATR-FTIR spectra of PP samples were recorded before and after treatment.

Figure 1. Images of water drops on the surface of PET film (a, b) and fabric (c, d) before (a, c) and after (c, d) plasma treatment at the *P*=50 Pa, *i*=80 mA, *t*=5 min.

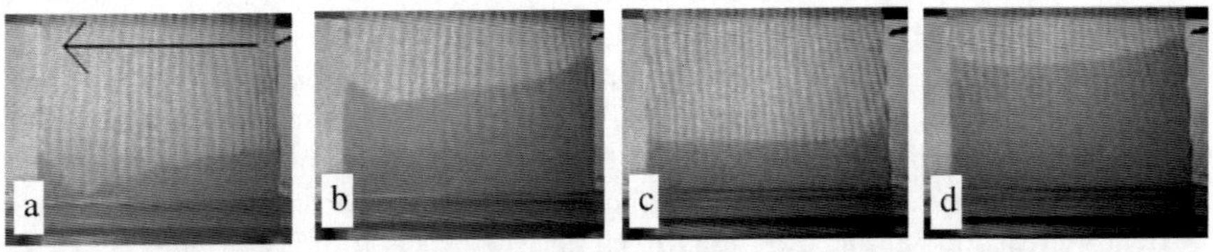

Figure 2. Influence the treatment time on the height of capillary rise: a, b – treatment time 0.5 min; c, d – treatment time 5 min; capillary rise time – 15 s (a, c) and 3 min (b, d). Arrow shows the direction of gas flow in plasma, width of fiber – 12 cm.

Table 1. Contact angle and surface energy of plasma treated PET films

Sample area (cm^2)	θ (degree)		Surface Energy ($mJ \cdot m^{-2}$)		
	Water	Glycerol	γ_d	γ_p	γ
Untreated sample	78.2	64.5	25.5	10.2	35.7
20	25	28	20	46	66
41	26	26	22	44	66
82	26	25	23	44	67
123	23	24	22	46	68

Table 2. Height of capillary rise (*h*, cm) after plasma treatment of PET fabric for wetting time of 15 s and 3 min [a].

Sample area, cm^2	Treatment time 0,5 min		Treatment time 1 min		Treatment time 5 min	
	15 s	3 min	15 s	3 min	15 s	3 min
20	2.20	5.58	2.14	6.40	2.70	7.36
41	2.23	5.80	1.97	5.67	2.60	7.23
82	2.15	5.68	2.21	6.12	2.59	7.12
123	2.01	5.96	2.31	5.92	2.51	6.85

[a] *P*=50 Pa, *i*=80 mA. *h*= 0.6 cm for untreated sample after 15 s wetting.

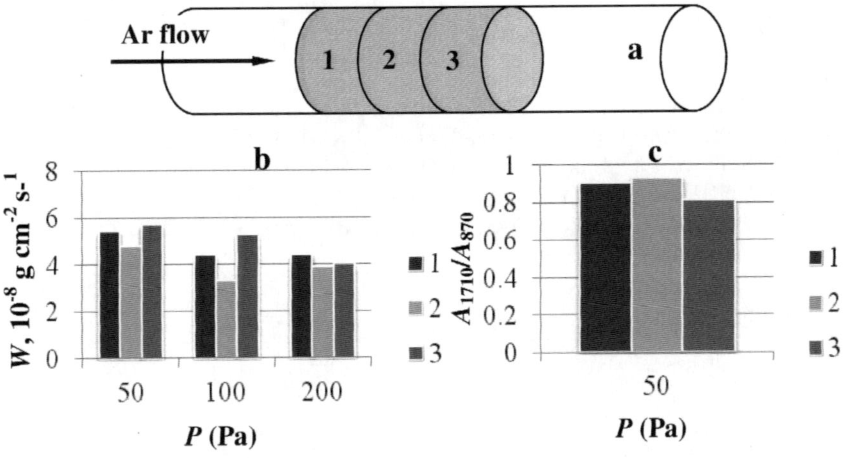

Figure 3. Scheme of sample placement in a reactor (a), etching rate (b) and reduced absorbance of C=O groups (ν=1710 cm^{-1}) at the treatment of PET film in argon plasma. Discharge current – 80 mA, treatment time – 5 min.

Contact angles and surface energy of PP films after treatment are represented in table 3. ATR-FTIR spectra show an increase of absorbance at ν=1572-1809 cm^{-1} (C=O groups) after treatment of PP films in remote plasma. When PET film is placed in argon plasma the absorbance is higher (table 3). It can be due to oxidative reactions initiated by long-lived active species forming in plasma. These species (oxygen atoms and OH radicals) can be the products of dissociation of PET gaseous destruction products. Oxidation of PP surface leads to increase in hydrophilic properties: contact angles decrease and surface energy increase. It has to be noted that presence of PET in the plasma leads to higher values of PP surface energy.

Table 3. Integral absorption at ν=1573-1809 cm^{-1} ($S_{c=o}$), contact angles (θ) and surface energy of PP films fter treatment in remote argon plasma [a]

Sample area of PET film in plasma (cm^{20})	$A_{c=o}$ (relative units)	θ, degree		Surface Energy ($mJ \cdot m^{-2}$)		
		Water	Glycerol	γ_d	γ_p	γ
Untreated PP	1.0	88	83	11.7	10.3	22
0	1.9	67	62	17.2	20.4	37.6
20.4	1.9	60	59	14.2	27.2	41.4
40.8	2.1	59	61	12.3	29.3	41.6
81.6	2.0	55	61	9.8	34.5	44.3
122.4	2.2	61	56	19.1	23.8	42.9

[a] Argon pressure 50 Pa, discharge current 80 mA, treatment time 5 min.

4. Conclusions

Treatment of PET films and fabrics results in increase of hydrophilic properties. Modification of samples with large area shows spatial non-homogeneity, which can be caused by the influence of gaseous etching products on plasma properties and kinetics of plasma active species formation. Experiments show the possible reactions of oxygen-containing active species originated from gas products of poly(ethylene terepthalate) etching in argon plasma

Acknowledgements

This work was supported by the Russian Foundation for Basic Research and Government of Ivanovo Region (grant No. 15-42-03124).

5. References

[1] Kutepov A M, Zakharov A G, Maximov A I 2004 Vacuumno-plazmennoe i plazmenno-rastvornoe modificirovanie polimernikh materialov M.: Nauka Publishers p 496 (In Russian)

[2] Titov V A, Shikova T G, Smirnov S A, Ovtsyn A A, Kuzmicheva L A, Chlustova A V 2016 *Izv. Vyssh. Uchebn. Zaved. Khim. Khim. Tekhnol.* 7 61 (In Russian)

[3] Smirnov S A, Titov V A, Shikova T G, Ovtsyn A A, Kadnikov D V 2016 *Applied Physics* **4** 43 (In Russian)

[4] Rybkin V V, Kuvaldina E V, Smirnov S A, Titov V A, Ivanov A N 1999 *High Energy Chemistry* **33** 409

IOP Conf. Series: Journal of Physics: Conf. Series **789** (2017) 012069 doi:10.1088/1742-6596/789/1/012069

Heat transfer in a longitudinal glow discharge

R F Yunusov
Kazan National Research Technical University named by A.N. Tupolev,420111, K. Marx str., 10, Kazan, Russian Federation

E-mail: optanir@mail.ru

Abstract. This article is devoted to the experimental study of heat transfer in a longitudinal glow discharge. The discharge was ignited in the discharge chamber (DC), consisting of a glass tube 10 mm in diameter and two electrodes. Copper electrodes were placed in the side branches, so that the average distance between them was 9 cm. The discharge pressure was varied in the range of P = (2.5 – 8.5) kPa. The air flow rate was varied from zero to G = 0.06 g / s. Current was varied in the range of I = (30-80) mA. Current-voltage characteristics of the discharge had falling form, and the voltage was varied in the range of U = (1-2) kV. The temperature of neutral particles in the plasma glow discharge was measured by six thermocouples, which were insertedon radius of DC.The aim of the article was to compare achieved experiment data with theoretical studies: recombination and diffusion plasma models.

1. Introduction

A plasma glow discharge is widely used in gas-discharge light sources, as the active medium of gas lasers [1-9], modern technologies for production of various coatings, nanotechnology methods for obtaining unconventional materials [10-24]. The study and research of glow discharge plasma parameters is of great scientific and practical interest. An important feature of glow discharge plasma is its thermal non-equilibrium: the electron temperature can be several times higher than the temperature of neutral molecules. For many applications it is important to know the heat transfer processes that occur in the gas discharge. Heat transfer in the gas discharge can be regarded as heat in the gas stream with internal heat sources. For theoretical analysis of these tasks, it's required to obtain the spatial distribution of the power heat sources. In the above works the potential distribution, the electron density, electric field strength, temperature of neural particles in the case of longitudinal glow discharges were obtained theoretically. The purpose of this article is a further study of heat transfer processes in a longitudinal glow discharge using experimental methods.

2. Experiment

Measurements of the neutral gas temperature in glow discharge plasma has certain difficulties related to the lack of balance between the neutral and charged particles, the possibility of chemical processes, etc. Definition of methodological error of temperature measurement is part of the general problem of the heat exchange system of bodies of research that are in contact with the environment or bodies. In case of radial thermocouple temperature measurements, the error sources are: the heat exchange by radiation between the thermocouple and the wall of the DC, the heat transfer by conduction from the sensor through the wires and thermocouple protection shield, heating of the thermocouple due to heterogeneous processes. Taking into account all the factors, the neutral gas temperature measurement error does not exceed 8%, which is consistent with the available data.

The circuit design and the DC for studying the processes of heat exchange is shown in Figure 1. Copper electrodes located at a distance ofl = (0.08 – 0.1) m on the side branches of the glass tube diameter of 0.01m. On the outer surface of the DCthere are nine capillaries for directing the thermocouple insertion and pressure measurement.The DC is connected to the pumping system using vacuum hoses. Current and discharge voltage were controlled by pointer-type instruments.Current-voltage characteristics were recorded using plotter model H-306.The temperature of the outer surface

IOP Conf. Series: Journal of Physics: Conf. Series **789** (2017) 012069 doi:10.1088/1742-6596/789/1/012069

of the DC was measured at eight points chromel-alumel thermocouples of 0.2 mm diameter. The working ends of thermocouple were pressed against the surface of the DC by means of four thin rings. The average temperature of the surface of the DC was determined by the measurements of thermocouples. The temperature of neutral particles was measured after the establishment of steady state, as determined by the stability of the input and output parameters. In each mode of measurements, the following parameters maintained constant: pressure in the DC, the gas flow rate, current and voltage discharge. Measurements in every mode were carried out 5 - 10 times. Thermo-emfthermocouple was measured using the compensation method by potentiometer model PP-63 (accuracy class of 0.05). The thermocouples shared a cold junctioninserted into constant-temperature ($0^{o}C$/ 32F) container.

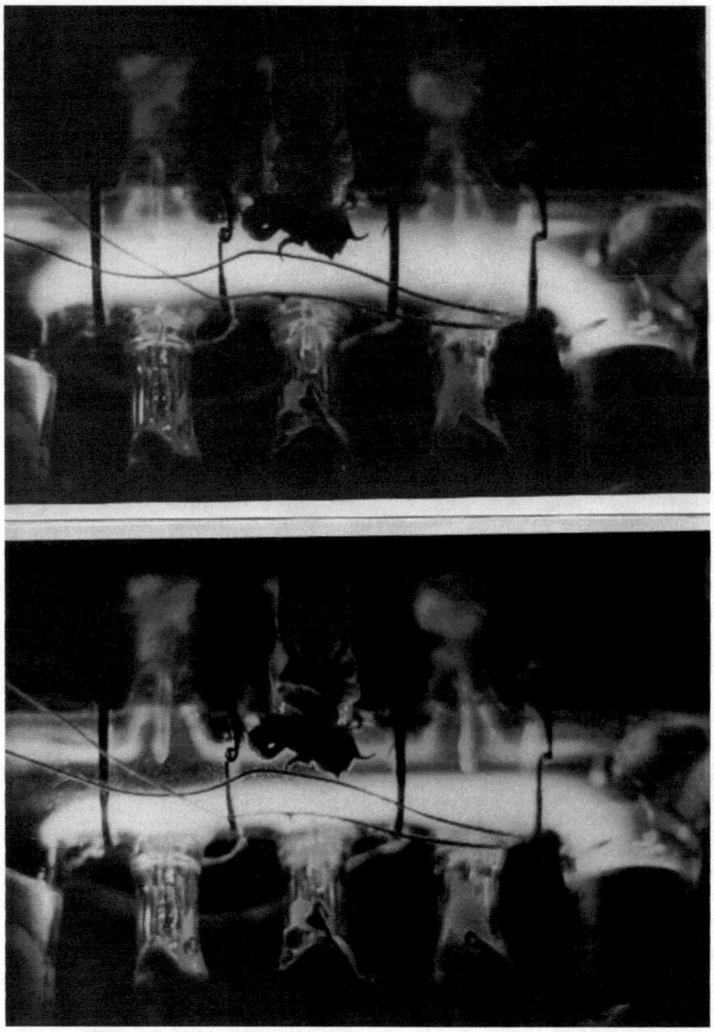

Figure1. Photographs of the discharge in the air flow

3. Results

The experiments showed that the air temperature in the longitudinal glow discharge increases linearly with increasing current, discharge capacity and pressure. The temperature increases approximately linearly with speed 6.5K / W on axis and 4K / W near the wall of the DC,when discharge power increases.Detected that temperature field of neutral particles essentially depends on the form of luminous discharge. A glow discharge glow between the two electrodes is shown on Figure 1. It also shows six capillaries, through which the vitrified thermocoupleswere inserted inside of the DC. The top photo corresponds to a small air flow rate G = 0.007 g/s, when the glow fills the entire cross-section of the discharge tube. The bottom photo corresponds to a higher air flow rate G = 0.035 g / s, in which the discharge is contracted to the bottom of the discharge tube, and can be clearly seen that the upper ends of the thermocouples are in the non-luminous discharge.Measured distributions of temperature along the length of the DC are shown in Figure 2.

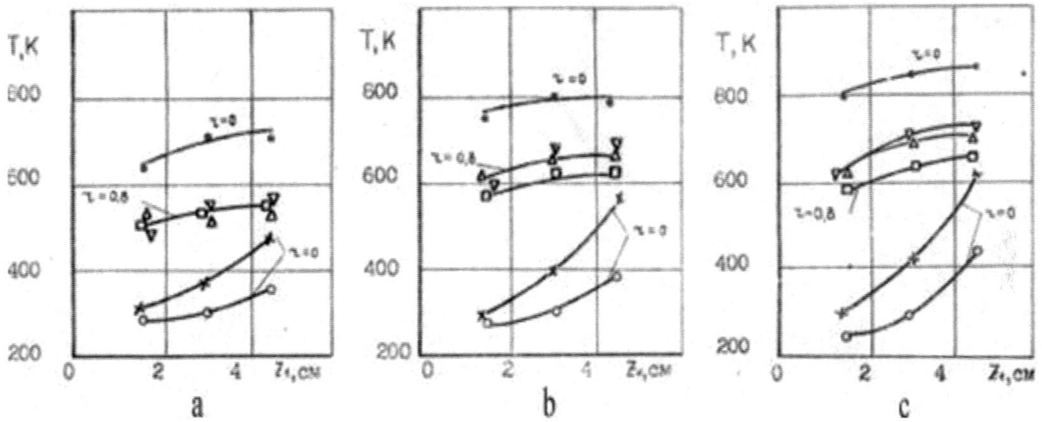

Figure2.The neutral particle temperature distribution along the length of the DC at P = 2.5 kPa: •, □ - G = 0.007 g/s; Δ, × - G = 0.017 g/s; ∇, ○ - G = 0.035 g/s; a - I = 30mA, b- I = 50mA, c- I = 70 mA.

The gas temperature at the outlet of the discharge chamber is increased by (50 - 400) Kon the axis of the discharge at a current I = 30 mA airflow range G = (0.007 -0.017) g/s. Under the same conditions of temperature increase of the gas at the radius r = 0.8 cm is 250K and weakly depends on the flow rate. When current increases, the gas temperature also increases, but the nature of its distribution along the length of the discharge chamber is not changed. Influence of air flow is particularly noticeable for the gas temperature at the discharge axis. When the air flow rate changes in the range G = (0.017 – 0.035) g/s (figure 2.a) on the discharge axis, the temperature decreases nearly to a room temperature. It happens because the flow of air causes a contraction of the discharge to the inner wall of the DC.

4. Discussion

In theoretical works [1-9] formulas were derived for calculating the temperature field of neutral particlesby solving the system of equations consisting of the charge continuity equation, Ohm's law in integral form and the energy equation. Calculations using these formulas were compared with the experimental temperature measurements, the results are shown in Figure 3.The recombination model has best agreement with experiment on the discharge axis, while the diffusion modelshows more accurate results near the wall of DC.

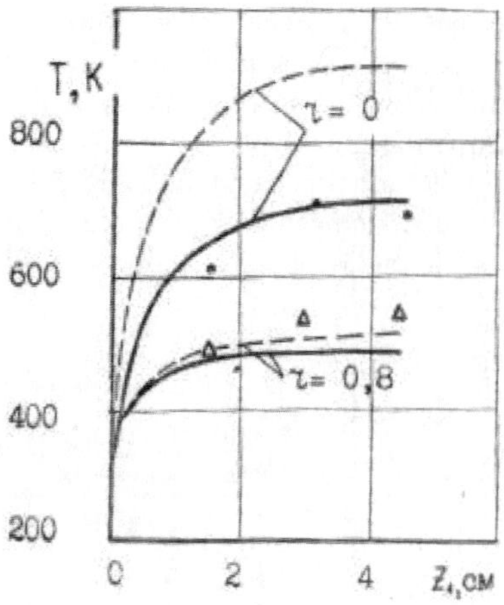

Figure3.The neutral particle temperature distribution along the length of the DC at P = 2.5 kPa, G = 0.007 g/s, I = 30 mA, ● - r = 0, Δ - r = 0, 8 cm. Dashed line – diffusion model, solid line – recombination model.

5. Conclusions

The aim of the article was experimental research of the temperature field of neutral particles in the plasma of a longitudinal glow discharge.The discharge was observed visually and photographically.Detected dependence of the temperature of the discharge current and gas flow rate. The experiment showed thatincrease of discharge power produces linear increase of temperature with speed 6.5K / W on axis and 4K / W near the wall of the DC.It has been found: the glow discharge is compressed to the boundary of the DC, when the air flow rate reaches a certain value equal to G = 0.035 g/s. Comparison of experimental and theoretical data was performed. It showed that the experimental temperature measurement results are in better agreement with the plasma recombination model on the discharge axis, while data near the wall of DC better fit in diffusion model.

References

[1] Raizer Y P 1992 *Gas Discharge Physics* (Moscow: "Science") p 536
[2] Yunusov R F1982*Journal of Engineering Physics and Thermophysics* **43** 1100-03
[3] Yunusov R F1985 *Journal of Engineering Physics and Thermophysics* **48**591-592
[4] Yunusov R F 1985 *Journal of Engineering Physics and Thermophysics* **48** 214-219
[5] Yunusov R F 1988*Journal of Engineering Physics and Thermophysics* **54** 76- 80
[6] Yunusov R F 1990*Journal of Engineering Physics and Thermophysics* **59** 990-994
[7] Fayrushin I, Kashapov N and Dautov I 2014 *J. Phys.: Conf. Ser.* **567** 012009
[8] Saifutdinov A I, Fairushin I I and Kashapov N F 2016 *JETP Lett.* **104** 180–185
[9] Israfilov Z K and Kashapov N F 1991 *Journal of Engineering Physics* **60** 364–368
[10] Yunusov F S and Yunusov R F 2005 *Vestnikmashinostroenija* **10** 52-56 (In Russian)
[11] Yunusov F S and Yunusov R F 2008 *Vestnikmashinostroenija* **8** 45-50 (In Russian)
[12] Yunusov F S and Yunusov R F 2008 *Vestnikmashinostroenija* **12** 35-40 (In Russian)
[13] Yunusov F S, Khisamutdinov R M andYunusov RF 2008 *Russian Engineering Research* **28** 965-973
[14] Yunusov F S andYunusov RF 2011*Russian Engineering Research* **31** 15-21
[15] Yunusov FS andYunusov RF 2011 *Russian Engineering Research* **31** 660-665
[16] Yunusov F S andYunusov RF 2011*Russian Engineering Research* **31** 951-959
[17] Yunusov F S and Yunusov R F 2012 *Vestnikmashinostroenija* **9** 52-57 (In Russian)
[18] Yunusov F S and Yunusov R F 2012 *Vestnikmashinostroenija* **10** 46-50 (In Russian)

[19] Kashapov L N, Kashapov N F and Kashapov R N 2013 *J. Phys.: Conf. Ser.* **479** 012011
[20] Kashapov L N, Kashapov N F and Kashapov R N 2014 *J. Phys.: Conf. Ser.* **567** 012025
[21] Zaripov R G, Kashapov N F, Tkachenko L A and Shaydullin L R 2016 *J. Phys.: Conf. Ser.* **669** 012053
[22] Denisov D G, Kashapov N F and Kashapov R N 2015 *IOP Conference Series: Materials Science and Engineering* **86** 012005
[23] Galyautdinov R T and Kashapov N F 2003 *Svarochnoe Proizvodstvo* (**3**) 27–31
[24] Abdullin I Sh, Galyautdinov R T and Kashapov N F 2001 *Inzhenerno-Fizicheskii Zhurnal* **74** 104–107

Influence of grinding on service properties of VT-22 powder applied in additive technologies

M N Zakharov, O F Rybalko, O V Romanova, B R Gelchinskiy, S A Il'inykh, V A Krashaninin
Institute of Metallurgy of Ural Branch of Russian Academy of Sciences, Amundsen 101, Ekaterinburg, 620016, Russia

E-mail: mr.mizani@mail.ru

Abstract. Powder of titanium alloy (VT-22) produced by plasma-spraying was subjected to grinding to obtain powder with size less 100 microns. These powders were sprayed by plasma unit using two types of gases, namely, air and air with methane (spraying in water and sputtering of coating on steel support). Influence of grinding time on yield of powder of required fraction was studied. Morphology and phase composition of the grinded powder and plasma sprayed one were under investigation. In the result of experiments, it appears that the grinding time genuinely influences the chemical and phase compositions, but there is no effect on physical-processing properties. For powders after plasma spraying some changes of non-metal elements content were detected by chemical analysis. Using gaseous mixture of air and methane in plasma spraying unit leads to formation of a new phase in the powder according X-ray diffraction data.

1. Introduction

Additive technologies e.g. selective laser melting and sintering, plasma spraying, and other are the production methods of 3D objects with predetermined mechanical properties using metal powders. Such technologies provide a possibility to manufacture different components having complex geometry in accord with technical requirements for product and mechanical properties of material.

The key requirements for powders applied in additive technologies are particles sphericity and dimensions. In references [1,2] the required particles dimensions are within the range from 20 to 100 microns.

In the present work the powder of VT-22 titanium alloy (Ti-5Al-5V-5Mo-Fe-Cr) were chosen as study subject. This titanium alloy is the widely used structural material. The powder was produced by plasma spraying of rotating electrode.

The dimensions of obtained particles were larger than required 100 μm. This raises the question of possibility to grind the particles down to target dimensions with further influence study of grinding duration on powder chemical and physical-processing properties.

Aim of the work was to study the influence of mechanical grinding process and plasma-forming gas composition on chemical and phase compositions of titanium-containing powder applied in additive technologies.

2. Research methods

The powder obtained by plasma spraying was subjected to grinding in IV-micro vibro-grinder during 15, 30, 45, and 60 minutes. Depending on grinding time the yield of powder with required fraction (from 20 to -100 microns) was determined by sieve analysis in accord with 18318-94 State Standard. After grinding, a degree of sphericity and symmetry of powder particles was estimated using Camsizer XT (Retsch Technology Gmbh, Germany). Carl Zeiss EVO40 XVP scanning electron microscope was used for morphology studying of VT-22 powder before and after grinding.

Physical-processing properties of initial powder and after grinding were measured according to the following State Standards: 19440-94 - the tap density, 25279-93 - the density after shake down, 20899-98 - the fluidity.

Phase composition was determined using X-ray diffractometer D8 ADVANCE (Bruker AXS, Germany).

The obtained powders were treated by Corvette 4 (Russia) plasma spraying unit using two types of plasma-forming gases: air and air+methane.

3. Results and discussion

For investigation the powders of two fractions (+100) and (-100) were used. Powder samples (+100) of 100 g were subjected to grinding in vibro-grinder. After grinding during different intervals of time, the amount of powder of required fraction was determined by sieve analysis, than the average values were calculated. The results are shown in Fig. 1.

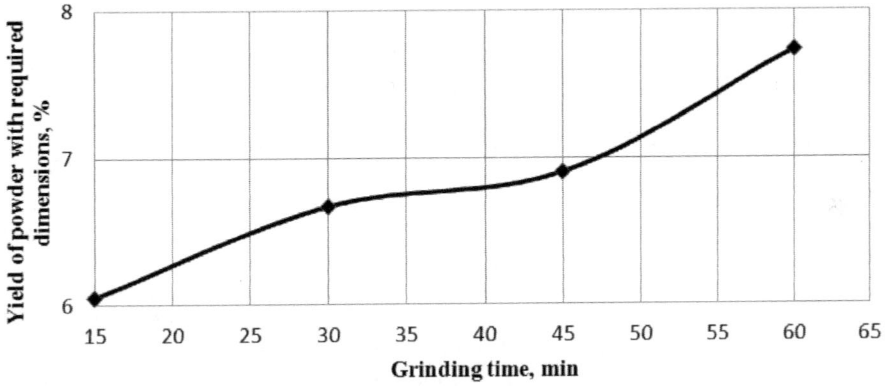

Fig. 1. Yield of powder of required fraction versus grinding time.

The longer time of grinding is the higher powder yield of required powder is.

Assessment of sphericity and symmetry of powder particles after grinding was carried out using Camsizer XT, besides the comparison with the initial powder of (-100) fraction.

The longer time of grinding is, the slightly higher the degree of powder particles sphericity is, but the degree of symmetry is almost independent on grinding time. The SEM images prove the results concerning the sphericity of particles before and after grinding (Fig. 2).

The values of physical-processing properties of initial sieved powder (-100) and after grinding (-100) are similar (Table 1).

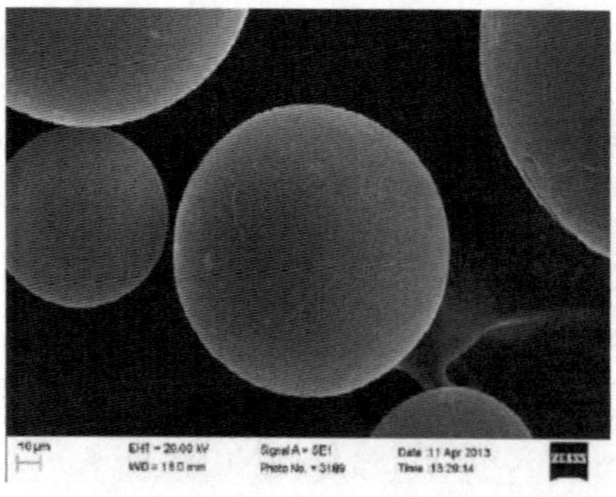

a

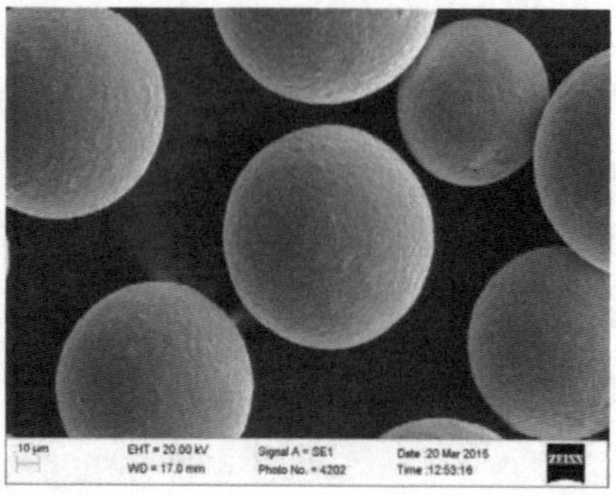

b

Figure 2. Surface morphology of VT-22 powder particles.
(a) - before grinding, (b) - after grinding. Magnification – 1000.

Table 1. Physical-processing properties of VT-22 powder

	Apparent density, g/cm^3	Fluidity, g/s	Tap density, g/cm^3
Initial state after sowing, (-100)	2.72	2.94	3.02
After grinding, granulometric composition (-100)	2.74	2.91	3.04

It was defined by chemical analysis of initially sieved (-100) and grinded powders that content of metal constituents in alloy is independent on grinding time. While the longer grinding time is, the higher content of non-metal components is. This can be attributed to the fact that the longer grinding time is, the higher temperature of powder is. Hence, the content of non-metal constituents is increased due to ability of titanium alloys to adsorb gases at elevated temperatures.

It was established by X-ray diffraction analysis that the main phase of initial sieved (-100) and grinded powders is β-Ti solid solution. Whereas the longer grinding time is, the higher percentage of α-Ti solid solution is. Likewise the content of titanium nitride and carbide (Ti3N1.29 and TiC) is gradually increased, however most likely that after 60 min grinding the nitride and carbide change structure and Ti4N2.33 and TiC0.3N0.7 are formed.

Table 2. Phase composition of VT-22 powder in initial state and after grinding of (-100) fraction

Time of grinding, min	Phase content					
	β-Ti,%	α-Ti, %	TiC, %	$Ti_3N_{1.29}$, %	$Ti_4N_{2.33}$, %	Ti_2CN , %
Initial powder after sieving (-100)	the balance	4.0	1.5	2.0	-	-
15	the balance	6.0	1.0	4.0	-	-
30	the balance	8.0	2.5	5.0	-	-
45	the balance	9.0	2.0	6.0	-	-
60	the balance	10.0	-	-	5. 0	2. 5

The obtained powders were treated by Corvette 4 plasma spraying unit using two types of plasma-forming gases: air and air with methane. The powders were quenched in water. In the first case an insignificant increase of content of non-metal elements (carbon, nitrogen, oxygen) in powders treated in plasma spraying unit using air was detected by chemical analysis. In the second case there was detected some growth of carbon and nitrogen content, whereas the oxygen content was increased significantly. According data collected from X-ray diffraction there was no change in phase composition of powder (in comparison with initial one) after application in air formed plasma. While there was detected a new phase, mostly comprised of titanium oxide mixture, when gaseous mixture of air and methane was applied.

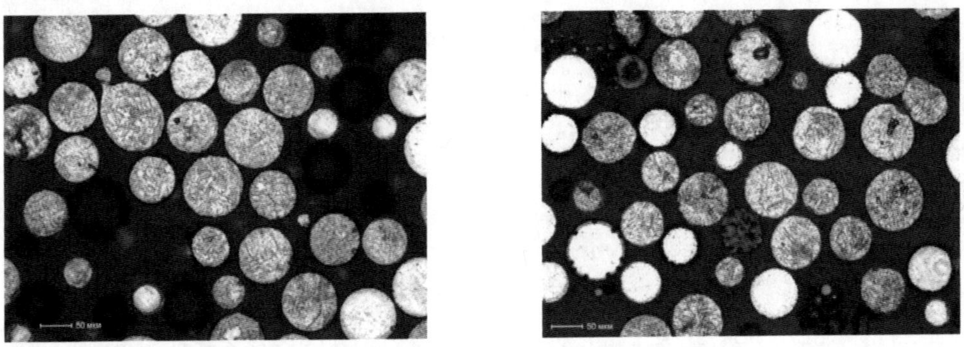

a b

Figure 3. Microstructure of VT-22 powder particles after plasma spraying using (a) - air, (b) - air + methane. Magnification – 200.

Coatings of the VT-22 powder sprayed by plasma on steel substrates are going to be investigated thoroughly. Now they are subjected to hardness, micro hardness, adhesion (three-point bending), corrosion-resistance (in chamber of salt fog) tests.

4. Conclusions

1. The longer time of grinding is, the higher powder yield of required powder is.
2. The longer time of grinding is, the slightly higher the degree of powder particles sphericity is, but the degree of symmetry is almost independent on grinding time.
3. There are differences in chemical and phase compositions of initial and grinded powders. The increase of oxygen and nitrogen occurs depending on the grinding time. Hence, it is recommended to grind the powder in a protective atmosphere.
4. Using air in plasma spraying unit the phase composition of powder is stable, but using gaseous plasma-forming mixture of air and methane leads to significant oxidizing of powder, so a new oxide phase is formed.

Acknowledgement
The authors acknowledge Ural Branch of Russian Academy of Sciences for financial support (research grant № -15-17-3-41). The study was carried out using equipment of the Ural-M Shared Research Center.

References

[1] Krashaninin V A, Il'inykh S A, Chusov S A and Gelchinskiy B R 2015 Sub- and supersonic plasma spraying of metal powders and production of combined protective coatings //

Proceedings of 11th International Scientific-Technical Conference devoted to State-of-the-art metal materials and technologies (SMMT'15) (Saint-Petersburg: Polytechnic university) P.1103-1114.

[2] Zlenko M A, Nagaycev M V and Dovbysh V M 2015 Additive technologies in machine-building (Moscow: NAMI), 220p. (in Russian)

[3] Leontiev L I and Gelchinsky B R 2014 *Proc. VI Int. Conf. on New perspective materials and technologies for production* (Volgograd: State technical University) p 278 (in Russian)

[4] Illarionov A G 2014 *Technological and service properties of titanium alloys* A G Illarionov and A A Popov (Ekaterinburg: Ural State University) 137(in Russian)

[5] Froes F H 2015 Titanium: Physical Metallurgy, Processing and Applications (Ohio: ASM International®Materials Park) 416

[6] Chumakov D M 2014 *Electronic journal Trudy MAI* 78 1 (in Russian)

[7] INTRODUCTION TO ADDITIVE MANUFACTURING TECHNOLOGY/A guide for Designers and Engineers/ European Powder Metallurgy Association/ www.epma.com/am

[8] Exploring Chemical and Thermal Non-equilibrium in Nitrogen Arcs. S Ghorui and A.K. Das, Journal of Physics: Conference Series 406 (2012) 012012

[9] M N Zakharov O F Rybalko S A Il'inykh S A Chusov and A V Dolmatov 2016 *XX Mendeleev congress on general and applied chemistry The XX Mendeleev Congress is held under the auspices of the International Union of Pure and Applied Chemistry* (Volgograd: Volgograd State Technical University) Vol 2 pp 214-215

Simulation of rarefied low pressure RF plasma flow around the sample

V S Zheltukhin[1], **A Yu Shemakhin**[2]

[1]Kazan National Research Technological University, Kazan, Russia
[2]Kazan Federal University, Kazan, Russia

Abstract. The paper describes a mathematical model of the flow of radio frequency plasma at low pressure. The hybrid mathematical model includes the Boltzmann equation for the neutral component of the RF plasma, the continuity and the thermal equations for the charged component. Initial and boundary conditions for the corresponding equations are described. The electron temperature in the calculations is 1-4 eV, atoms temperature in the plasma clot is (3-4) • 10^3 K, in the plasma jet is (3.2-10) • 10^2 K, the degree of ionization is 10^{-7}-10^{-5}, electron density is 10^{15}-10^{19} m^{-3}. For calculations plasma parameters is developed soft package on C++ program language, that uses the OpenFOAM library package. Simulations for the vacuum chamber in the presence of a sample and the free jet flow were carried out.

1. Introduction

RF plasma discharges at low pressures (P = 13, 3 − 133 Pa) is used successfully for the modification of various materials: dielectric, conducting, semiconducting. The plasma has the following properties: ionization degree is 10^{-7}–10^{-5}, electron density n_e=10^{15}–10^{19} м$^{-3}$, electron temperature T_e=1–4 эВ, temperature of atoms and ions in the plasma bunch T_a=(3-4)$\cdot 10^3$ K, in a plasma stream. T_a=(3.2-10)$\cdot 10^2$ K [1,2,3,9].

2. Mathematical model of low pressure RF plasmas stream

The hybrid mathematical model, combining a kinetic model for the carrier gas and the continuous model for charged particles is developed.
The model includes:

1) the Boltzmann's transport equation for neutral atoms:

$$\frac{\partial f}{\partial t} + \mathbf{c}\cdot\frac{\partial f}{\partial \mathbf{r}} + \tilde{\mathbf{F}}\cdot\frac{\partial f}{\partial \mathbf{c}} = S(f),$$
$$f(\mathbf{c}, \mathbf{r}, 0) = f_0(\mathbf{c}, \mathbf{r}),\ \tilde{\mathbf{F}} = -(1/m_a)\ \mathrm{grad}\ W_T \tag{1}$$

2) the equation of the electron continuity:

$$\frac{\partial n_e}{\partial t} - \mathrm{div}(D_a \mathrm{grad} n_e - \mathbf{v}_a n_e) = \nu_i n_e,$$
$$n_e\big|_{inlet} = n_{einlet}, n_e\big|_{outlet} = 0, n_e\big|_{walls} = 0, \quad n_e\big|_{body} = 0, n_e\big|_{t=0} = n_{e0} \tag{2}$$

3) the equation of the electron heating:

$$c_p\rho_e\frac{\partial T_e}{\partial t}-\mathrm{div}(\lambda_e\,\mathrm{grad}T_e-\frac{5}{2}k_Bn_eT_e\mathbf{v}_e)+\frac{3}{2}k_B\delta\nu_cn_e(T_e-T_a)=\sigma E^2-\nu_in_eE_I,$$

$$T_e\Big|_{inlet}=T_{einlet},T_e\Big|_{outlet}=T_{eroom},T_e\Big|_{walls}=T_{eroom},\frac{\partial T_e}{\partial\mathbf{n}}\Big|_{body}=0,T_e\Big|_{t=0}=T_{e0},\qquad(3)$$

4) the equation of the metastable atoms:

$$\frac{\partial n_m}{\partial t}-\mathrm{div}(D_m\,\mathrm{grad}(\mathrm{n}_m))=Rn_mn_e,$$

$$n_m\Big|_{inlet,outlet,sides}=0\qquad(4)$$

and closing equations:

$$p_a=n_ak_BT_a,\quad \mathbf{v}_a(\mathbf{r},t)=\int_{-\infty}^{\infty}\mathbf{c}f(\mathbf{c},\mathbf{r},t)d\mathbf{c},\quad \mathbf{v}_e=\mathbf{v}_a-(D_a/n_e)\mathrm{grad}\,n_e.\qquad(5)$$

Here **c** and **r** are vectors of velocity and coordinates of atoms, $f(\mathbf{c}, \mathbf{r}, t)$ are function of distribution of neutral atoms by velocities, f_0 – Maxwell function of distribution by velocities, $S(f)$ is collisions integral, $\tilde{\mathbf{F}}$ is force, which influence on neutral atoms in elastic collision process, W_T is energy, which transferred to atoms in elastic collision process , n_e is density of electrons, D_a is ambipolar diffusion coefficient, l_e is coefficient of thermal conductivity of electron gas, n_i is ionization frequency , c_p is heat capacity of the electron gas, R is rate coefficient of step-wise ionization, n_c is the frequency of elastic collisions of electrons with atoms and ions, s is plasma conductivity, **E** is the electric field vector, $E=|\mathbf{E}|$, E_i is ionization potential, k_B is Boltzmann constant, $d = m_e/2m_a$, m_a is atom mass, m_e is electron mass, R_{vk} is radius of the cylindrical vacuum chamber, L_{vk} is length, R_{rk} is the radius of the plasma torch outlet, subscripts *inlet*, *outlet*, *body*, *walls* devoted parameter value on inlet and outlet of the chamber, on the walls of the sample, and the vacuum chamber, respectively, $W_T = \mathrm{T}\, E_c dVdt$, dV is volume element, $E_c = \frac{3}{2}k_B\delta\nu_cn_e\left(T_e-T_a\right)$ is energy of elastic collisions transport. Coefficients D_a, n_i , l_e are functions of electron temperature T_e.

3. Numerical method and program

Bird's method [6] is modified by introducing the area heat source power W_T. A three-step iterative process is constructed for solving the problem (1) - (5). Software package that allows to find the spatial distribution of the main plasma flow characteristics is developed. OpenFOAM [10] library environment is used by the software program under the Linux operating system.

4. Calculations of low pressure RF plasma flow characteristics

The gas-dynamic characteristics of low pressure RF plasma undisturbed flow as well as stream with overflowing sample in the vacuum chamber of R_{vk}=0,2 m, L_{vk}=0,5 m and R_{gk}=0,012 m, on the center of the base plate is carried out. A cylindrical sample of radius R_b=0,03 m and a height L_b=0.02 m located in the plasma jet at a distance L_{tb}=0,02 m. Flow input parameters are the following: the plasma forming gas is argon, gas flow rate G=0,12–0,24 g/s, pressure P_{inlet} = 35–85 Pa, the temperature T_{inlet} = 400–600 K, the degree of ionization δ_i=10^{-4}. The initial pressure in the vacuum chamber

IOP Conf. Series: Journal of Physics: Conf. Series **789** (2017) 012071 doi:10.1088/1742-6596/789/1/012071

P_0 =3,5–8,5 Pa. Distributions of temperature, velocity, and pressure of carrier gas as well as electron temperature and density are obtained.

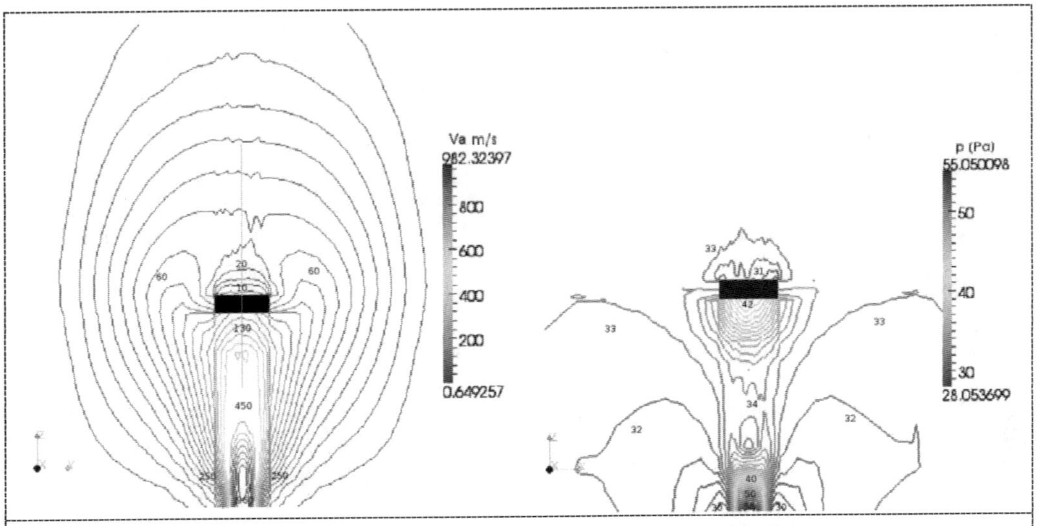

Figure 1. Pressure (left side) and neutral's velocity (right side) distributions.

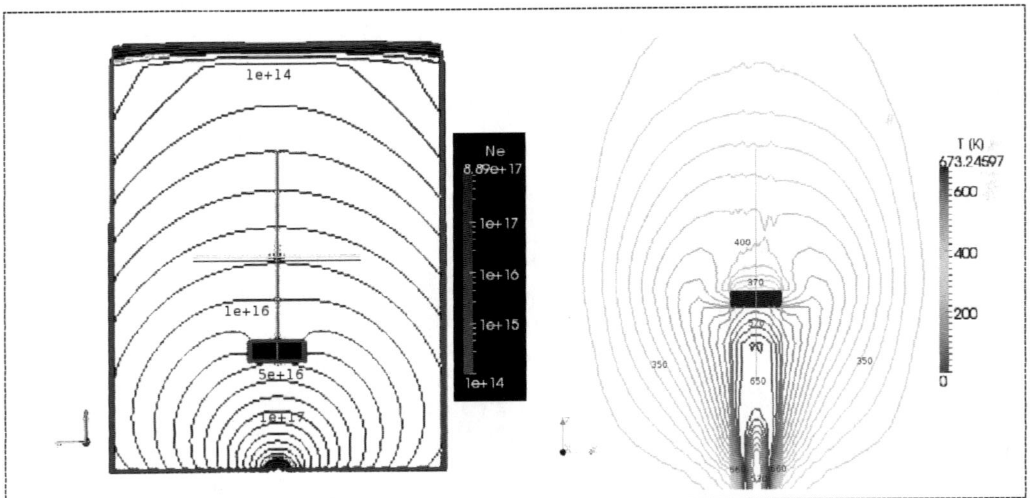

Figure 2. Electron density (left side) and neutral's temperature (right side) distributions.

Figure 1 illustrates the velocity and pressure contour lines of neutral gas. It can be noted that the pressure increases near the sample because of stream stagnation. Pressure is straightening around 30 Pa in the area behind the sample.

Figure 2 shows electron density and temperature of carrier gas contour lines. The maximum value of electron density equal to $9 \cdot 10^{17}$ $1/m^3$ in the region of the inlet hole. The density goes down to zero near the boundaries, because of electron-ion recombination. Overheating of neutral atoms is found on the jet edge. The temperature maximum is 650 - 670 K in this area. The reason is abrupt deceleration of stream atoms on statical atoms in chamber. Neuman's conditions is laid on the sample surface and thus the temperature behind the sample is about 370 K. The temperature in the jet middle is in range of 550 – 650 K. The calculations results are in satisfactory agreement with experimental data [1,5,9].

Acknowledgments
The work was supported by RFBR (grant № 16-31-60081, theoretical part) and the Ministry of Education of the Russian Federation (the basic part of the state task number 2196, the experimental part of the work).

References
[1] Abdullin I S, Zheltukhin V S, Sagbiyev I R and Shayekhov M F 2007 Modifikaciya Nanosloev v Vysokochastotnoj Plazme Ponizhennogo Davleniya (Kazan: Izdatel'stvo Kazanskogo Tekhnologicheskogo Universiteta)
[2] Khubatkhuzin A A, Abdullin I S, Gatina E B, Zheltukhin V S and Shemakhin A Y 2012 Vestn. Kazan. Tekhnol. Univ. 15 72
[3] Khubatkhuzin A A, Abdullin I S, Bashkirtsev A A and Gatina E B 2012 Vestn. Kazan. Tekhnol. Univ. 15 71
[4] Dulov V G and Lukyanov G A 1984 Gazodinamika Processov Istecheniya (Moscow: Nauka)
[5] Mitchner M and Kruger C H 1973 Partially Ionized Gases (New York: John Wiley and Sons)
[6] Bird G A 1976 Molecular Gas Dynamics and the Direct Simulation of Gas Flows (Clarendon Press: Oxford)
[7] Zheltukhin V S and Shemakhin A Y 2011 Uch. Zap. Kazan. Univ., Ser. Fiz.-Mat. Nauki 4 135
[8] Zheltukhin V S and Shemakhin A Y 2014 Math. Model. Comput. Simul. 6 101
[9] Abdullin I S, Zheltukhin V S, Khubatkhuzin A A and Shemakhin A Y 2014 Matematicheskoe Modelirovanie Gazodinamiki Strujnykh Techenij Vysokochastotnoj Plazmy Ponizhennogo Davleniya (Kazan: Izdatel'stvo Kazanskogo Tekhnologicheskogo Universiteta)
[10] OpenFOAM OpenFOAM Foundation. Free Open Source CFD 2011–2016 http://www.openfoam.org

Magnetic and Mössbauer effect studies of ZnO thin film implanted with iron ions to high fluence

A L Zinnatullin[1], A I Gumarov[1], I F Gilmutdinov[1], V F Valeev[2], R I Khaibullin[2] and F G Vagizov[1]

[1] Kazan (Volga Region) Federal University, Kremlevskaya 18, 420008 Kazan, Russia
[2] Zavoysky Physical-Technical Institute, Sibirsky Trakt 10/7, 420029 Kazan, Russia

E-mail: almaz.zinnatullin@gmail.com

Abstract. We present the results of magnetic and Mössbauer effect studies of zinc oxide thin film obtained by RF magnetron sputtering and implanted with 40 keV iron ions to a fluence of $1.5\cdot10^{17}$ ion/cm^2. As-implanted and post-annealed sample shows ferromagnetic properties at room temperature and consists of paramagnetic and ferromagnetic phases according to Mössbauer spectroscopy.

1. Introduction

There has been considerable interest to materials that show semiconducting and magnetic properties simultaneously. Materials with this combination of properties can be used to create spintronic devices such as spin field effect transistor (spin-FET). Such materials should have the relative long spin coherence length of electrons, otherwise, information carried by each spin will be distorted. Therefore, these materials should have homogeneous crystal structure. On the other hand, it is necessary to have the ferromagnetic properties of these materials at room temperature for creating the functional devices [1]. T. Dietl et al. [2] predicted that few semiconductors doped with "magnetic" ions can show room temperature ferromagnetic behavior. The prediction stimulated intensive research devoted to the theoretical and experimental investigations of materials called as diluted magnetic semiconductors (DMS). Iron doped zinc oxide (ZnO) is one of such materials.

We investigated the ZnO thin film implanted with iron ions. Magnetic properties were measured by vibrating sample magnetometer (VSM). Phase composition and valence state of implanted iron ions were studied by room-temperature conversion electron Mössbauer spectroscopy (CEMS).

2. Sample preparation

ZnO thin film with the thickness of 130 nm was deposited by RF magnetron sputtering on monocrystalline silicon (100) substrate at room temperature. The film was grown under the pressure of $5\cdot10^{-3}$Torr in the *BESTEC* vacuum system. Magnetron power was about 100 W. Then, the ZnO film was implanted with 40 keV iron ions (enriched with Fe^{57} isotopes up to 40 %) to a fluence of $1.5\cdot10^{17}$ ions/cm^2 in the *ILU-3* ion accelerator (Zavoysky Physical-Technical Institute). The ion implantation was carried out at room temperature and ion beam current density was 8 μA/cm^2. The subsequent annealing of the sample was performed in high vacuum (~10^{-8} Torr) at 500 °C for half an hour.

3. Results and discussion

CEMS measurements of ^{57}Fe nuclei were performed using constant acceleration spectrometer (WissEl) employing ^{57}Co embedded in Rh matrix source. All measurements were carried out at room temperature. CEMS spectra of the sample before and after annealing are presented on figure 1. They consist of one sextet with wide distribution of hyperfine magnetic field on iron nuclei and two doublets. The best fits of experimental data were achieved under assumption of the non-symmetric Gauss distribution of hyperfine magnetic field on ^{57}Fe nuclei. The obtained non-symmetric distributions are wider on the side of lower fields. The hyperfine parameters of the ascertained components are presented on table 1.

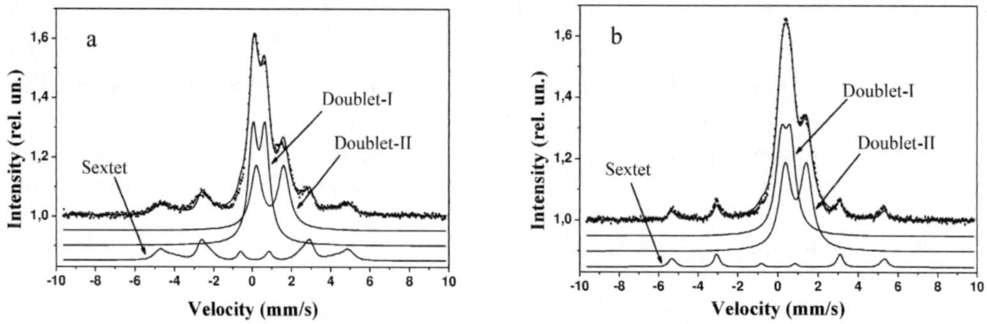

Figure 1. Room-temperature CEMS spectra of ZnO thin film implanted with iron ions with energy 40 keV to the fluence of $1.5 \cdot 10^{17}$ ions/cm^2 (a) - before annealing and (b) - after annealing at 500 °C for 30 minutes.

The sextets in Mössbauer spectra correspond to the magnetically ordered phases of the sample. The hyperfine parameters of the sextet for as-implanted sample (δ = 0.11 mm/s, ΔE_Q =- 0.06mm/s, B_{hf}= 290 kOe) are a slightly different from the well known parameters for iron-based bulk phases. However, they are close enough to the parameters which characterize the ensemble of small metallic iron nanoparticles with a wide particle size distribution. In this case, an appreciable amount of iron atoms can be located on the shell of nanoparticles and between these magnetically ordered nanoparticles. These circumstances lead to a decrease of the average hyperfine magnetic field, $\langle B_{hf} \rangle$, and to a positive value of the average isomer shift δ of the ensemble of atoms relative to the bulk α-Fe.

Table 1. Hyperfine parameters of identified phases.

Sample	Component of the spectrum	Valence state of iron	$\langle B_{hf} \rangle$, kOe	δ, mm/s	ΔE_Q, mm/s	Relative content, %
$1.5 \cdot 10^{17}$ions/cm^2 as-implanted	Sextet	Fe^{+3} or Fe^0	290	0.11	-0.06	32.8
	Doublet-I	Fe^{+3}	-	0.36	0.61	45.2
	Doublet-II	Fe^{+2}	-	0.91	1.39	22
$1.5 \cdot 10^{17}$ions/cm^2 post-annealed	Sextet	Fe^0	331	0	-0.02	15.8
	Doublet-I	Fe^{+3}	-	0.39	0.46	49.6
	Doublet-II	Fe^{+2}	-	0.89	1.05	34.6

$\langle B_{hf} \rangle$ - average hyperfine magnetic fields on iron nucleus, δ –isomer shift relative to α-Fe, ΔE_Q–quadrupole splitting

The hyperfine parameters of the sextet for the annealed sample are in good agreement with the parameters of the metallic iron. This is probably due to the fact that an increase of thermal energy leads to the intensification of the diffusion processes, which causes the aggregation of dispersed metallic iron as well as the growth of magnetic clusters in size. The distribution of hyperfine fields becomes significantly narrower after subsequent annealing (figure 1). It should be noted that the relation of intensities for 1 and 2 lines (as well as for 5 and 6) of sextet is equal to 3:4. In our experiment the propagation direction of the gamma photons is perpendicular to the sample plane, so this ratio is possible only when the magnetic moments of the iron atoms are in plane of the sample.

The presence of doublets in the spectrum is usually associated with the formation of paramagnetic iron phases. In earlier papers [4, 5] was shown that the implantation of Fe^{+} ions into ZnO single crystal leads to the formation of metallic iron and zinc ferrite ($ZnFe_2O_4$) nanoparticles. The hyperfine parameters of the weakly split doublet ($\delta = 0.36$ mm/s, $\Delta E_Q = 0.61$ mm/s) indicated in the spectrum as "Doublet-I", are very close to the parameters of the $ZnFe_2O_4$ particles [4]. Subsequent vacuum annealing increases the content of this phase for 4.4%.

The vacuum annealing caused more pronounced effect on the relative content of the magnetically ordered phase and paramagnetic phase of the iron atoms in the divalent state. The relative content of magnetically ordered phase decreases about 17%, whereas the content of the paramagnetic phase (Doublet-II) increases up to 12.6%.. Last one may be associated with the formation of the Fe_xO nanoparticles during the implantation process [6]. The revealed growth of its content after annealing can be explained by the oxidation of the metallic iron nanoparticles. The origin of these paramagnetic doublets is not clear and requires further detailed studies of the symmetry and the local environment of iron atoms.

Magnetic studies were performed with a vibrating sample magnetometer (PPMS-9, Quantum Design). Magnetization curve versus magnetic field are obtained. Measurements were carried out at 10 K and room temperature (figure 2). Parameters of hysteresis loops are shown on the table 2.

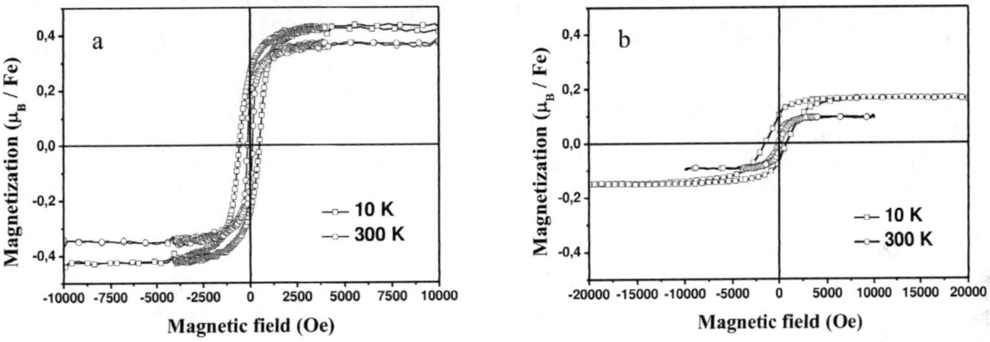

Figure 2. Magnetic hysteresis loops of ZnO thin film implanted with the 40 keV iron ions to the fluence of $1.5\cdot10^{17}$ ions/cm^2 (a) - before annealing and (b) - after annealing at 500 °C for 30 minutes.

It is seen that the sample under study exhibit the ferromagnetic properties before and after annealing even at room temperature. The saturation magnetization and the coercive field values of hysteresis loops increase while the temperature decrease. The annealed sample has a smaller value of the saturation magnetization.

Table 2. Magnetic hysteresis loops parameters.

Sample	Measuring temperature, K	Coercive field, Oe	Saturation magnetization, μ_B/Fe	Remnant magnetization, μ_B/Fe
$1.5\cdot10^{17}$ions/cm^2 as-implanted	10	504	0.43	0.26
	300	133	0.37	0.17
$1.5\cdot10^{17}$ions/cm^2 post-annealed	10	810 (-1487)	0.17	0.10
	300	109	0.10	0.02

Hysteresis loop of the annealed sample at 10 K is shifted to smaller magnetic fields. This shift may arise due to exchange bias. Exchange bias occurs in the case of the tight magnetic contact between the ferromagnetic and antiferromagnetic phases. The exchange interaction at the boundary of two types of magnets may lead to appearance of the preferred directions of the magnetization vector at low temperatures. Typically exchange anisotropy arises in "core-shell" nanoparticles, where the pair of two types of magnets is formed due to the oxidation of the surface of the nanoparticles [7]. In our case, a surface of the ferromagnetic iron nanoparticles can be oxidized to antiferromagnetic FeO phase.(For the bulk FeO the Neel temperature is about 198 K[6]).

4. Conclusion

Zinc oxide thin film prepared by RF magnetron sputtering technique and post-implanted with 40 keV iron ions to fluence of $1.5 \cdot 10^{17}$ ions/cm^2 has been studied. The implanted sample exhibits the ferromagnetic properties at room temperature. The iron nanoparticles are primary source of room temperature ferromagnetism. Composition of iron-containing phases in Fe-implanted ZnO film consists of magnetically ordered and paramagnetic phases. Vacuum annealing reduces the content of the magnetically ordered phase. Magnetic properties of iron nanoparticles embedded inside ZnO can be tuned by vacuum annealing.

Acknowledgments
The authors acknowledge support by the Program of Competitive Growth of Kazan Federal University. The Mössbauer and magnetic measurements were carried out at the PCR Federal Center of Shared Facilities of KFU.

References

[1] Wolf S A, Awschalom D D, Buhrman R A, Daughton J M, von Molnár S, Roukes M L, Chtchelkanova A Y and Treger D M 2001 *Science* **294** 1488
[2] Dietl T, Ohno H, Matsukura F, Cibert J and Ferrand D 2000 *Science* **287** 1019
[3] Lyadov N M, Bazarov V V, Vagizov F G, Vakhitov I R, Dulov E N, Kashapov R N, Noskov A I, Khaibullin R I, Shustov V A and Faizrakhmanov I A 2016 *Applied Surface Science* **378** 114
[4] Zhou S, Potzger K, Reuther H, Talut G, Eichhorn F, von Borany J, Skorupa W, Helm M and Fassbender J 2007 *Journal of Physics D: Applied Physics* **40** 964
[5] Potzger K, Zhou S, Reuther H, Muecklich A, Eichhorn F, Schell N, Skorupa W, Helm M, Fassbender J, Herrmannsdoerfer T and Papageorgiou T P 2006 *Applied Physics Letters* **88** 052508
[6] Kozioł-Rachwał A, Ślęzak T, Nozaki T, Yuasa S and Korecki J 2016 *Applied Physics Letters* **108** 041606
[7] Noguésa J and Schuller I 1999 *Journal of Magnetism and Magnetic Materials* **192** 203

IOP Conf. Series: Journal of Physics: Conf. Series **789** (2017) 012073 doi:10.1088/1742-6596/789/1/012073

Synthesis and properties of the molybdenum and tungsten disulfide thin films

A L Zinnatullin, I R Vakhitov, A I Gumarov, S I Nikitin, S S Kharintsev, I V Yanilkin and R V Yusupov

Kazan Federal University, Kremlevskaya 18, 420008 Kazan, Russia

E-mail: almaz.zinnatullin@gmail.com

Abstract. In the paper results of thin film synthesis and primary characterization of molybdenum and tungsten disulfides are presented. The synthesis is performed by means of the high-temperature sulfurization of the metallic Mo and W films in the flow of argon. Chemical and phase compositions and morphology of thin films are reported. Optical transmission and Raman spectra are presented and discussed.

1. Introduction

Synthesis of the single-layer modification of carbon – the graphene – and studies of its extraordinary properties [1] have triggered an increased interest to two-dimensional (2D) materials. These compounds in the single layer (SL) form reveal unique properties that cannot be achieved or are even impossible in the bulk. Transition metal dichalcogenides (TMD) attract a special interest among two-dimensional materials. These compounds with the general formula MX_2, where M is a transition metal (Mo, W, Nb etc.) and X is the element of chalcogen group (S, Se, Te etc.), are layered crystals where the layers are bound to each other by Van der Waals forces. Each monolayer is in turn a "sandwich", where the transition metal atoms are located between two chalcogen layers. Large difference between the binding energies within a layer and between the layers determines the anisotropy of TMDs.

These compounds in a single layer form are the direct bandgap semiconductors while their bulk analogues are indirect bandgap ones. These materials reveal huge potential in such fields of science and technology as spintronics, valley-tronics, voltaics [2]. Therefore the development of reliable and reproducible method of TMD ultrathin film synthesis and studies of these materials are very important.

In the paper results of the synthesis and studies of MoS_2 and WS_2 thin films are presented. Films were prepared by thermal annealing in the sulfur atmosphere of the molybdenum or tungsten thin films pre-deposited by either the magnetron sputtering or molecular beam epitaxy (MBE). Chemical and phase compositions were studied by X-ray photoelectron (XPS) and Raman spectroscopies, surface morphology was investigated with scanning electron microscopy (SEM). Optical transmission spectra of ultrathin films of MoS_2 were also studied.

2. Experimental details

Initial metallic Mo and W films with the thicknesses of 3.5 nm and 10.8 nm, respectively, were obtained by magnetron sputtering onto the silicon (Si) substrates at room temperature with the sputtering power of 120 W. Molybdenum films with the thickness of 1, 2 and 4 nm on the fused silica (SiO_2) substrates and 5.7 nm film on the R-cut sapphire have been obtained with MBE. Pre-deposited metallic films were annealed then in an atmosphere of sulfur at a temperature of 800°C. A weak flow of ultrapure argon (~15 sccm) was used to transport gas for sublimated sulfur. After the annealing samples were cooled down to room temperature also in the flow of Ar (figure 1).

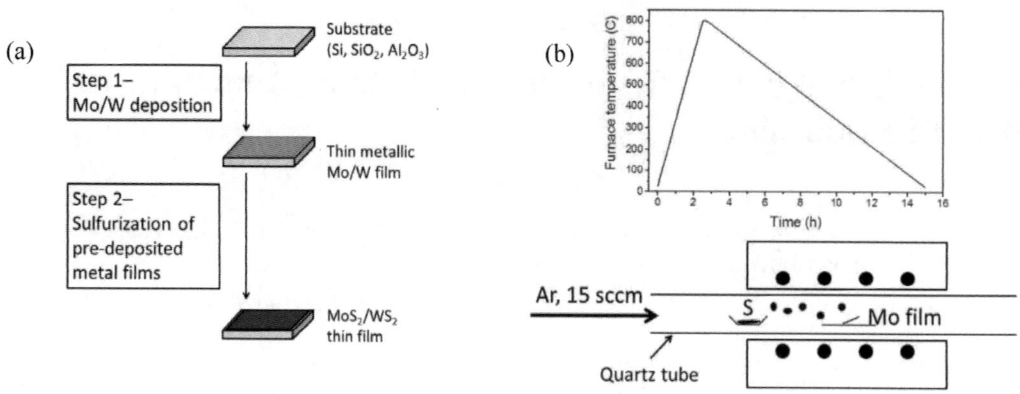

Figure 1. Disulfide film synthesis sequence (a); diagram of a setup for sulfurization of Mo and W thin films (b). Upper graph in (b) illustrates approximately the dependence of the furnace temperature on time.

3. Results and discussion

XPS measurements were performed in the ultrahigh vacuum chamber (base pressure ~ 3×10^{-10} mbar) equipped with the Mg K_α X-ray source operated at 12.5 kV and 250 W, and a Phoibos 150 hemispherical energy analyzer (all from SPECS).

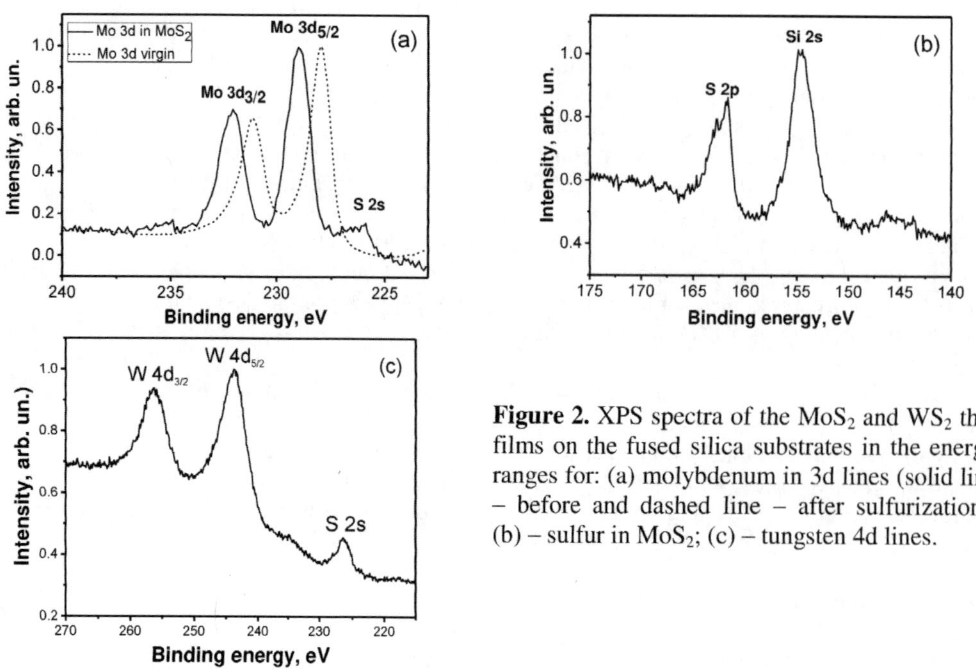

Figure 2. XPS spectra of the MoS_2 and WS_2 thin films on the fused silica substrates in the energy ranges for: (a) molybdenum in 3d lines (solid line – before and dashed line – after sulfurization); (b) – sulfur in MoS_2; (c) – tungsten 4d lines.

XPS measurements (figure 2) show that the initially metallic films of molybdenum and tungsten have been successfully transformed into the disulfides of these elements during sulfurization [3]. This is confirmed first of all by the energy shift of molybdenum and tungsten lines towards the higher binding energy range, as well as the appearance of sulfur line in the spectra. The change in energies of the Mo and W lines is the chemical shift that is associated with the displacement of the electron density in the sulfur direction during the formation of a compound. Positive charge on a metal atom creates additional Coulomb attraction of the photoelectron emitted due to the interaction with X-rays [4].

Raman spectra of molybdenum and tungsten disulfides were studied with the NTEGRA SPECTRA setup (NT-MDT). Laser emitting at a wavelength of 532 nm was used for excitation of phonons in the samples. The measurements were performed at room temperature. Matching of well-known vibrational modes [5, 6] with the peaks in experimental spectra indicates that indeed as a result of sulfurization molybdenum and tungsten disulfide films have been obtained (figure 3). Furthermore, it is well-known that phonon frequency splitting between the A_{1g} and E^1_{2g} modes depends on the number of layers for MoS_2 [7]. Splitting of these lines in our experimental spectra is ~ 25 cm^{-1}, which corresponds to 6-8 molecular layers. Rich Raman spectrum for WS_2 is due to the known fact that under excitation with the 532 nm wavelength laser light, not only the first order Raman modes but also longitudinal acoustic modes of the second order are excited. In addition, there are several peaks associated with multiphonon processes [6].

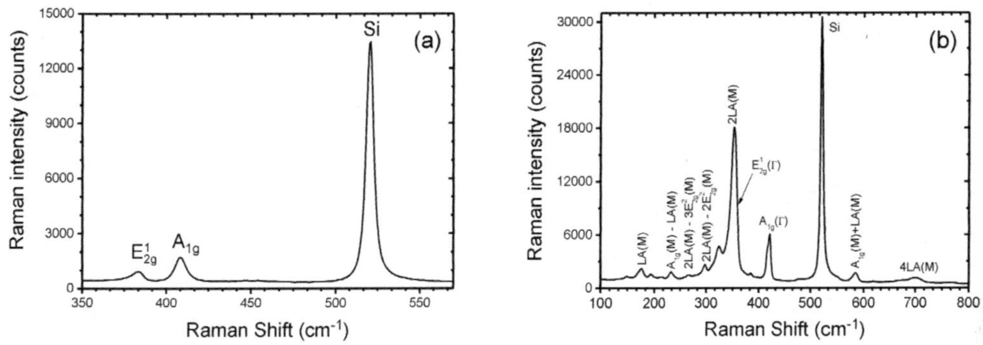

Figure 3. Raman spectra of thin films for: (a) MoS_2 and (b) WS_2. Raman modes corresponding to the peaks are marked in spectra.

Morphology of the surface was studied with the universal scanning electron microscope Merlin (Carl Zeiss). SEM results (figure 4a) show that the investigated samples represent the nanogranular films, wherein each granule is a platelet with the thickness of several nanometers in-plane dimensions of about a hundred of nanometers. Granules have different orientations with respect to the the substrate, as it also can be seen from figure 4a. Moreover, the granules have different shapes: triangular, truncated triangular, and near-round ones. Different shapes of these granules are related in the published data to the varying conditions inside the oven during the growth [8]. In particular, the sulfur concentration may vary.

Optical transmission spectra of molybdenum disulfide samples were studied using dual beam spectrophotometer SPECORD M40 (Carl Zeiss). In the spectra (figure 4b) two absorption peaks are revealed for MoS_2 both on the sapphire and fused silica substrates. These peaks are associated with the excitonic states and are labelled traditionally as A and B [9]. Figure 4b shows that the position of these peaks depends on the film thickness. Namely, excitonic absorption bands shift to the higher energies on decreasing of the film thickness. Such a change in transmission spectrum is related to the band structure modification of semiconducting molybdenum disulfide with the decrease of the molecular layer number [10]. The peak position dependence on the material and structure of substrate remains poorly studied.

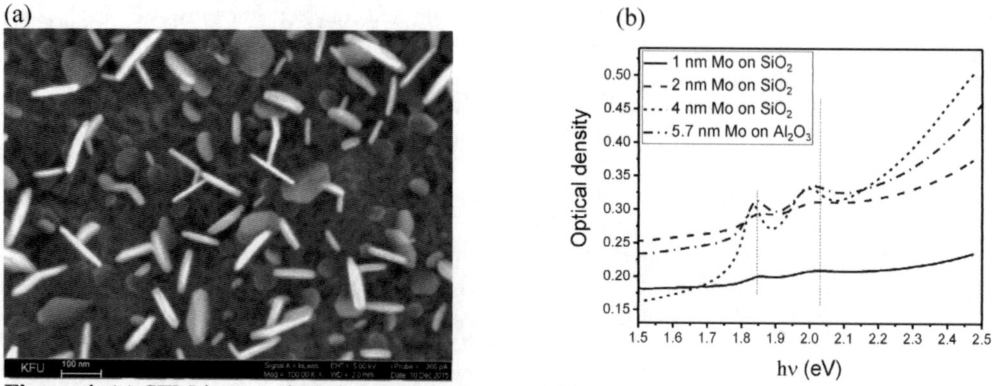

Figure 4. (a) SEM image of MoS_2 on a single crystal Si substrate; (b) optical transmission spectra of MoS_2 on the fused silica and sapphire substrates.

4. Conclusions

The paper presents the results of the synthesis and investigations of thin and ultrathin molybdenum disulfide and tungsten disulfide films. It was shown that applied synthesis method leads to the formation of multilayered TMD in the form of nanogranular thin films. The transmission spectra of the samples contain two excitonic absorption bands. These bands are shifted to higher energies by decreasing film thickness.

Acknowledgements

Vakhitov I.R., Gumarov A.I., Yanilkin I.V., Tagirov L.R. were supported by the Program of Competitive Growth of Kazan Federal University. The work of A. L . Zinnatullin and R.V. Yusupov was funded by the subsidy allocated to Kazan Federal University for the state assignment in the sphere of scientific activities. Synthesis of films and XPS measurements were carried out at the PCR Federal Center of Shared Facilities of KFU. SEM measurements were performed at the Interdisciplinary Center for Analytical Microscopy.

References

[1] Novoselov K S, Geim A K, Morozov S V, Jiang D, Zhang Y, Dubonos S V, Grigorieva I V, and Firsov A A 2004 *Science* **306** 666
[2] Ganatra R and Zhang Q 2014 *ACS Nano* **8** 4074
[3] Wanger C D, Riggs W M, Davis L E, Moulder J F and Muilenberg G E 1979 *Handbook of X-ray Photoelectron Spectroscopy* (Eden Prairie: Perkin-Elmer Corp.)
[4] Winograd N 1980 X-Ray Photoelectron Spectroscopy *Physical Methods in Modern Chemical Analysis* ed. T Kuwana. (London: Academic Press) volume 2 p 411
[5] Ye M, Winslow D, Zhang D, Pandey R and Yap Y K 2015 *Photonics* **2** 288
[6] Berkdemir A *et al.* 2013 *Scientific Reports* **3** 1755
[7] Lee C, Yan H, Brus L E, Heinz T F, Hone J and Ryu S 2010 *ACS Nano* **4** 2695
[8] Tongay S, Varnoosfaderani S S, Appleton B R, Wu J and Hebard A F 2012 *Applied Physics Letters* **101** 123105
[9] Pogna E A A *et al.* 2016 *ACS Nano* **10** 1182
[10] Splendiani A, Sun L, Zhang Y, Li T, Kim J, Chim C-Y, Galli G and Wang F 2010 *ACS Nano* **10** 1271

IOP Conf. Series: Journal of Physics: Conf. Series **789** (2017) 012074 doi:10.1088/1742-6596/789/1/012074

Plasma technologies application for building materials surface modification

A.G. Luchkin , Hakki A., Rahimov N.F., Sadikov K.G., Luchkin G.S.
Kazan (Volga Region) Federal University, 18 Kremljovskaja str., Kazan 420008, Russian Federation

E-mail: luchkin_a.g@mail.ru

Abstract. Low-temperature plasma modification of $LiYF_4$ crystal surface in Helium atmosphere caused microhardness decreasing and increasing of roughness of crystal surface. The change of microhardness and morphology is a possible result of Fluorine outgoing from material structure due to heating of surface and plasma chemical reactions and ingoing of Oxygen. As a result of exchange and diffusion processes crystal surface structure become more crumbly, its morphology and mechanical properties change.

1. Introduction

Low-temperature has wide application in different kinds of industry [1-4], including optics, where it is used for surface cleaning, deposition and assisting with deposition of thin functional coatings [5]. Ion's low energy allows using low-temperature plasma also for modification of some crystals surface, i.e. $LiYF_4$. That crystal is used in laser systems [6], it has relatively high surface microhardness, that prevents its high quality finishing.

The aim of this article is low-temperature plasma modification of $LiYF_4$ crystal surface for increasing of quality of this crystal finishing. To achieve this goal crystal surface was processed by Helium low-temperature RF plasma at atmosphere pressure in the presence of water vapour.

2. Experiments

Experimental unit consists of chamber, RF poser supply, gas supply system. Functional scheme of experimental unit is shown on Fig. 1. The upper and bottom electrodes are situated in chamber vertically and coaxially. The upper electrode is RF power supplied (frequency 13.56 MHz, power up to 50 W) and bottom electrode is grounded. Distance between electrodes was 5 mm. Both electrodes are made from stainless steel and have cylinder form for evenly plasma processing of crystal. The crystal LiYF4 (Fig. 2) was put on the bottom electrode. There was a barrier layer between crystal and bottom electrode. Passive Aluminum radiators was used for electrode cooling and situated outside the chamber. Between RF power supply ("Magniterm LCL", Russia) and upper electrode there was a matcher. Helium with maximum humidity was supplied through gas inlet.

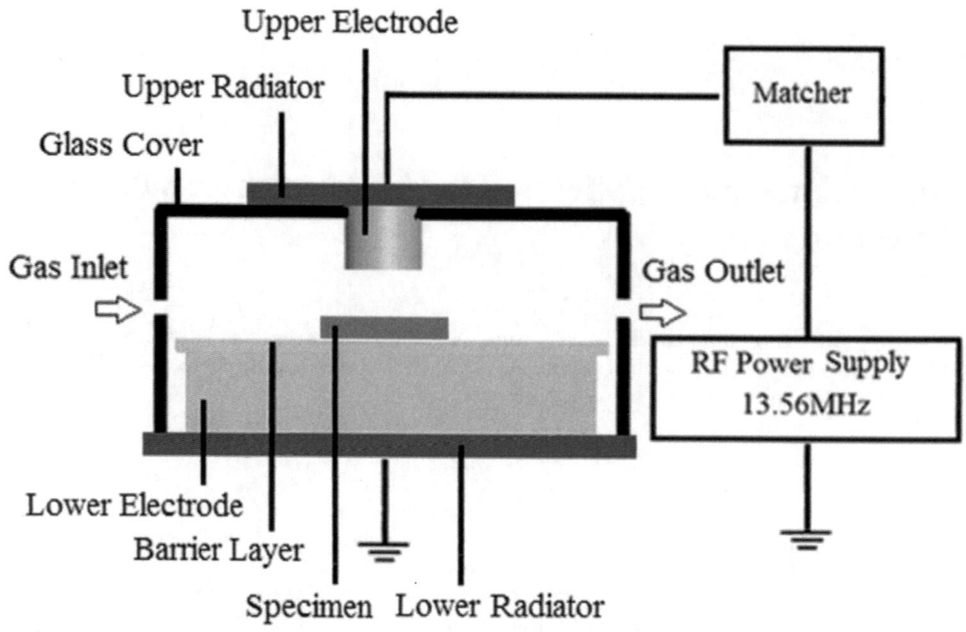

Figure 1. Functional scheme of experimental unit.

Atom force microscope NT-MDT Solver P47H was used for measurements of morphology. Microhardness was measured by hardness measuring unit PMT-3. Parameters of experiments of low-temperature plasma processing of $LiYF_4$ crystals are shown in Table 1.

Table 1. Experiment parameters

Parameter	Value
Sample	
Material	monocrystalline $LiYF_4$
Diameter	6 mm
Thickness	2 mm
Gas	
Composition	Helium 100%
Humidity	1%
Rate	1 l/min
Supply	
Type	RF
Discharge power	20 W
Frequency	13.56 MHz
Current density (upper electrode)	7.5 mA/mm^2
Processing time	15 min, 30 min, 60 min
Chamber pressure	atmospheric

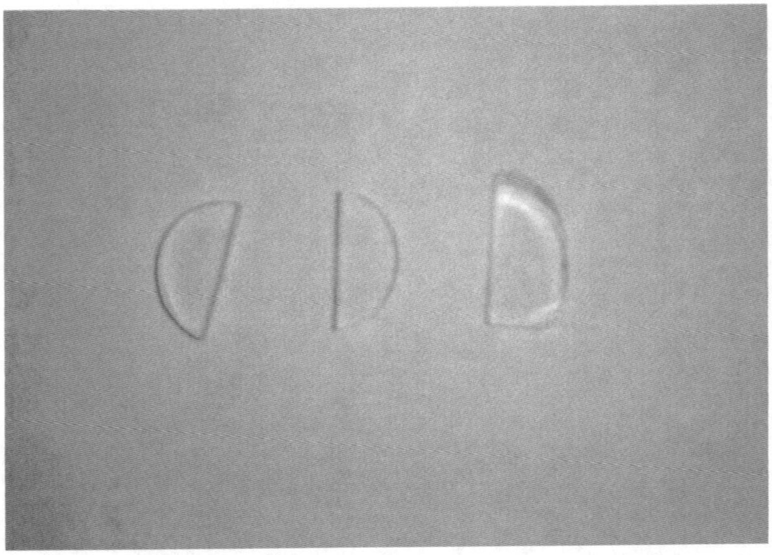

Figure 2. $LiYF_4$ crystals

3. Results and discussions

The view of low-temperature plasma within crystal processing is shown in Fig. 3. Wide part of plasma zone is near bottom electrode and sample. Results of AFM measuring is shown in Fig. 4 and 5. Changes of microhardness and roughness of crystal surface are shown in Fig.6. The microhardness of sample surface decreases with increasing of processing time. The surface roughness increases with increasing of processing time. It could be caused by surface structure changes. Fluorine from crystal material under influence of low-temperature plasma diffuses to the surface and desorbs from it. As a result the stoichiometry of crystal is broken.

Also there is Oxygen in water vapor in Helium and the possibility of plasma chemical treatment of crystal surface is not small. As result in crystal appear Yttrium and Lithium oxides, not only stoichiometry is broken, but chemical composition is changed.

Figure 3. The view of low-temperature plasma.

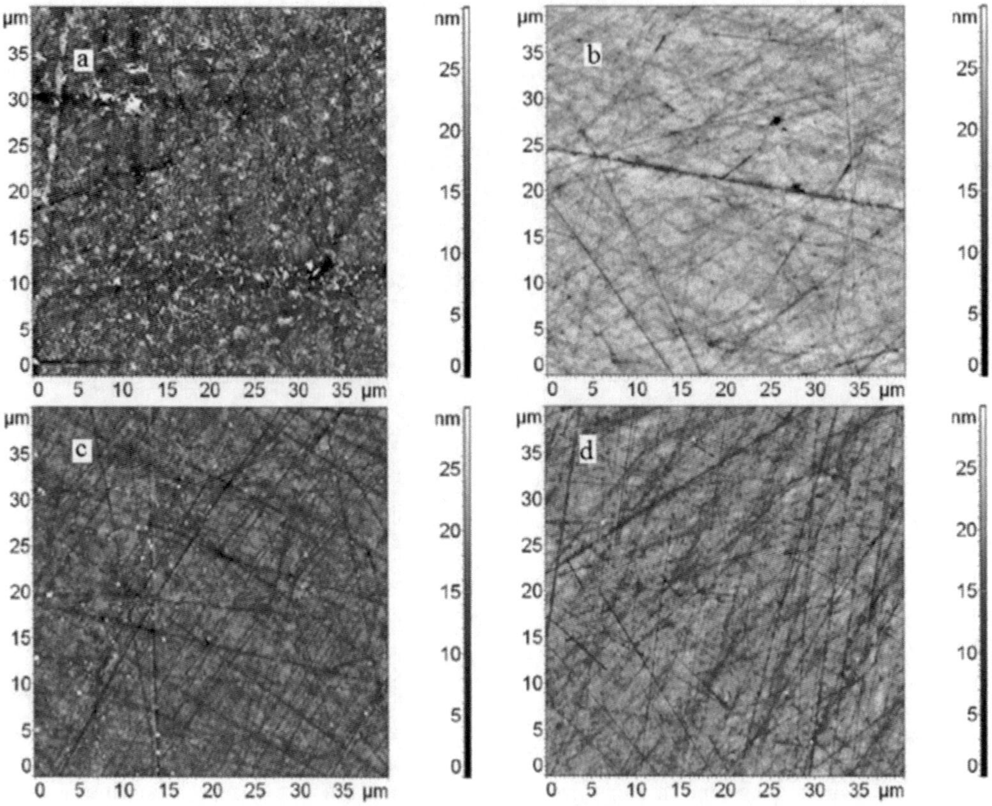

Figure 4. AFM images of the surfaces treated by the helium plasma (a) exposition times 60 min, (b) 30 min, (c) 15 min; (d) is the typical AFM image of the reference virgin sample's surface

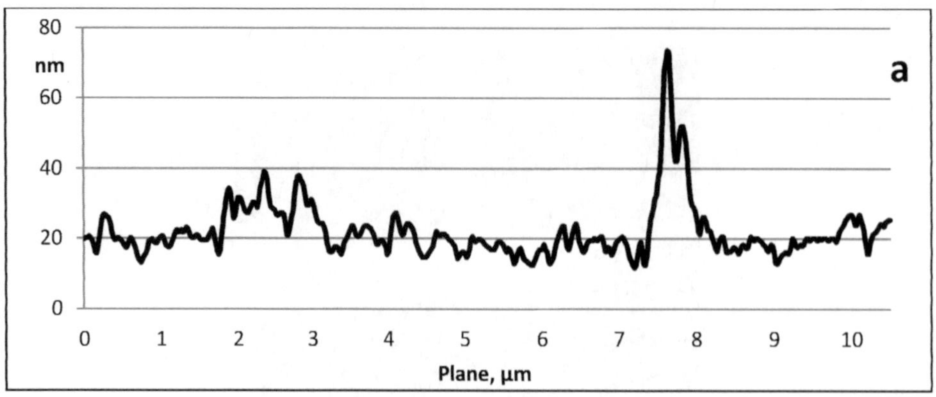

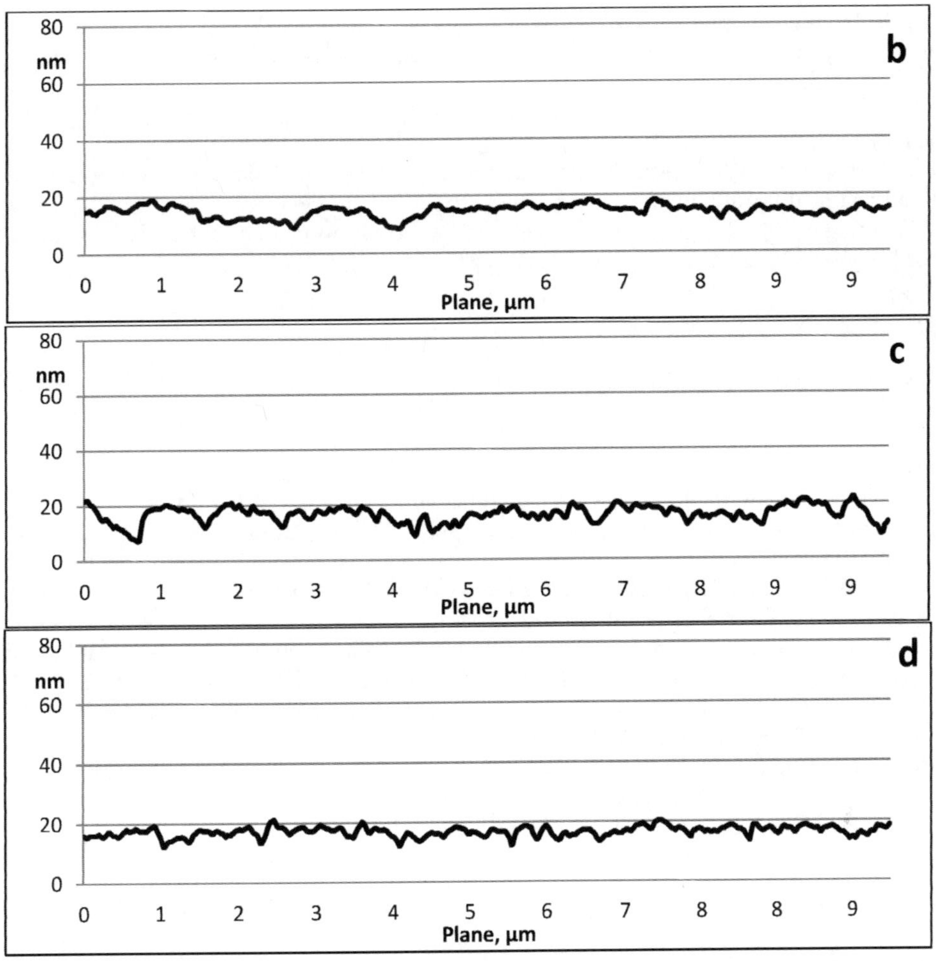

Figure 5. The roughness of the surfaces processed by helium plasma. AFM image: (a) 60 min, (b) 30 min, (c) 15 min, (d) non-treated.

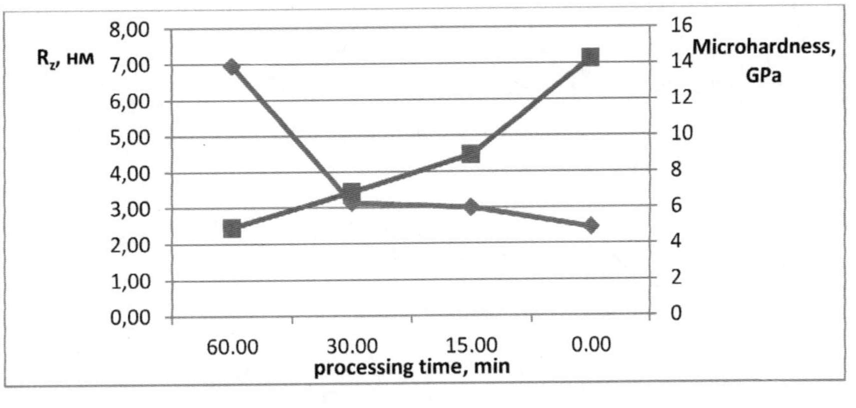

Figure 6. Dependence of microhardness (■) and roughness (♦) of crystal surface from processing duration.

The difference of crystal surface microhardness from crystal bulk microhardness gives opportunity for next mechanical, chemical or mechanic-chemical polishing of crystal surface and for increasing of finishing quality.

4. Conclusion

The microhardness of $LiYF_4$ crystal surface decreased in more than 2 times due to low-temperature Helium plasma modification within 1 hour. This result gives an opportunity for increasing of crystal surface finishing quality. Increasing of surface roughness could be caused by fact that crystal surface structure become more crumbly.

Acknowledgements

A.G. Luchkin, A. Hakki, N.F. Rahimov, K.G. Sadikov, G.S. Luchkin were supported by the Program of Competitive Growth of Kazan Federal University.

References

[1] Volokitin G.G., Skripnikova N.K., Volokitin O.G., Shehovtzov V.V., Kashapov N.F. Plasma technologies application for building materials surface modification, J. Phys. Conf. Ser. 669 (2016) 012065

[2] Galyautdinov R.T., Luchkin A.G., Luchkin G.S. Low-temperature plasma in processes of deposition of functional coatings. collection of articles IV Republican scientific-technical conference: collection of articles - Kazan : KSTU Publishing house, - 2013, - P. 155-161. (rus)

[3] Galyautdinov R.T., Kashapov N.F., Luchkin G. S. // Applied physics. . - 2005.- № 6.- P. 88-93.

[4] Luchkin A.G., Kashapov N.F. Reactive magnetron sputtering model at making Ti-TiOx coatings, J. Phys. Conf. Ser. 567 (2014) 012027

[5] Spravochnik optika tehnologa / M.A. Okatov. – 2-d edition – SPb.: Polytehnika, 2004. – 679 p.

[6] Laser performance of in-band pumped Er : LiYF4 and Er : LiLuF4 crystals/ Semashko, V.V. et. al/ Quantum Electronics, Volume 46, Issue 2, 2016, Pages 95-99.